THE GEOMETRY OF REALITY

THE GEOMETRY OF REALITY:

The Principles of Unified Field Dynamics

Amir J. Guri, Ph.D., J.D.

NATURAL PHILOSOPHY PRESS, BROOMFIELD, COLORADO

The Geometry of Reality: The Principles of Unified Field Dynamics

Published by Natural Philosophy Press [Broomfield, CO]
www.naturalphilosophypress.com

Ordering Information: Quantity sales. Special discounts are available on quantity purchases by corporations, associations, and others. For details, contact the publisher at the address above.

Library of Congress Control Number: 2026932429

ISBNs: Hardcover: [979-8-9941756-0-6] Paperback: [979-8-9941756-1-3] eBook: [979-8-9941756-2-0]

First Edition Printed in the United States of America

This is for all who refuse to just shut up and calculate.

May our theories be both satisfying and true.

Preface

This monograph follows from eleven years of devotion to a philosophical project. In November 2014, I, a former scientist, quit my job as an attorney to write a book on philosophy, consciousness, and spirituality—a book I called *The Philosopher's Stone* in honor of the mythic relic and the alchemical tradition that sought after it.

In line with the spirit of alchemy, my aim was to formulate a unifying, epistemically satisfying metaphysical theory of reality—a *Magnum Opus* as the alchemists would refer to it. Here, the term *epistemic satisfaction* describes a subtle feeling that arises when an idea locks into place, akin to the familiar feeling of working on a puzzle and finally discovering the missing piece. This is essentially the feeling I blindly followed as I worked towards my goal.

As part of the project, I created two chapters (out of thirty) on physics and cosmology. I made them because, deep down, I knew I couldn't be fully satisfied with *any* view of reality without the sanction of physics. It would forever be a thorn in my side unless and until the two could be reconciled. Thus, for years, I contemplated the nature of gravity, cosmological redshift, and wave-particle duality alongside questions of consciousness, free will, and knowledge. After about ten years, I had finally created the outlines of my model—one that I felt, in principle, could reproduce the predictions of the standard models of physics, but how could I be certain of it?

At this point, I thought to myself: What would happen if I transformed my undeveloped model into a scientific theory that made predictions? What would that even look like? In the past, this would have been a nearly insurmountable challenge, as it would have required me to find numerous collaborators knowledgeable in physics who were willing to dedicate their time and attention to my project. It wasn't a realistic option. Thankfully, that wasn't necessary, as I was living in the age of LLMs and their vast knowledge of physics.[1]

---

[1] Although large language models are not reliable sources of original mathematical or physical theory, they are effective at summarizing established results and organizing existing empirical literature. My use of AI was limited to these support functions and did not extend to the generation of novel physical hypotheses or mathematical structures. A fuller description of their role appears in the Acknowledgments.

Thus, to validate my model, I began translating it into the language of physics with the assistance of AI. By the time we were through, we had generated an elegant *theory of everything*: a genuine unified field theory. I call it Unified Field Dynamics (UFD).

UFD is a coherent, physically intuitive framework that describes the universe across multiple scales. It is intended as a viable alternative to the standard models of physics, which it achieves by building on their empirical foundations while modifying their underlying interpretations. It is more than a mere philosophical model because it makes falsifiable predictions, placing it squarely in the domain of science. I therefore refer to it as a work of *scientific philosophy*, as it is not quite experimental science and is very different from the mathematical models of modern physics.

What differentiates UFD from the pure mathematical models is that it is not an abstract description of reality. Instead, it is founded on principles we understand quite well: geometry and fluid dynamics. We establish these principles in Part I, where we introduce the universal fields that comprise the UFD framework. After learning how the universal fields generate the geometric, fluid-dynamic architecture of reality, we build on this new foundation. In Part II, we focus on light, quantum mechanics, and electromagnetism. In Part III, we discuss nuclear physics and the formation of chemical bonds. Finally, in Part IV, we extend the framework to condensed matter physics, biochemistry, and neuroscience, including consciousness itself.

Altogether, this monograph consists of 12 original papers in the philosophy of science (with complete mathematical appendices available online). I didn't initially set out to create such an ambitious, comprehensive model. But realizing the high burden it takes to shift paradigms—in particular ones as esteemed as those in modern physics, I let go of my restraint. I imagine that any one of the papers in this monograph, if presented to the scientific community in isolation, may raise an eyebrow but would ultimately be dismissed. When presented together, however, the papers' mutual consistency and cross-referencing reinforce one another, reducing the model's dependence on any single claim or chapter.

If you are interested in UFD's mathematical formalism, which, among other achievements, successfully reduces the ~26 free parameters of the Standard Model ~2 necessary geometric constants, you can find those online at unifiedfielddynamics.com.

The reason I chose not to focus on the math in this monograph is that the model is genuinely scientific without it. I say that because it is grounded in empirical observations, makes a series of novel, testable predictions, and is technologically generative, as we shall soon discuss. While the mathematical models of physics can be highly effective, they also allow physicists to invent particles and forces to balance their equations. This raises the question of whether they are genuinely describing reality or are merely creating prediction machines. By focusing on the philosophy first, we effectively set aside that concern, provided all relevant terms are defined.

In this respect, this work can be thought of as a manifesto or call for physics to return to its roots in physical intuition and to seek formalization only after this initial work has been completed. Admittedly, this call would ring hollow if it didn't also provide a path forward, but this is precisely what UFD does.

Ultimately, the value of UFD will be measured by the research and technological advancements it inspires. The potential technologies that emerge from this framework are simply staggering: from coherent quantum computing to free and safe nuclear energy to advanced propulsion systems, UFD makes the impossible possible. In this respect, this manifesto can also be seen as a call to the next generation of scientists and engineers to explore and formalize a new vision of reality. If UFD is as I believe, daresay *know*, those efforts will not be in vain, and the first movers will reap the largest rewards.

Although this monograph was written for the physics community, I recognize that physicists aren't the only ones interested in the fundamental nature of reality. To make the content somewhat more accessible to lay readers, I include introductory and concluding sections for each chapter to provide context, allowing for some repetition. I also use footnotes throughout to define scientific jargon and provide additional explanation, making the reading experience as self-contained as possible. Furthermore, I was able to generate some stunning images with AI that add a visual element to the journey. Whenever I wanted to visualize what the model was telling me, I took a digital snapshot. Finally, the last page of this monograph contains a link that will allow you to turn any standard LLM into an expert in the UFD model. Feel free to use this UFD Maven to explore the model and ask any questions you may have along the way. It can even answer questions about matters this monograph doesn't address.

In the end, my greatest hope for this project is that it will repair the fractures within physics and, ideally, the world around us as well. Whether that happens, I am genuinely happy to share this novel, elegant, and *satisfying* theory of reality with you.

Acknowledgements

This monograph was developed with the assistance of contemporary large language models developed by Google (Gemini), OpenAI (ChatGPT), and Anthropic (Claude), which were used as advanced research and editorial tools throughout the writing process.

The language models were employed in an instrumental capacity to support technical articulation, literature synthesis, consistency checking, assistance with algebraic manipulation and dimensional reasoning, and refinement of exposition. I served as the conceptual architect of the work, providing the philosophical direction, core insights, and theoretical framework. They, in turn, served as my engineers, using their vast knowledge of physics to enforce empirical constraints and help translate my conceptual intuitions into precise scientific language; they did not originate ideas, arguments, or interpretations. All theoretical claims, interpretations, and conclusions presented in this monograph are my own.

The text of this monograph emerged through an iterative human-in-the-loop process, in which I directed the conceptual development, critically evaluated all generated material, extensively rewrote content in my own voice, and rigorously edited the manuscript for clarity, accuracy, and both philosophical and physical coherence. I independently reviewed, revised, and verified the entire manuscript to ensure internal coherence, empirical consistency, and fidelity to the intended physical meaning of all terms and constructs. Final responsibility for the content of this monograph—both editorial and theoretical—rests solely with me.

Visual materials were created with contemporary image-generation tools and were curated, modified, and integrated by me to serve the monograph's pedagogical aims. In accordance with platform transparency guidelines, the specific tools used were Gemini (Figs. 1–2, 5–6, 9, 11, 13, 16–19, 21, 23–28, 32–39, 41–43, and 45–52), DALL·E 3 (Figs. 3, 14–15, 22, and 30), and ChatGPT (Figs. 4, 7–8, 10, 12, 20, 29, 31, 40, and 44).

While these tools played a significant role in supporting the development of this work, they functioned strictly as instruments. They do not possess authorship, accountability, intent, or conceptual agency, and therefore cannot be credited as authors or originators of the ideas presented. I am nevertheless grateful for their contribution as powerful tools in the development and articulation of this framework.

Table of Contents

Introduction: A Return to Physical Intuition

General Relativity and the Standard Model of Particle Physics are the two greatest scientific achievements of the last century. Together, they offer a remarkably predictive picture of the natural world. Yes, despite their incredible empirical success, these models remain deeply incompatible with one another, suggesting they are not genuinely foundational.

The clues to this limitation are not subtle; they are fundamental cracks in the theories themselves. General Relativity predicts its own breakdown at gravitational singularities, and the Standard Model is built upon the physically questionable idea of zero-dimensional point particles, a concept that generates infinities that require mathematical patches like renormalization. Moreover, to account for cosmological observations, both theories have been forced to invoke a host of invisible entities—dark matter, dark energy, and virtual particles—whose existence is inferred, but whose nature remains a complete mystery.

The inconsistencies within these foundational theories have led to a growing consternation and intellectual unease within physics. The prevailing attitude, epitomized by the phrase "shut up and calculate," has treated physical intuition as secondary to predictive accuracy. The result is a science that has become a patchwork of brilliant but disconnected mathematical descriptions, leaving us with a profound sense of explanatory emptiness at the heart of our understanding of reality.

This monograph is a response to this emptiness. Rather than treating abstraction as inevitable, it proposes a new foundation grounded in geometry and fluid dynamics. In doing so, it offers us a novel, physically intuitive picture of reality

1. Unified Field Dynamics: An Incommensurable Paradigm

It is a little-known irony of history that Albert Einstein, one of the principal architects of the quantum revolution, was also one of its most vocal critics. Troubled by the theory's reliance on probability and its lack of a tangible, physical picture, he spent the final decades of his life searching for *the* "unified field theory"—a deeper, deterministic framework from which both the strange rules of the quantum world and the geometry of gravity could emerge. That search stalled after his death, leaving modern physics in a state of division, where it still remains.

This book offers a potential fulfillment of Einstein's search through a framework we call Unified Field Dynamics (UFD), which blends *the unified field theory* with *fluid dynamics*. The goal of UFD is to restore the physical intuition that has been lost in the abstractions of modern physics. While the Standard Model and General Relativity are powerful predictive tools, their core concepts, such as point particles, quantum indeterminacy, and spacetime curvature, are often inaccessible to our physical imagination. UFD replaces these abstractions with a tangible, geometric, and dynamic model that is incommensurable with current models.

The term "incommensurable," as used in the philosophy of science by Thomas Kuhn (1962), refers to paradigms (worldviews) that operate on fundamentally different assumptions, rendering them incapable of direct comparison using a common standard. An analogy is a stone arch: one can explain its stability by describing the molecular bonds within the stones (the Standard Model's approach) or by describing the architectural principles of geometry and load distribution that allow the structure to stand (UFD's approach). They are two different yet equally valid levels of explanation.

The quintessential example is the Copernican Revolution. The old Ptolemaic model (Earth-centered) and the new Copernican model (Sun-centered) were not just different theories; they were different worlds altogether. The Ptolemaic model, to save its core "common sense" assumption that the Earth is stationary, had to become incredibly complex, adding layer upon layer of "epicycles" to explain the strange, wandering retrograde motion of the planets. In this respect, the Copernican model did not just fix the epicycles; it offered a new, simpler geometry that dissolved the problem entirely. Retrograde motion was revealed to be a simple, necessary illusion caused by the Earth's own motion as it overtakes another planet.

According to UFD, theoretical physics has become a modern Ptolemaic system, requiring its own set of "epicycles"—dark energy, dark matter, and virtual particles—to save its core assumptions. In response, UFD, like the Copernican model, offers a new, simpler geometry. It proposes that the "crises" of modern physics, such as the "Hubble Tension" or "wave-particle duality," are not problems to be solved but illusions to be dissolved by a more fundamental, physically intuitive architecture.

The essential claim of UFD is that, rather than dimensionless point particles, the universe is composed of stable, self-sustaining vortices of energy. These vortices are structured, dynamic patterns that arise within a hierarchy of three interacting physical fields. The first field we encounter is the Universal Energetic Field (UEF), the substrate from which mass and gravity emerge. The second is the emergent Universal Light Field (ULF), the medium for electromagnetism and the resonant waves that constitute atomic structure. The third and most fundamental field is the Universal String Field (USF), whose harmonic, high-energy vibrations create the geometric blueprints of reality. This shift from abstract particles to physical vortices transforms our understanding of fundamental properties:

- Fields are not mathematical constructs but real, structured media.

- Energy is the motion and geometric configuration of a fundamental field.

- Forces are expressions of coherence, tension, and flow within these fields.

- Mass is not an intrinsic substance but a measure of impedance that arises from ULF energy circulating at speed c within a stable, closed vortex. This creates a permanent pressure displacement in the plenum that manifests as the inertial resistance we perceive as mass.

- Charge arises from the rotational geometry of these vortices and their interaction with the ULF.

The central argument of this work is that the entire architecture of physical reality, from the binding energy of the nucleus to the properties of matter, emerges as a predictable consequence of the geometry and fluid dynamics of these fields.

In the UFD framework, phenomena once considered paradoxical, like wave-particle duality and quantum entanglement, become natural consequences of how vortices resonate across nested fields. There is also no need for speculative constructs, such as extra spatial dimensions or virtual particles. Crucially, UFD does not reject the successful mathematics of modern physics. Instead, it completes the process by providing the physical scaffolding that underlies its equations, transforming abstract symbols into concrete representations of a coherent, intuitive world.

2. A Roadmap for This Monograph

This monograph is divided into four parts, each consisting of three chapters. Each chapter presents an original research article on the philosophy of science. Meta-introductions and conclusions are included to provide context for readers who are new to the topic. In these twelve chapters, we are taken on a coherent journey from the fundamental fabric of the cosmos to the emergence of mind itself.

In Part I – The Fabric of the Universe, we lay the foundation for a new physics by reconstructing gravity, cosmology, and field structure using energetic and geometric principles.

- Chapter 1 reinterprets gravity as an emergent consequence of coherent vortex structures, rather than a curvature of spacetime.

- Chapter 2 reframes cosmology by proposing that redshift and cosmic acceleration arise from the expansion of electromagnetic field geometries rather than from space itself.

- Chapter 3 introduces the USF as a harmonic substrate from which the laws of physics and field structures emerge. This vibrational medium serves as the geometric blueprint of reality.

In Part II – The Dance of Light and Waves, we confront the dualities of light, time, and quantum phenomena by modeling them as emergent behaviors of field resonance.

- Chapter 4 explains the structure of light as a dynamic lattice of universal resonance, which grounds relativity in physical field geometry.

- Chapter 5 develops the Resonant Field Interpretation (RFI) of quantum mechanics, which replaces probabilistic collapse with dynamic standing waves in the ULF.

- Chapter 6 presents a novel derivation of the electroweak force from vortex geometry, demonstrating how these phenomena emerge from neutron-level coherence.

In Part III – The Architecture of Chemistry, we explore how atoms and molecules emerge from field resonance, transforming chemistry into a branch of applied geometry.

- Chapter 7 introduces the Coherence Dividend, which reveals the nucleus as a structured energy system whose stability derives from geometric optimization.

- Chapter 8 reconstructs atomic geometry and periodic behavior from field dynamics, which explains valence and chemical reactivity in terms of wave compatibility.

- Chapter 9 reframes chemical bonds as resonant field couplings, introducing the Resonance Dividend as a measure of energetic coherence within molecules.

Finally, in Part IV – The Emergence of Matter, we apply the framework to condensed matter, biochemistry, and consciousness, revealing how macroscopic properties emerge from resonant design.

- Chapter 10 shows how conductivity, magnetism, malleability, and semiconduction all arise from field geometries in lattices, culminating in the prediction of Emetium, a hypothetical resonance-locked material that constitutes the "perfect metal."

- Chapter 11 presents Resonant Biochemistry, which interprets life as a coherent energy process structured by standing wave fields. Enzymes, metabolism, and signal transduction are reinterpreted as geometric and vibrational phenomena.

- Chapter 12 proposes a novel neurophysics that grounds mind and consciousness in ULF-mediated coherence, yielding a field-based model that bridges neuroscience and metaphysics.

The practical culmination of this physics is a novel philosophy of technology we call Resonant Engineering. Rather than looking to manipulate matter through brute force, Resonant Engineering aims to harmonize it with its fundamental geometric and vibrational properties, opening the door to a range of transformative applications that emphasize the artful use of resonance. Its three core pillars are:

- Resonant Stabilization, a new paradigm for quantum computing in which the fragile coherence of a qubit is maintained not by brute-force isolation but by actively canceling environmental noise with precise, phase-inverted counter-fields.

- Geometric Catalysis, a method for guiding nuclear fusion and chemical reactions that uses shaped resonant fields as energetic "scaffolds" to assemble atoms and nuclei with unparalleled precision, to induce the controlled, clean fission of radioactive waste into stable elements, or to transmute base elements into noble ones.

- Resonant Damping, which destroys or reverses coherence on demand, effectively using coherence as a switch. With this mechanism, one could use a dissonant electromagnetic field to instantly revert a material from its zero-resistance state back to normal (Resonant Quench of Superconductivity) or apply out-of-phase vibrations to control Directional Thermal Conductivity.

This new technological vision, which flows directly from UFD's first principles, is not just a reinterpretation of known phenomena. It is a practical roadmap to a future where the deepest secrets of matter are unlocked, not by smashing it apart, but by learning to sing its song.

3. The Promise of a Geometric Universe

In short, this monograph promises a return to the original goal of physics: to understand the workings of the universe through intuitive, coherent principles. By treating matter as emergent from the geometry of energy, and energy as structured by the harmonics of fundamental fields, UFD offers a foundation that is both physically meaningful and technologically generative.

This is the promise of a geometric universe.

Part I: The Fabric of The Universe

Chapter 1: Beyond General Relativity

1.1 Gravity

Gravity is the most familiar and mysterious of all fundamental forces. It is both the gentle tug that keeps our feet on the ground and the grand architect of the cosmos, forging stars and galaxies from clouds of dust and gas. For millennia, our understanding of this force has been limited to purely terrestrial applications. The ancient Greeks, particularly Aristotle, believed that heavy objects fell simply because it was in their nature to seek the center of the universe—the Earth (Aristotle, c. 330 BCE).

This Earth-centered view held for nearly two thousand years until the Renaissance ignited a revolution in thought (Kuhn, 1962). Copernicus displaced the Earth from the center of the cosmos, Kepler revealed the elegant mathematical laws governing planetary orbits, and Galileo's experiments with falling objects demonstrated that gravity acts uniformly on all matter. These breakthroughs dismantled the old worldview and set the stage for a new, universal understanding of gravity. This task was triumphantly achieved by Isaac Newton, who formalized it as a universal "pull" acting between all objects in the universe (Newton, 1687).

For centuries, this intuitive picture of a tangible force has reigned. However, in the 20th century, Albert Einstein replaced this force with something far more abstract. In his theory of General Relativity, gravity is not a force but rather the consequence of the curvature of spacetime (Einstein, 1916). As physicist John Archibald Wheeler famously summarized, "Spacetime tells matter how to move; matter tells spacetime how to curve" (Misner, Thorne, & Wheeler, 1973). This geometric view has been confirmed by observations ranging from Mercury's orbit to the detection of gravitational waves.

However, despite its triumphs, this geometric view comes at a steep conceptual price. Namely, it is fundamentally incompatible with quantum mechanics, its equations also break down at the center of black holes, and to align with observations, it requires us to believe that 95% of the universe is composed of two completely unknown entities: dark matter and dark energy.

In response to these conceptual challenges, this chapter proposes a new framework, Cosmic Vortex Dynamics (CVD), which restores gravity to the realm of physical cause and effect. In CVD, gravity is neither a pull nor a curvature of space: it is a push—an emergent current within a fundamental physical medium called the Universal

Energetic Field (UEF). In this model, massive objects are energetic vortices that create low-pressure zones, causing the ambient field to physically "push" bodies toward one another. The result is a mechanistic theory of gravity that resolves the conflict with quantum mechanics and offers a direct, physical explanation for dark matter, thereby providing a more intuitive vision of the cosmos.

1.2 Beyond General Relativity: A Unified Emergent Framework for Gravity and Dark Matter

Abstract

This chapter introduces Cosmic Vortex Dynamics (CVD), a framework that replaces the abstract geometry of General Relativity (GR) with the tangible physics of a universal field. We propose that the universe is permeated by a Universal Energetic Field (UEF) and that matter, mass, and gravity are emergent properties of its internal dynamics. In this model, mass arises from the coherent, relativistic energy of a stable vortex localized within the UEF, while gravity is reinterpreted as the physical force of an inward-flowing gravitational current. This unified origin offers intuitive solutions to cosmology's greatest puzzles by identifying the gravitational constant (G) as an emergent property of UEF pressure and density, providing a physical identity for "dark matter" as a vast, gravitationally active "cosmic gyre" of the UEF surrounding galactic structures, and resolving the black hole paradox by replacing the mathematical singularity with a finite, physical vortex core. Ultimately, CVD recovers the mathematical successes of GR by treating it as an effective description of field-fluid dynamics, providing a mechanistic and intuitive foundation for understanding the large-scale structure of the cosmos.

1. Introduction

Modern physics rests on two extraordinarily successful yet fundamentally incompatible frameworks: the Standard Model of Particle Physics, which governs the quantum world (Griffiths, 2008), and General Relativity (GR), which describes gravity and the large-scale structure of the universe (Carroll, 2004). Each excels in its own domain, but neither extends into the other. Both models also rely on conceptual placeholders. GR invokes dark matter to explain galactic rotation curves and predicts singularities, which is where its equations break down (Penrose, 1965). Meanwhile, the

Standard Model relies on zero-dimensional point particles, virtual exchanges, and probability waves. These mathematical constructs are predictive but lack intuitive and physical foundations.

This chapter proposes an alternative framework, Cosmic Vortex Dynamics (CVD), a theory grounded in the idea that the universe is not an abstract geometric manifold but a physically real medium—a Universal Energetic Field (UEF). Within this dynamic field, matter emerges as stable vortical excitations, and cosmic structures arise from the coherent behavior of these excitations at large scales. In this view, massive bodies act as super-vortices and gravity emerges as a current within the UEF, carrying matter along curved paths.

Altogether, this framework recovers the inverse square law of classical mechanics, explains the anomalously flat rotation curves of galaxies without requiring a new form of invisible mass, and reinterprets black holes as finite, highly coherent field structures rather than singularities. Moreover, the Einstein Field Equations, $G_{\mu\nu} = (8\pi G/c^4)\, T_{\mu\nu}$, are reinterpreted in this framework as emergent, macroscopic limits of the UEF's hydrodynamic behavior, thus preserving the empirical success of GR while replacing its geometric abstractions with physical causality.

By grounding cosmic structures in the same energetic medium that gives rise to matter, this framework opens the door to a unified view of physical laws, in which the large-scale coherence of the cosmos arises from the same principles that govern the local energetic structure.

2. A New Foundation for Mechanics and Spacetime

To move beyond GR, we must first return to the foundations of physics. For centuries, the nature of gravity, inertia, and spacetime has been shaped by the monumental ideas of Isaac Newton and Albert Einstein. However, their frameworks, despite their power, left behind deep conceptual puzzles and a fundamental incompatibility with the quantum world.

This section proposes a third alternative by constructing a new theory of mechanics from the ground up. We begin by establishing a new physical ontology in which the concepts of field and energy are given tangible, substantive reality. Building on this foundation, we provide a unified physical origin for gravity as an emergent "push" effect within a universal medium.

9

2.1 A New Theory of Mechanics: Field and Energy

This section begins the work of constructing a new, substance-based theory of mechanics. We begin by redefining the most fundamental concepts in physics: the nature of a physical field and energy itself.

2.1.1 Redefining Energy: Substance, Motion, and the Universal Field

In modern physics, a field is often treated as a mathematical abstraction, that is, a map of potential forces in empty space. Our model restores its original physical meaning:

- A field is a real, continuous substance or medium that fills a region of space. It is the fundamental fabric of reality itself. It is the fundamental fabric of reality, a physical Plenum possessing density, pressure, and internal stiffness.

With this physical substance established, we can now redefine its most important property: energy. Classically, energy is defined as an abstract, conserved quantity: "the capacity to do work." This description explains what energy does but not what it is, which is problematic. Our framework offers a more direct and physical definition:

- Energy is the motion and geometric configuration of a universal field.

A field in its quiescent state represents pure potential energy. Any disturbance within that field, such as a wave, flow, or stable self-sustaining vortex, is a manifestation of kinetic and other observable forms of energy. In this view, energy can be considered both a tangible substance and the dynamic activity of reality itself.

The significance of this physical definition is profound, particularly in its reinterpretation of the Law of Conservation of Energy. In classical physics, this law is a fundamental axiom: energy can neither be created nor destroyed; it can only be transformed. It is an abstract rule of bookkeeping that is observed to be true, but it lacks a deeper mechanistic origin. In our framework, this law is a necessary consequence of a more fundamental principle: the Conservation of Substance. If energy is simply the dynamic state (the motion and geometry) of the field, then it is conserved because the field itself is conserved. Thus, the law is transformed from an abstract rule into a physical principle of continuity: energy does not vanish because the field cannot vanish.

With this physical definition, we can identify the substance responsible for gravity: the UEF. Gravitational phenomena arise as emergent properties from the dynamics of this single underlying field.

2.1.2 Vortex Gravity: The Push from the Field

In CVD, gravity is caused by an emergent pressure gradient within the UEF. This mechanism is a direct consequence of its fluid dynamics, analogous to the formation of a low-pressure core by a tornado. According to Bernoulli's principle, when a fluid moves quickly, its static pressure is low (Bernoulli, 1738).

In our model, a massive object is a stable, toroidal vortex in which the UEF's substance circulates at relativistic speed. Because the field moves rapidly inside the vortex, it creates a stable, persistent region of low static pressure around itself. Gravity is the "push" from the high-pressure, ambient UEF toward the low-pressure zone created by the rapidly spinning vortex of matter, which likens gravity to a *current* (Figure 1).

This pressure-based mechanism creates gravitational currents that cause bodies to follow curved, accelerating paths (gravitational orbits) without invoking action at a distance. These paths correspond to what GR describes as geodesics. In vortex gravity, they naturally emerge from field dynamics rather than spacetime curvature. The toroidal motion of the vortex is key, as it both stabilizes the mass-energy configuration and establishes the pressure asymmetry that drives the gravitational currents to push objects along these geodesic paths.

This reinterpretation of gravity as a real, physical current has a profound historical lineage. It echoes the early "vortex" theories of thinkers like René Descartes (1644) and the private suspicions of Isaac Newton, who, despite his law of universal gravitation, found the idea of "action at a distance" to be philosophically absurd (Newton, 1693). Even in modern physics, the most powerful analogy for GR is the "river model," which describes objects as carried along by the flow of spacetime (Visser, 1998). The crucial difference is that in these previous models, the river was either a flawed mechanical aether or a pure mathematical analogy. Our model of vortex gravity provides the missing piece: the river is real, and it is the UEF.

This model of gravity as a physical current succeeds where older "push gravity" models have failed for several reasons. First, it avoids the historical problem of drag. Classical models, such as the Le Sage model, relied on a flux of corpuscles that would have produced significant drag on celestial bodies (Le Sage, 1784). Vortex gravity avoids

11

this because an object in free fall is not fighting a barrage of particles; it is carried along *with* the UEF's gravitational current. Because there is no relative motion between the object and the field that carries it, there is no drag. This is possible because the UEF is treated as a superfluid-like medium capable of supporting coherent motion without dissipative loss (Barceló, Liberati, & Visser, 2005) (*see* Chapters 3 and 7).

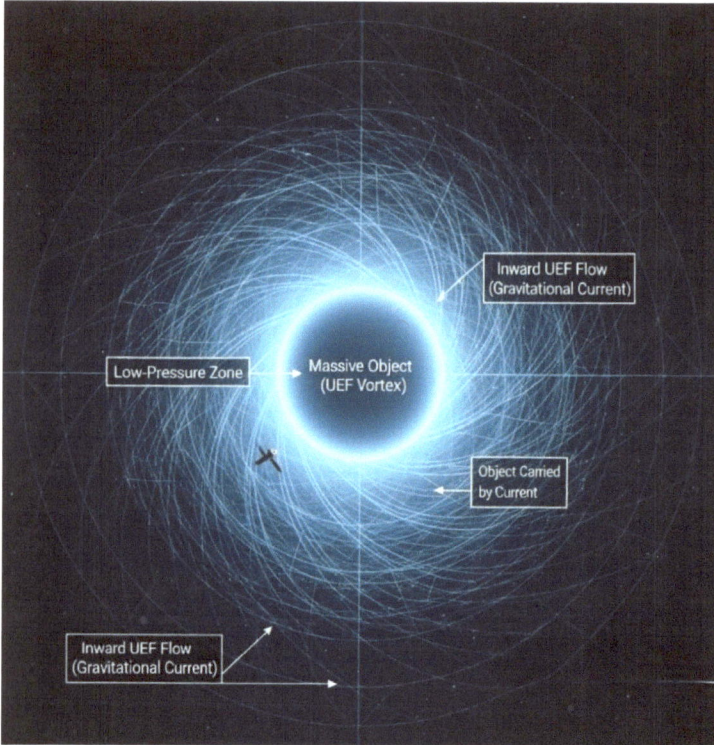

Figure 1: Vortex Gravity. This illustration shows gravity as a real, physical flow of the UEF. A stable spinning vortex of matter (such as a planet or star) creates a persistent low-pressure zone in the field. This pressure difference causes the surrounding, higher-pressure UEF to flow inward, creating the "gravitational current." This inward flow is what we experience as gravity, which carries objects along. The curved streamlines indicate the stabilizing rotation of the central vortex. In this model, gravity is the tangible, fluid-dynamic flow of the universe's energetic substance.

Second, vortex gravity resolves the "unresolvable" thermal problem. In older models, a body that is constantly bombarded by energetic particles heats up to catastrophic temperatures. Our framework avoids this by proposing that a stable vortex of matter is in a state of perfect resonant coherence with the ambient UEF. Similar to a tuning fork that vibrates sympathetically with an incoming sound wave of the same frequency, a particle does not chaotically absorb the field's energy as heat; instead, it responds coherently to it. The gravitational current is a coherent, non-thermal flow of the field, and the particle is effectively carried by this flow without absorbing its energy as random heat, thereby resolving the thermal paradox.

By treating gravity as a current caused by a pressure gradient, vortex gravity reintroduces causal intuition into gravitational theory and restores the possibility of unifying gravity with the other forces through a common substrate: energy in motion within a resonant, continuous field.

2.2 Recovering Newtonian Gravity

Having provided a causal, substance-based origin for the gravitational force in the UEF, the model must now demonstrate that it can recover the established mathematics of classical mechanics. The goal of this section is to prove that the simple fluid dynamics of the gravitational current naturally lead to the two core principles of Newtonian physics: the inverse square law and the derivation of the gravitational constant (G).

2.2.1 Recovering the Inverse Square Law

The mechanism of vortex gravity also provides a direct, causal derivation for the most fundamental observation of gravity: the inverse square law (F $\propto$ $1/r^2$). In vortex gravity, this principle is not a random mathematical rule; it is an expression of the geometric conservation of flow (flux) in a three-dimensional field.

1. The Source: A massive object (the UEF vortex) acts as a low-pressure sink, creating a constant, uniform, and stable gravitational current that flows inward from all directions. This pressure field is not arbitrary; it is the unique energetic solution that minimizes the UEF's stored energy while containing the vortex's mass deficit. This energetic constraint forces the pressure field to adhere to the $P(r) \propto 1/r$ form (the solution to Poisson's Equation).

2. Conservation of Fluid Flux: In fluid dynamics, if a substance is flowing uniformly inward toward a center point, the total volume of fluid (Q) passing through any sphere surrounding that center must be conserved, regardless of the sphere's radius (r).

$$Q = \text{Velocity} \times \text{Area}_{\text{sphere}} = \text{Constant}$$

3. The Geometric Constraint: The surface area of any sphere in 3D space is proportional to the square of its radius ($A = 4\pi r^2$)

4. The Inverse Law: To keep the total flux (Q) constant, the velocity of the gravitational current (v_g) must be inversely proportional to the area it is spread over:

$$v_g \propto \frac{1}{\text{Area}} \propto \frac{1}{r^2}$$

Since, in the CVD framework, the acceleration of a falling body is determined by the speed and acceleration of this gravitational current, Newton's Law of Universal Gravitation is a necessary consequence of the conservation of fluid flow in a three-dimensional medium. The inverse-square law is thus transformed from a mysterious universal rule into a simple statement of how fluid must distribute itself in space.

2.2.2 A Physical Origin for the Gravitational Constant (G)

In both Newtonian mechanics and Einstein's GR, the gravitational constant (G) is treated as a fixed, fundamental quantity, that is, a universal conversion factor that relates mass to gravitational force or spacetime curvature. It appears in Newton's law of universal gravitation as:

$$F = \frac{G m_1 m_2}{r^2}$$

and in Einstein's field equations as:

$$G_{\mu\nu} = (8\pi G / c^4)\, T_{\mu\nu}$$

However, in both paradigms, the origin of G remains unexplained: it is empirically measured rather than theoretically derived. Its numerical value:

$$G = 6.67340 \times 10^{-11} \text{m}^3 \text{kg}^{-1} \text{s}^{-2}$$

is inserted into the equations as given. This has long been a source of conceptual discomfort for physicists, some of whom, such as Dirac (1937), have speculated that G might vary over cosmic time. The dimensionality suggests that it mediates the relationships among energy (mass), space, and time. In CVD, this dimensional interplay arises naturally from the structure of the field:

- Mass is interpreted as a coherent energy vortex in a universal field.

- Distance measures the spatial separation between these vortices.

- Time represents the internal oscillatory phase of the field itself.

Thus, G is not imposed externally; it emerges as a consequence of the field's physical configuration, vibrational coherence, and resistance to compression under energetic strain. It is essentially a macroscopic measure of the intrinsic properties of the UEF that arises from two field-based quantities:

1. The Energy Density (ρ_{UEF}) of the UEF, which determines how much "inward pressure" the field stores per unit volume, and

2. The Compressibility or Elastic Modulus (K_{UEF}) of the field, which sets the rate at which pressure gradients can propagate through the medium.

In this view, G is understood as a measure of the field's effective gravitational stiffness—that is, how efficiently a mass-induced pressure gradient can accelerate other masses.

$$G \propto K_{UEF}$$

This parallels the measurement of the speed of sound in a fluid:

$$v_{sound} = \sqrt{\frac{K}{\rho}}$$

In this analogy, just as sound propagates through a medium based on its compressibility and density, gravitational effects propagate through the UEF as pressure disturbances. A "stiffer" UEF transmits gravitational effects more strongly, resulting in a higher effective G. This relationship can be expressed conceptually as:

$$G \propto f\left(\frac{K_{UEF}}{\rho_{UEF}}\right)$$

where G is shown to be a function of the UEF's intrinsic properties, not a fundamental constant of nature.

3. A Substance-Based Ontology: Reinterpreting General Relativity

Having established that gravity emerges from a universal field, we can now address the broader ontological divide between GR and CVD. GR is a purely geometric theory: spacetime itself is the actor, and matter merely follows its curvature. In contrast, CVD is a substance-based theory in which space serves as a static coordinate system. Physical phenomena arise from the dynamics of the UEF, a physical energetic field. This shift from geometry to substance reshapes the interpretation of all gravitational phenomena.

3.1 Force vs. Geometry: A New Look at the Equivalence Principle

In GR, gravity is treated as an "apparent" force; an object in free fall simply follows the straightest possible path (a geodesic) through curved spacetime. In CVD, gravity is a real, physical force arising from pressure gradients in the UEF. This leads to a profound conceptual shift. In GR, free fall is inertial motion. In CVD, it is continuous acceleration caused by a real pressure gradient.

Our framework is therefore in direct tension with the philosophical interpretation of the Einstein Equivalence Principle (EEP). The EEP states that the laws of non-gravitational physics in a small, freely falling laboratory are indistinguishable from those in an inertial frame far from any gravitational source. While a powerful and experimentally verified principle, GR uses it to make a profound philosophical leap: if the effects of gravity can be locally eliminated by the choice of a reference frame, then gravity itself must not be a fundamental force but an artifact of the geometry of spacetime.

Our framework rejects this philosophical leap. We contend that the local undetectability of a force does not equate to its absence. In the classic example of a person in a free-falling elevator, the person feels weightless because the UEF's gravitational field carries both the person and the elevator at the same rate, akin to two boats being carried by the same current.

This same principle explains why all objects fall at the same rate: the current pushes harder on a more massive object, such as a boulder, but this stronger push is perfectly counteracted by the boulder's greater resistance to motion (its inertia) (*see* Chapter 4). For a feather, the gentle push is perfectly matched by its tiny resistance, resulting in the exact same acceleration for both.

3.2 The Nature of Spacetime and Spacetime Curvature

In GR, spacetime is a dynamic, four-dimensional manifold that can be bent and warped. As physicist John Wheeler famously summarized, "Spacetime tells matter how to move; matter tells spacetime how to curve" (Misner, Thorne, & Wheeler, 1973). From this perspective, space and time are active players in the unfolding of the cosmos.

In CVD, we return to a more classical view by separating these two roles. We posit that space is a static, absolute background that serves only as a passive coordinate system—a fixed stage or arena. Within this arena lies the UEF, a dynamic and energetic entity that is the true actor. All physical phenomena, from the motion of particles to the slowing of time, arise from interactions within this field, not from the geometry of space itself. In this model, the properties associated with spacetime, such as the "stretching" of distance or the "slowing" of time, are physical effects caused by the gravitational currents of the UEF.

Thus, what we perceive as the curvature of spacetime is an effective mathematical description of the pressure and flow gradients within the UEF. To visualize this, consider a whirlpool in a lake. An observer focused solely on the surface geometry might describe the whirlpool as a static, curved depression. CVD, however, describes the underlying cause—the depression is being actively created and sustained by the spinning flow of the fluid. What GR calls "curvature" is simply the visible, geometric consequence of the UEF's underlying hydrodynamics.

3.3 Gravitational Time Dilation

In GR, time is said to slow down in a strong gravitational field, a phenomenon first confirmed with high precision by the Pound-Rebka experiment (1959). CVD interprets this not as a warping of time but as a direct physical effect on the processes we use to measure it.

As the UEF flows inward to form a gravitational current, its substance becomes highly compressed and denser in the vicinity of a massive vortex. This high-density field provides greater resistance to processes occurring within it, physically slowing all rates of change. The situation is analogous to running a race underwater versus in the air: the runner's fundamental ability is the same, but the denser medium slows the pace. Therefore, in CVD, time dilation is a direct, measurable consequence of the increased density of the field.

3.4 Gravitational Lensing: Bending Light in a Physical Medium

If the UEF becomes denser near a massive vortex, it acts as a medium with a variable refractive index, where the paths of light and matter are bent by the changing density of the field through which they travel. This principle provides a direct, physical mechanism for another famous confirmation of GR: gravitational lensing. First famously observed during a solar eclipse by Sir Arthur Eddington (1919), gravitational lensing is the effect in which the gravitational field of a massive object bends light from a distant source. In GR, this is explained as light simply following the straightest possible path—a geodesic—through the spacetime that the massive object has curved.

To account for gravitational lensing, the CVD framework applies the physical mechanism we just established for gravitational time dilation. When a photon from a distant source passes through a dense region of the UEF, its path is bent or refracted. This is analogous to how light is bent when it passes from a thin medium, such as air, into a denser medium, such as water, causing a straw to appear "bent."

Thus, gravitational lensing is a tangible and predictable phenomenon of light refraction. The "lens" is not a warped coordinate system; it is a real, physical region of a denser universal field—the very same denser field responsible for slowing time.

3.5 Bridging the Models: From Fluid Dynamics to an Emergent Metric

The critical test for our model is whether it can recover the mathematical predictions of GR. We propose that this is possible, in principle, by reinterpreting GR's core mathematical concepts as effective descriptions of the UEF's fluid dynamics. This section provides a "dictionary" for translating the abstract geometry of GR into the tangible physics of our framework.

3.5.1 Geodesics

In GR, a geodesic is the straightest possible path that an object can take through curved spacetime. In our model, a geodesic is simply the path of least resistance for an object being propelled by the UEF's gravitational currents. The situation is analogous to a small boat being guided by a river's currents. The path it takes—the geodesic—is a direct reflection of the river's dynamic currents.

3.5.2 The Metric Tensor as a Field Descriptor

The mathematical machine of GR is the metric tensor, which defines the geometry of a system by providing the rules for measuring distance and time.

$$ds^2 = g_{00}(cdt)^2 + \sum_{i=1}^{3}\sum_{j=1}^{3} g_{ij}dx^i dx^j$$

In GR, the metric tensor is a 4x4 matrix of numbers at every point of spacetime. It consists of a one-time component and three space components. The time component (g_{00}) reveals the rate of time, where changes signify time dilation. The space components (g_{ij}) reveals the geometry of space. If those values change, it means that space is curved. To intuitively understand this, we can imagine space as a flexible rubber sheet printed with a perfect grid, similar to graph paper. In an empty, flat space, every square on the grid is identical. The "metric" is simply the rule that this uniform grid uses to measure the distance between any two points. In GR, a massive object, such as a star, warps this sheet, stretching and distorting the grid around it. In this respect, the metric tensor is akin to a tiny instruction manual attached to every point on the sheet. It describes the exact shape and size of the local grid squares and explains how curvature has altered the measurement rules.

In CVD, we propose that GR's metric tensor can be understood as an effective mathematical tool for describing the local pressure, density, and flow of the UEF. The time component of the metric (g_{00}) corresponds to the local pressure and density. It governs the rate at which physical processes unfold, an effect known as gravitational time dilation. The space components (g_{ij}) correspond to the spatial pressure gradients and flow patterns of the gravitational currents. They describe the direction and intensity of inward flow at every point. These components govern the deflection of an object's trajectory—the "bending" of its path through space—as it is carried along by the gravitational currents. Thus, the metric tensor is transformed from a map of warped spacetime geometry into a dynamic "weather map" of the UEF, showing its pressure, density, and current.

3.5.3 The Einstein Field Equations as Emergent Hydrodynamics

The Einstein Field Equations (EFEs), $G_{\mu\nu} = (8\pi G/c^4)\, T_{\mu\nu}$, form the mathematical heart of GR. In the standard interpretation, they describe exactly how the distribution of mass and energy dictates the curvature of spacetime. The right side of the equation, the stress-energy tensor ($T_{\mu\nu}$), represents the matter and energy content of the system. The left side, the Einstein tensor ($G_{\mu\nu}$), represents the resulting curvature of the geometric fabric of spacetime.

CVD proposes a new physical interpretation of this celebrated equation: the EFEs provide an effective macroscopic description of UEF hydrodynamics, detailing how energetic vortices generate gravitational currents. In this new physical picture, the two sides of the equation represent a direct, physical cause-and-effect relationship. The right side ($T_{\mu\nu}$) describes the source of the disturbance: the density, pressure, and motion of the energetic vortices that constitute matter. The left side ($G_{\mu\nu}$) describes the field's collective response: the resulting large-scale currents—flow patterns and intensities— that we observe as gravity.

This cause-and-effect relationship is perfectly analogous to the Ideal Gas Law ($PV = nRT$). The gas law is an elegant macroscopic equation that relates simple properties such as pressure and volume. It emerges from the complex, chaotic, and untraceable statistical mechanics of numerous microscopic collisions. To predict the gas's overall behavior, it is not necessary to track every particle. Similarly, we propose that the elegant geometric laws of the EFEs emerge as a stable, large-scale description of the UEF's underlying and likely very complex hydrodynamics. In this view, the EFEs are the Ideal Gas Law for the UEF, providing an accurate macroscopic recipe for gravity without revealing the microscopic, fluid-dynamic reality from which it arises.

Now that we have established the fundamental "rules of the road" that govern how matter interacts, we are now equipped to explore the origin of the road itself: the grand top-down cosmogenesis that gives birth to universal structure.

4. Cosmogenesis: A Top-Down Alternative

The standard cosmological model describes a bottom-up universe: a hot, dense, nearly uniform beginning, seeded by quantum fluctuations amplified by inflation, with gravity patiently assembling structure over billions of years (Guth, 1981). While it successfully reproduces many observations, this picture relies on speculative components, such as inflationary fields and cold dark matter particles, whose physical nature remains uncertain.

CVD offers a radically different, top-down narrative grounded in the dynamics of the UEF (described in greater detail in Chapter 3). In this model, the fundamental substrate of the universe is the UEF itself. Dark matter, rather than being a new particle species, is the UEF organized into a vast, stable, and slowly rotating whirlpool or "cosmic gyre."

4.1 Phase One: Birth of the Supermassive Black Hole

The primordial UEF begins in a high-energy, unstable "false vacuum" state.[2]

---

[2] One can understand the concept of a "false vacuum" through the analogy of being in a valley high up in the mountains. Imagine a ball nestled in a valley on a mountain range. Though locally stable, a movement in any direction would cause it to roll down the mountain. Because the state is locally stable, it is referred to as a "vacuum." However, because there is a more stable, true valley below it, we refer to it as a "false vacuum."

Local instabilities trigger a first-order phase transition,[3] nucleating bubbles of a lower-energy "true vacuum" protected super-vortices—supermassive black holes (SMBHs)—whose formation releases vast amounts of energy. The rapid growth and outward thrust of these bubbles provide a physical driver for inflation, generating the large-scale smoothness of the cosmos without invoking a separate inflaton field.[4]

4.2 Phase Two: Pair Production and the Neutron Epoch

After the initial collapse of the UEF into an SMBH, the surrounding field becomes a turbulent, high-energy environment. This is the crucible in which matter is forged. The intense energy triggers massive pair production of proton ("sources") and antiproton ("sinks") trefoil vortices (*see* Chapter 7), creating a perfect balance of matter and antimatter.[5] At this extreme pressure and energy density, the volatile antimatter is not annihilated but is instead converted into the colossal energy of the Coherence Dividend, a process that drives the formation of the stable, composite neutron (*see* Chapters 3, 5, and 7). This immense energy release, derived from the mass of antimatter, powers the quasar that marks the birth of a galaxy.[6]

It is in this formative moment that a new conservation law emerges: the Standard Model's Law of Conservation of Baryon Number. In CVD, this conservation law is not yet applicable. Instead, our framework proposes that it is an emergent property of

---

[3] All phase transitions describe how a substance changes state. The first-order phase transition is a familiar, abrupt change that involves "latent heat," like boiling water turning into steam. Second-order phase transitions, in contrast, are more gradual, like a block of iron transforming into a magnet. When iron is hot, it is not magnetic. It is only when it cools below a critical temperature that its magnetic qualities re-emerge smoothly and continuously. In the context of our model, we claim that the energy of cosmic vortices does not abruptly shatter into nucleons but instead condenses or settles into these more stable states as local energy conditions become favorable.

[4] While we do not endorse the view that a runaway first-order phase transition led to cosmic inflation (*see* Chapter 3), cosmic inflation is conceptually consistent with our top-down cosmogenesis model.

[5] In our fluid-dynamic model, the terms "source" and "sink" describe the fundamental flow patterns induced by charged vortices. A source (positive charge, like a proton) is a vortex whose geometry perpetually generates an outward flow of the field, analogous to a spring continuously emitting water. A sink (negative charge, like an antiproton or electron) is a vortex whose geometry perpetually generates an inward flow, analogous to a drain continuously drawing water in. These enduring flow patterns are the physical reality that we measure as an electric field, as described more fully in Chapter 6.

[6] The composite nature of the neutron immediately raises two critical questions about its mass: why it is not twice the mass of a proton, and why it is slightly heavier than a proton. These questions are answered by the principles of the Coherence Dividend and environmental stability. A full, detailed explanation is provided in Chapter 6.

neutrons. The proton and antiproton vortices, generated from the same UEF substrate, interlock to form a new, stable topology. From this point onward, the baryon number emerges as a conserved topological law of the UEF, with the neutron setting the baseline.

This model echoes but reinterprets earlier speculations about the origin of baryon asymmetry. Dirac's "sea" of particles and antiparticles envisioned that the vacuum itself could generate matter–antimatter pairs (Dirac, 1930), while Sakharov's baryogenesis conditions posited that CP violation, nonequilibrium dynamics, and baryon number violation were necessary to explain the predominance of matter (Sakharov, 1967). In contrast, CVD proposes that baryon conservation is an emergent property of neutrons. In this respect, the neutron plays the role of a "keystone particle": its formation marks the moment when the baryon conservation law crystallizes into existence.

However, this "Neutron Epoch" was a transient state. As this sea of matter expanded and cooled, moving away from the high-pressure zone of the central SMBH, the free neutrons were no longer in the environment that sustained them. Like a deep-sea creature brought to the surface, a free neutron becomes unstable in the low-pressure vacuum of empty space. Its decay is analogous to the spontaneous crystallization of a supercooled liquid, in which a high-energy, unstable state suddenly snaps into a lower-energy, more ordered state, releasing energy in the process.[7]

This decay initiated a crucial "race against time" that determined the ultimate composition of the galaxy:

- A significant fraction of the free neutrons decayed back into protons, contributing to the vast clouds of hydrogen that permeate the universe.

- The remaining neutrons were "saved" from decay by being captured by protons to form stable nuclei, primarily Helium-4 and deuterons. Within these nuclei,

---

[7] The Stefan problem is a classic mathematical framework in physics, first formulated by Josef Stefan (1889) to describe the position of a moving boundary between two different phases of a substance, such as the surface of an ice cube melting in water. While melting requires an input of energy (it is endothermic), a better analogy for the neutron's energy-releasing decay is the opposite process: spontaneous crystallization. A "supercooled" liquid is a high-energy, metastable state that can be triggered to suddenly "freeze" into a more ordered, lower-energy solid. This process is exothermic—it releases a burst of energy (latent heat) as it finds its more stable configuration. In CVD, the decay of a neutron is modeled as this type of Stefan problem, where the unstable, "supercooled liquid" state of the free neutron "crystallizes" into the more stable, "solid-like" geometry of the proton, releasing its stored strain energy as the Coherence Dividend.

they could finally relax into their true, non-strained, and perfectly stable ground state.

This process naturally forges a universe composed of stable nuclei (containing the "hidden" antimatter) and a vast surplus of free protons (hydrogen). It is a complete, "top-down" cosmogenesis in which the visible galaxy, its central black hole, and its dark matter halo are all born together in a single formative event. Having established a model for the formation of these cosmic structures, we can now explain their observed dynamic behavior, beginning with the puzzle of galaxy rotation curves.

5. Solving the Galaxy Rotation Curve Problem

One of the most significant observational challenges for GR is the measurement of galaxy rotation curves. GR predicts that the orbital velocities of stars should decrease with distance from the galactic center. However, observations have consistently shown a flat rotation curve, in which stars in the outer regions orbit at the same speed as those in the inner regions (Rubin & Ford, 1970). The standard cosmological model resolves this discrepancy by postulating that every galaxy is embedded in a massive, invisible halo of dark matter, whose extra gravity holds the outer stars in their rapid orbits.

The "dark matter halo" is now a cornerstone of the Lambda-Cold Dark Matter (ΛCDM) model of cosmology. The term is somewhat misleading, as it is not merely a "halo" or "ring" surrounding the galaxy. Instead, it is envisioned as a vast, roughly spherical cloud of invisible matter that permeates and envelops the entire visible galaxy, extending far beyond its luminous edge. The visible galaxy is understood to be a small, dense concentration of baryonic matter (stars, gas, etc.) that has condensed *within the center* of a much larger, more massive dark matter structure.

This cloud is believed to outweigh the visible galaxy by a factor of five to one; however, its nature remains a complete mystery, as it does not interact with light in any way. Physicists theorize that it is composed of a new, undiscovered type of subatomic particle. This hypothesis is not just an *ad hoc* fix for galaxy rotation, as the gravitational scaffolding provided by these halos is now considered essential for explaining the large-scale structure of the universe and patterns in the Cosmic Microwave Background (CMB). CVD provides a physical explanation for the dark matter halo.

In CVD, the dark matter halo is not a cloud of undiscovered particles. It is the vast, coherent region of the primordial UEF from which the visible galaxy condensed, stabilized by a finite galactic boundary (Figure 2).

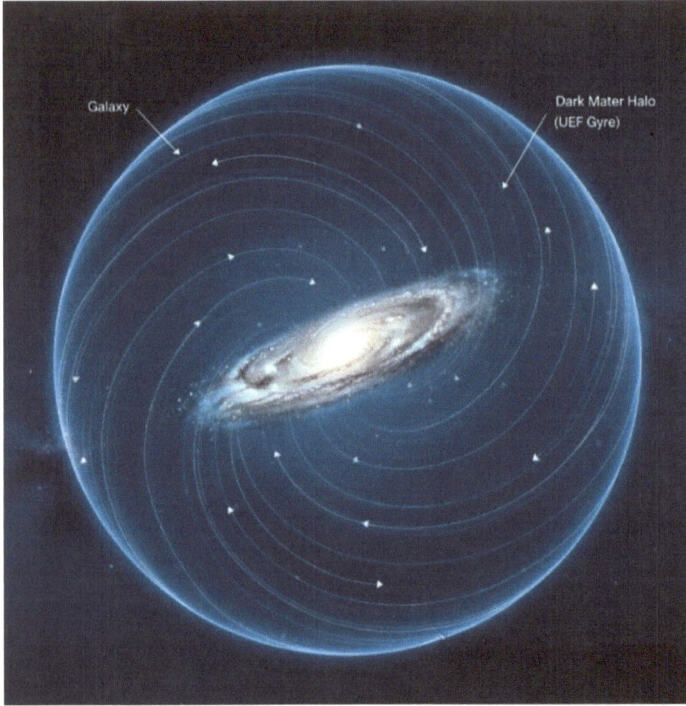

Figure 2: The Dark Matter Halo. This figure illustrates the CVD's "top-down" model of a galaxy, where all components emerge as different states of the UEF. The visible spiral galaxy precipitates in a thin disk around a central supermassive black hole. The entire system is enveloped by a dark matter halo, which represents a vast, gravitationally active region of the UEF itself, organized into a slowly rotating "cosmic gyre" by the Genesis Event. The flattened shape of the halo is a direct consequence of this rotation, and its immense systemic gravitational current holds the visible galaxy together.

We propose that the total gravitational influence on a star is the sum of two distinct gravitational currents arising from different parts of this UEF system. The total gravitational force, F_{total}, experienced by a star at any given distance (r) is the sum of the force from the visible matter, $F_{local}(r)$, and the force from the surrounding halo, $F_{systemic}(r)$:

$$F_{total}(r) = F_{local}(r) + F_{sytemic}(r)$$

Here, the systemic gravitational current is directly generated by the boundary-stabilized pressure geometry of the UEF, which forms a vast, slowly rotating vortex or "cosmic gyre." While this systemic current is more diffuse, its scale is immense, and its strength does not diminish rapidly with distance across the galactic disk.[8]

---

[8]The reason for this difference in how the force diminishes lies in the geometry of the source. Local gravity, generated by a concentrated object like a star, radiates its influence outward in all directions. This influence must spread out over the surface of an expanding sphere, causing its strength to decrease rapidly according to the inverse-square law ($1/r^2$). This is analogous to how the light from a single light bulb gets dimmer very quickly as you move away from it. Systemic gravity, in contrast, is not generated by a central point. It is the collective inward pressure exerted by the entire surrounding halo of the UEF. An object in the outer galaxy is not just being pulled by the center; it is being pushed inward by the immense, distributed mass-energy of the halo that surrounds it. The geometry of this interaction is

At large radii, the local gravitational current from visible matter becomes negligible; however, the force from the massive halo's systemic current remains strong and nearly constant. This immense and persistent contribution is what carries the outer stars in their rapid orbits. The observed flat rotation curves are thus the direct observational signature of these two nested gravitational currents.

6. The Nature of Black Holes: Resolving the Three Great Paradoxes

In GR, the black hole is a nexus of profound conceptual paradoxes in which the theory itself seems to fail. Our framework resolves these paradoxes by replacing the abstract, mathematical black hole of GR with a tangible, physical object: a finite, dynamic, maximally coherent super-vortex of the UEF.

6.1 The Singularity Problem

According to GR, the gravitational collapse of a sufficiently massive star is predicted to inevitably lead to a singularity: a point of zero volume and infinite density where spacetime curvature diverges and the known laws of physics cease to apply (Penrose, 1965). In physics, such infinities are not physical predictions but indicators of an incomplete model. They signal that an extrapolation has been pushed beyond its domain of validity.

CVD resolves this problem by eliminating the assumption of arbitrary compressibility. In this framework, the UEF is not an abstract continuum but a structured, elastic medium possessing a finite coherence scale (*see* Chapter 7). As gravitational collapse proceeds and shear and curvature increase, the system encounters a coherence floor beyond which further compression is dynamically forbidden. At this threshold, the effective impedance of the medium diverges, preventing infinite inflow velocities or unbounded energy density.

Instead of collapsing to a mathematical point, the core of the black hole settles into a phase-stabilized state: a maximally coherent, finite-density vortex configuration determined by the intrinsic stiffness and topology of the UEF. Rather than being trapped indefinitely, excess energy is redirected through boundary-mediated phase transitions

---

more like the pressure one feels deep in the ocean, which depends on the entire column of water above and around, not just a single point. This results in a force that diminishes much more slowly with distance, remaining nearly constant across the outer regions of the galaxy.

into subsidiary field degrees of freedom. Thus, the GR singularity is replaced by a well-posed physical object with finite extent and regulated internal dynamics.

In this view, the singularity is not a feature of nature but a symptom of neglecting the coherence structure of the underlying medium. Once this structure is accounted for, gravitational collapse terminates naturally, and the black hole interior remains globally regular.

6.2 The Information Loss Paradox

The information loss paradox is one of the most challenging problems in modern physics, as it represents a direct contradiction between two of our most successful theoretical frameworks. Quantum mechanics demands that information is never fundamentally destroyed (the principle of unitarity). GR, however, implies that anything crossing a black hole's event horizon is permanently removed from the observable universe. Hawking's discovery that black holes emit thermal radiation and slowly evaporate over immense timescales appeared to sharpen this contradiction: if the radiation is purely thermal and uncorrelated with the infalling matter, then the information is lost forever (Hawking, 1975).

In CVD, this paradox does not arise. Information is not an abstract bookkeeping quantity but a physical property of geometry and coherence. A black hole is a maximally coherent super-vortex of the UEF, and all information associated with infalling matter is encoded in the precise geometric, topological, and resonant state of this vortex and its surrounding boundary layers. Crucially, there is no absolute separation between "inside" and "outside" the black hole in the GR sense. This is because in CVD, the event horizon is not a metaphysical boundary. Instead, it is a dynamical surface defined by competing field propagation speeds. The vortex, its boundary, and the surrounding field constitute a single, continuous physical system. As a result, information is neither destroyed nor sequestered; it is redistributed.

The outgoing Hawking radiation—interpreted here as a form of boundary dissipation or field friction arising from the interaction between the rotating super-vortex and the surrounding field—is subtly modulated by the internal geometric state of the system. Over cosmological timescales, this modulation allows information to be gradually imprinted back into the external field environment. Information preservation is therefore not miraculous or acausal; it is a direct consequence of the continuity and coherence of the underlying medium.

Thus, the information loss paradox dissolves once black holes are treated as physical objects embedded in a real field substrate rather than as abstract regions excised from spacetime.

6.3 The No-Hair Theorem

The classical "no-hair theorem," famously summarized by Wheeler (Misner, Thorne, & Wheeler, 1973), asserts that a stable black hole is completely characterized by only three externally observable parameters: mass, spin, and charge. All other information about the matter that formed the black hole—its composition, structure, or history—is presumed to be erased. This conclusion reinforces the information-loss problem by portraying black holes as ultimate information shredders.

CVD predicts the opposite. As the most extreme expression of a stable, coherent vortex in the UEF, a black hole cannot be geometrically featureless. Stability at maximal impedance requires high-order geometric organization, not simplicity. Rather than being "bald," black holes possess rich internal and boundary structure encoded in their topology, symmetry, and coherence patterns.

While the precise symmetry class of these structures remains an empirical question, UFD predicts that black hole vortices should preferentially adopt geometrically optimal configurations, consistent with Platonic or quasi-crystalline symmetries. These structures are not arbitrary adornments but the physical carriers of information, determining how energy, coherence, and phase are exchanged with the surrounding field.

This prediction leads to clear and falsifiable consequences. Deviations from pure GR ringdown behavior, anisotropies in horizon-scale dynamics, and structure-dependent modulation of emitted radiation are all natural outcomes of a geometrically structured super-vortex. The absence of such effects would place direct constraints on the UFD/CVD framework.

Taken together, the postulate of a physical, geometric super-vortex resolves not just one but all three of the great paradoxes of black hole physics. The singularity problem, the information loss paradox, and the no-hair theorem are revealed to be artifacts of treating spacetime as an abstract geometry rather than as a structured, coherent medium.

7. Falsifiable Predictions

To be considered scientific, a framework must make novel, falsifiable predictions that distinguish it from the existing paradigm. Here, the CVD framework leads to the following unique, testable predictions.

7.1 The Existence of Longitudinal Gravitational Modes

In GR, gravitational waves are strictly transverse tensor perturbations of spacetime. Because spacetime is not a physical medium and admits no pressure, flow, or internal structure, longitudinal modes—disturbances aligned with the direction of propagation—are fundamentally forbidden. The absence of longitudinal gravitational modes is therefore not an empirical result but a direct structural consequence of GR's ontology.

UFD leads to a different conclusion. In the UFD framework, gravity is not a manifestation of spacetime curvature but arises from pressure geometry and coherent flow constraints in a real physical medium: the UEF. Gravitational radiation, however, does not propagate through the UEF itself. Because the UEF is incompressible, it supports no propagating compression waves. Instead, gravitational waves propagate through the ULF (*see* Chapter 2) at the speed of light, carrying energy and radiation consistent with relativistic causality.

Although the UEF does not support propagating waves, it imposes geometric and topological constraints on the surrounding fields. Coherent vortical structures in the UEF generate flow-aligned pressure geometries that act as boundary conditions on ULF dynamics. As a result, gravitational waves propagating in the ULF need not exhibit purely transverse tensor polarization. Instead, they may acquire longitudinally correlated components—not as independent radiative degrees of freedom, but as constraint-induced projections of UEF geometry onto ULF wave propagation.

These longitudinally correlated components do not involve oscillations in mass density or sound-like compression waves. Instead, they arise from coordinated pressure–velocity structure imposed by the incompressibility constraint of the UEF and encoded in the polarization and phase structure of ULF gravitational radiation. Such effects are strictly excluded in general relativity, where no physical medium exists to impose pressure geometry or flow constraints.

This leads to a clear, falsifiable prediction: gravitational-wave signals should exhibit measurable deviations from purely transverse tensor polarization, including flow-aligned or longitudinally correlated components tied to source geometry. These components are expected to be:

- weaker than the dominant transverse tensor modes,

- most prominent near highly coherent or rapidly evolving sources (such as black-hole mergers or intense mass-current reconfigurations),

- detectable as systematic phase or amplitude correlations inconsistent with GR's tensor-only polarization structure.

The experimental detection of such longitudinally correlated gravitational-wave components would constitute decisive evidence for gravity as a hydrodynamic, medium-based phenomenon rather than a purely geometric property of spacetime.

7.2 A Near-Luminal Speed for Gravitational Waves

While GR postulates that gravitational waves must travel at precisely the speed of light (c), CVD makes a different and more nuanced prediction. Current observations have confirmed that the speed of gravity is indeed almost identical to the speed of light. Our model, based on its universal scaling principle, provides a direct physical explanation for this finding.

The speed of any wave is determined by the ratio of its medium's stiffness to its density ($v = \sqrt{K/\rho}$). Our model posits that the UEF is as stiff as it is dense relative to the medium through which light travels. As a result, the scaling factors cancel out, and the CVD framework predicts that the speed of a gravitational wave (vg) should be nearly identical to the speed of light.

However, an important distinction remains. A wave traveling through a dense, energetic medium will constantly interact with the medium's background energy. This field self-interaction creates a minuscule time delay, thereby slowing the wave's effective propagation speed. This is analogous to how light travels more slowly through a dense medium, such as glass, than through a vacuum; the constant interaction of the photon with the electromagnetic field of the glass atoms slows its effective speed.

This leads to a falsifiable prediction: because a gravitational wave is a disturbance in the incredibly dense UEF, it must be subject to this self-interaction. Therefore, future, more precise measurements will reveal that gravitational waves travel infinitesimally slower than light. The discovery of any deviation from an exact 1:1 ratio would be a direct violation of GR and a confirmation of the CVD framework's view of gravity as a wave in a real, physical medium.

7.3 The Non-Constant Nature of G

In GR, the gravitational constant, G, is considered a fundamental, isotropic, and unchanging constant of nature. Our framework, which posits G to be an emergent property of the UEF, predicts that its value is not absolute. This leads to two distinct, testable consequences:

- Anisotropy: As our galaxy moves through the "cosmic currents" in the UEF, the background pressure of the field may be infinitesimally stronger from one direction. This should be detectable as a small dipole anisotropy in high-precision measurements of G.

- Environmental Dependence: Because G is a function of the UEF's gravitational stiffness; its effective value must therefore vary with local environmental pressure. We predict that measurements of G conducted within a dense galactic cluster will yield a slightly different value than measurements conducted in the near-emptiness of an intergalactic void (like the Boötes Void).

7.4 The Anisotropy of a Black Hole's Gravity

While the gravitational field of a planet or star appears perfectly isotropic (the same in all directions), our framework explains this as a statistical illusion—the averaged effect of trillions of randomly oriented nucleon vortices. However, a spinning black hole is a different kind of object. In CVD, it is a single, massive, and coherent UEF super-vortex. We therefore predict that its gravitational field should not be perfectly isotropic. Instead, it should reflect the underlying toroidal geometry of the vortex. This effect would be measurable as a specific quadrupole moment—a slight "flattening" of its gravitational field, making it effectively stronger at its poles than at its equator.

While this effect is related to the "frame-dragging" predicted by GR, our model provides a direct, intuitive, and physical mechanism for it, grounding it in the geometry of the vortex itself. High-precision measurements of the spacetime around a nearby spinning black hole, for instance, by observing the orbits of stars like those around Sagittarius A*, should reveal this specific geometric signature.

7.5 A New Astronomy: Gravitational Wave Spectroscopy

This final prediction directly tests our model of the black hole as a physical, geometric object. When two black holes merge, the event unfolds in two stages: the first is the merger, and the second is the ringdown. Both should leave unique, detectable signatures.

The initial merging of two immense UEF super-vortices would be a violent and messy fluid dynamic event. We predict that the gravitational waves from this phase would contain signatures of this turbulence, such as chaotic frequencies that deviate from the clean inspiral predicted by GR.

After the initial chaotic merger, the new, larger super-vortex would settle into its final, stable state—a perfect geometric object. As it did, it would "ring down" like a struck bell. GR's no-hair theorem predicts this ringdown will be a simple tone, but our model predicts it to be a complex chord of overtones and harmonics. The specific frequencies of this chord are a direct fingerprint of the black hole's underlying geometry (e.g., an icosahedron, *see* Chapter 3).

The detection of either the initial turbulence or its subsequent geometric chords would provide powerful evidence for our framework.

8. Conclusion

This chapter has introduced CVD, the cosmological-scale branch of the UFD framework, which provides a physical, substance-based foundation for gravity. By modeling gravity as an emergent current within the UEF, CVD offers a concrete, mechanistic alternative to the abstract geometric curvature of spacetime.

At the heart of this new foundation is a more physical definition of the universe's primary substrate. We began by redefining energy not as an abstract scalar, but as the tangible motion and configuration of a Universal Plenum. From this principle, a new theory of gravitational mechanics emerges: gravity is the physical force of an inward-flowing current driven by the pressure differentials created by localized vortex structures. Within this framework, the gravitational constant (G) is no longer treated as a fundamental, arbitrary axiom of nature. Instead, it emerges as a derived property of the UEF, representing the field's intrinsic energy density and elastic tension.

This deeper mechanical framework provides a cascade of solutions on the cosmological scale. It offers intuitive explanations for the origins of dark matter halos and the flat rotation curves of galaxies, reinterpreting them as large-scale "cosmic gyres" or rotations of the UEF. Furthermore, by replacing the mathematical singularity with a physical, maximally coherent super-vortex, CVD resolves the long-standing paradoxes of black hole physics, treating these objects as finite, high-energy phase transitions in the field.

The true significance of this new paradigm lies in its power to unify the disparate realms of physics. By demonstrating that the core mathematics of GR can be recovered as an effective description of the UEF's fluid dynamics, CVD places gravity on the same physical footing as the quantum world. Gravity (as a current) and localized energetic structures (as vortices) are no longer two different realities governed by incompatible rules but are simply different behaviors of the same underlying substance. This resolves the foundational conflict between them and offers a direct, physical path toward the realization of a unified theory of gravity.

Moving forward, the path to validating this framework is clear. The theoretical task is to further formalize the UEF's fluid dynamics, while the experimental task is to test the unique, falsifiable predictions outlined herein, from the existence of longitudinal gravitational waves to the non-constant nature of G over cosmic time.

Ultimately, CVD challenges the modern interpretation of the cosmos, suggesting that gravity is not a property of empty space, but a tangible manifestation of the energetic field from which all structures in the universe arise.

1.3 Vortex Gravity

In this chapter, we introduced Cosmic Vortex Dynamics (CVD), a physical, field-based framework that challenges the geometric interpretation of gravity established by General Relativity (GR). By providing a physical medium for gravity, the Universal Energetic Field (UEF), this chapter lays the groundwork for a more tangible, mechanistic, and unified understanding of the cosmos.

The central thesis of CVD is that gravity is not the curvature of spacetime but a current that flows within the UEF. This substance-based ontology provides a unified physical origin for a wide array of cosmological phenomena:

1. A Physical Definition for Energy and Mass: The framework begins by redefining energy as the physical motion and configuration of the UEF, transforming the Law of Conservation of Energy from an abstract axiom into a principle of

Conservation of Substance. From this, mass emerges as a measure of coherent, localized, relativistic energy, providing an intuitive, causal basis for the relationship.

2. A Novel Theory of "Push" Gravity: Gravity emerges as the force of the UEF's inward-flowing gravitational current.

3. A Bridge to GR: The framework shows how GR's mathematics can be recovered as an emergent, large-scale description of the UEF's hydrodynamics, preserving its predictive power while grounding it in a physical mechanism.

4. A Physical Origin of the Gravitational Constant: The model reinterprets the gravitational constant, G, as a composite value representing the intrinsic energy density and stiffness of the UEF. This transforms a "magic number" of physics into a predictable, physical property.

5. Top-Down Cosmogenesis: Galaxies form not through a bottom-up hierarchy but via the condensation of the UEF, resolving the matter–antimatter asymmetry by embedding antimatter within neutrons as stable composites.

6. A Physical Identity for Dark Matter: The "dark matter" halo is revealed as a cosmic gyre of the UEF, whose systemic gravitational current naturally explains flat galaxy rotation curves without invoking new particles.

7. Resolution of Black Hole Paradoxes: By replacing the singularity with a finite, maximally coherent UEF super-vortex, CVD resolves the singularity problem, the information loss problem, and the no-hair theorem.

8. Novel, Falsifiable Predictions: From longitudinal gravitational modes to geometric harmonics in black hole ringdowns, CVD presents clear experimental pathways by which it can be confirmed or refuted.

The significance of this framework extends beyond any single result. By grounding both quantum and gravitational phenomena in a single underlying field, CVD resolves the foundational conflict between the physics of the very small and the very large. Its testable predictions offer a roadmap for future experimental programs, while its substance-based ontology promises to restore causality and physical intuition to our picture of the cosmos.

If validated, CVD would mark a significant step toward a unified, coherent theory of the universe, with implications that extend beyond cosmology. For decades, the search for a theory of quantum gravity has been the central pursuit of fundamental physics. The failure to reconcile GR with quantum mechanics has led to increasingly abstract proposals, including loop quantum gravity and string theory, which attempt to quantize the geometric structure of spacetime or redefine matter at the Planck scale.

CVD offers us a new path forward. By reinterpreting gravity as the currents and flow dynamics of a universal energetic medium, the need to quantize space is bypassed. Instead, CVD points to a deeper substrate—one in which quantum phenomena and gravitational behavior emerge from the same field-based reality. This removes the false dichotomy between the smooth curves of relativity and the discrete quanta of particle physics. If gravity and quantum effects are simply two scales of a field's coherent behavior, their long-sought-after "unification" may not require a new mathematics but rather a new ontology.

Ultimately, CVD marks a return to physical intuition, to a world where the forces of nature are not abstract constructs but the consequences of the structure and flow of a universal medium. In doing so, it offers a viable path to move beyond the mathematical abstractions of GR.

References

1. Aristotle. (1999). *Physics*. (R. Waterfield, Trans.). Oxford University Press. (Original work c. 330 BCE).
2. Barceló, C., Liberati, S., & Visser, M. (2011). "Analogue gravity". *Living Reviews in Relativity*, 14(3).
3. Bernoulli, D. (1738). *Hydrodynamica, sive de viribus et motibus fluidorum commentarii* [Hydrodynamics, or commentaries on the forces and motions of fluids]. Johann Reinhold Dulsecker.
4. Carroll, S. M. (2004). *Spacetime and Geometry: An Introduction to General Relativity*. Addison-Wesley.
5. Descartes, R. (1644). *Principia Philosophiae* (Principles of Philosophy).
6. Dirac, P. A. M. (1930). A Theory of Electrons and Protons. *Proceedings of the Royal Society A: Mathematical, Physical and Engineering Sciences, 126*(801), 360–365.
7. Dirac, P. A. M. (1937). "The Cosmological Constants." *Nature, 139*(3512), 323.
8. Eddington, A. S. (1919). The deflection of light by gravitation and the Einstein theory. *The Observatory, 42*, 119-122.

9. Einstein, A. (1916). *Die Grundlage der allgemeinen Relativitätstheorie*. Annalen der Physik, 354(7), 769–822.

10. Griffiths, D. (2008). *Introduction to Elementary Particles* (2nd ed.). Wiley-VCH.

11. Hawking, S. W. (1975). Particle Creation by Black Holes. *Communications in Mathematical Physics*, 43(3), 199–220.

12. Krane, K. S. (1988). *Introductory nuclear physics*. Wiley.

13. Kuhn, T. S. (1962). *The Structure of Scientific Revolutions*. University of Chicago Press.

14. Le Sage, G. L. (1784). *Physique Mécanique de la Gravitation*. Geneva.

15. Misner, C. W., Thorne, K. S., & Wheeler, J. A. (1973). *Gravitation*. W. H. Freeman, p. 5.

16. Newton, I. (1693). Third Letter to Richard Bentley.

17. Penrose, R. (1965). "Gravitational Collapse and Spacetime Singularities". *Physical Review Letters*, 14(3), 57–59.

18. Pound, R. V., & Rebka, G. A. Jr. (1959). Gravitational Red-Shift in Nuclear Resonance. *Physical Review Letters*, 3(9), 439–441.

19. Rubin, V. C., & Ford, W. K. Jr. (1970). "Rotation of the Andromeda Nebula from a Spectroscopic Survey of Emission Regions". *The Astrophysical Journal*, 159, 379

20. Sakharov, A. D. (1967). Violation of CP Invariance, C asymmetry, and baryon asymmetry of the universe. *JETP Letters*, 5(1), 24–27.

21. Stefan, J. (1889). Über einige Probleme der Theorie der Wärmeleitung [On some problems of the theory of heat conduction]. *Sitzungsberichte der Kaiserlichen Akademie der Wissenschaften in Wien, Mathematisch-Naturwissenschaftliche Classe*, 98, 473–484.

22. Uzan, J.-P. (2003). "The Fundamental Constants and Their Variation: Observational and Theoretical Status." *Reviews of Modern Physics*, 75(2), 403–455.

23. Visser, M. (1998). Acoustic black holes: horizons, ergospheres, and Hawking radiation. *Classical and Quantum Gravity*, 15(6), 1767–1791.

24. Weinberg, S. (1995). *The quantum theory of fields* (Vol. 1). Cambridge University Press.

Chapter 2: Emergent Cosmology

2.1 Cosmology

Cosmology is the study of the origin, evolution, and ultimate fate of the universe. It is the scientific endeavor that asks the grandest of questions: Where did everything come from? Why is there something rather than nothing? And where is it all going? Since antiquity, humans have crafted creation myths and philosophical models to answer these questions. Over time, our answers have evolved from Earth-centric worldviews to the revolutionary Copernican principle that we do not occupy a special place in the universe. For centuries, the universe was imagined as a static, eternal, clockwork machine. This view held sway until the dawn of the 20th century, when it was radically transformed by modern cosmology.

The modern scientific era of cosmology began with a single discovery. In 1929, the astronomer Edwin Hubble observed that light from distant galaxies was systematically redshifted (Hubble, 1929), with redshift increasing in proportion to distance. Interpreted through the Doppler effect, this suggested that the universe was not static but was rather expanding uniformly in all directions. This discovery gave birth to the Big Bang theory: the idea that the universe began in an unimaginably hot, dense state and has been expanding and cooling ever since.

While the Big Bang theory successfully explained Hubble's observations and the existence of the Cosmic Microwave Background (CMB)—the faint afterglow of the initial fireball—it faced significant challenges. It could not, for instance, explain why the universe is so remarkably uniform on large scales (the "horizon problem"), why its geometry is so perfectly "flat" (the "flatness problem"), or why we do not observe the massive magnetic monopoles predicted to have been created in the early universe (the "monopole problem"). To resolve these issues, the theory of cosmic inflation was introduced in the 1980s (Guth, 1981). Inflationary cosmology proposes that in the first fraction of a second after the Big Bang, the universe underwent a period of hyper-accelerated, exponential expansion, which would have diluted the density of any monopoles to near zero, smoothed out initial irregularities, and set the stage for the universe we see today.

The final significant addition to this framework occurred in the late 1990s when observations of distant supernovae revealed that the expansion of the universe was not slowing down, as would be expected due to gravity, but was accelerating (Riess et al.,

1998). To account for this finding, cosmologists postulated the existence of a mysterious repulsive force called dark energy, which is now believed to comprise nearly 70% of the universe's total energy content. The resulting paradigm, known as the Lambda-Cold Dark Matter (ΛCDM) model, is the Standard Model of Cosmology. Its primary strength is its remarkable ability to explain a vast range of observations, from the CMB to the large-scale structure of galaxies, with a *relatively* simple, six-parameter model.

Notwithstanding its predictive success, ΛCDM is far from a perfect theory. For one, the model relies on the existence of dark matter and dark energy, two entities whose physical nature remains mysterious. Furthermore, recent, increasingly precise measurements of the universe's expansion rate have revealed a persistent discrepancy—the Hubble tension—that suggests a potential crack in the model's foundations. This tension represents a three-part contradiction between the primary, large-scale datasets:

1. "Early universe" measurements using data from the CMB (Planck Collaboration, 2018) infer a slower expansion rate of approximately 67-68 km/s/Mpc.

2. "Local late universe" measurements using distance ladders of stars and supernovae (e.g., the SH0ES team) consistently find faster rates of approximately 73-74 km/s/Mpc (Freedman et al., 2019).

3. "Global late universe" measurements that find the "best fit" for the *entire* 1701-point Pantheon supernova dataset (Scolnic et al., 2022) converge on a "systemic" rate of approximately 70 km/s/Mpc.

This three-part contradiction, in which the local, global, and primordial measurements all disagree, poses an intractable crisis for the Standard Model.

In response to the conceptual and empirical challenges of the Standard Model, this chapter proposes a radical alternative. We contend that the foundational observation of cosmology—the cosmological redshift—is not a consequence of the metric expansion of space; instead, it is an emergent phenomenon arising from the interaction of light with a physical, expanding medium that we call the Universal Light Field (ULF).

In the sections that follow, we will demonstrate how this physical, mechanistic framework can account for the observations that underpin the entire Standard Model of Cosmology, from the apparent acceleration of the universe to a definitive, three-part resolution of the Hubble tension. This leads us to the startling yet intuitive conclusion that the 13.8-billion-year story of our expanding universe is a grand illusion and that the universe is much more ancient than we had imagined.

2.2 Emergent Cosmology: A Unified Framework for Redshift, Dark Energy, and the Hubble Tension

Abstract

This chapter presents a new cosmological framework in which redshift, time dilation, and phenomena attributed to dark energy are emergent properties of a physical field, rather than the metric expansion of spacetime. Building on the previous chapter, which introduced the medium of mass and gravity—the Universal Energetic Field (UEF)—we propose that all matter perpetually radiates an expanding Universal Light Field (ULF), the medium through which light propagates and electromagnetism operates. We demonstrate that a photon's interaction with this expanding field leads to a single, exponential redshift-distance relationship: $z = e^{Hd} - 1$. This law naturally accounts for the apparent cosmic acceleration, provides a physical explanation for dark energy and time dilation, and offers a testable solution to the Hubble tension by employing a two-rate model that accounts for both local and systemic expansion. Furthermore, the framework provides a new cosmic distance ladder and reinterprets the Cosmic Microwave Background (CMB) as the thermal equilibrium of the ULF itself. Ultimately, this work presents a physically intuitive and testable alternative to the Lambda-Cold Dark Matter (ΛCDM) model.

1. Introduction

Edwin Hubble's 1929 observation that light from distant galaxies is systematically redshifted stands as the cornerstone of modern cosmology, as it shifted the scientific paradigm from a static universe to an expanding one. This concept underpins the Lambda-Cold Dark Matter (ΛCDM) model, the Standard Model of Cosmology. While ΛCDM has been remarkably successful at explaining phenomena such as the Cosmic Microwave Background (CMB) (Planck Collaboration, 2018), it faces persistent challenges. These include its reliance on unknown entities, such as dark energy (Peebles & Ratra, 2003), the growing Hubble tension between different cosmic measurements (Freedman et al., 2019), and the philosophical difficulty of an expanding spacetime itself.

This chapter proposes a radical alternative to the Standard Model. Building on the principle established in Chapter 1 — that matter and gravity are manifestations of a Universal Energetic Field (UEF) —we now introduce a secondary, emergent field into the Unified Field Dynamics (UFD) framework: the Universal Light Field (ULF). We contend that all matter perpetually radiates this ULF and that its systemic expansion is the direct physical cause of cosmological redshift, time dilation, and phenomena attributed to dark energy.

Here, we systematically develop this new cosmological paradigm. First, we review the standard interpretation of cosmological redshift. Second, we detail the origin and properties of the ULF, derive a mathematical relationship for cosmological redshift, and show how it accounts for the observed cosmic acceleration. Third, we demonstrate how our emergent model resolves the Hubble tension through a two-rate model that accounts for both local and systemic field expansion. Finally, we discuss the novel, testable predictions of this framework and explore its profound implications for our understanding of the universe.

2. The Standard Paradigm: An Expanding Universe

Modern cosmology is based on a single foundational postulate: the universe is expanding. This idea originated from Edwin Hubble's 1929 discovery that light from distant galaxies is systematically redshifted in proportion to their distance (Hubble, 1929). While initially interpreted as a simple Doppler shift of galaxies moving away from us, this concept evolved with the application of General Relativity into the modern paradigm, indicating that the expansion is not due to galaxies moving through space but rather to the expansion of the fabric of spacetime itself.

This model is mathematically described by the Friedmann-Lemaître-Robertson-Walker (FLRW) metric, which governs a homogeneous and isotropic universe. In this framework, as a photon travels through expanding space, its wavelength is stretched along with the cosmic fabric. This stretching is the cause of the observed redshift, meaning that the redshift we see from a distant galaxy is a direct measure of how much the universe has expanded since the light left that galaxy. The relationship between a galaxy's redshift (z) and the expansion of the universe is formally defined by the cosmic scale factor, $a(t)$:

$$1 + z = \frac{a(t_{obs})}{a(t_{emit})}$$

Here, $a(t_{emit})$ is the scale factor of the universe at the time the light was emitted, and $a(t_{obs})$ is the scale factor at the time it is observed (Carroll, 2004; Ellis et al., 2012). For relatively nearby galaxies, this relationship simplifies to the famous linear Hubble-Lemaître law:

$$z \approx \frac{H_0 d}{c}$$

where H_0 is the Hubble constant, d is the proper distance to the galaxy, and c is the speed of light.[9]

The true complexity of the Standard Model, however, lies in determining the evolution of the scale factor, $a(t)$, over cosmic time. This is governed by the Friedmann equations of GR, which describe how the expansion rate depends on the total energy density of the universe. This "cosmic recipe" is a complex balancing act between multiple, seemingly unrelated components, such as the gravitational "brake" from baryonic matter and dark matter as well as the repulsive "accelerator" of dark energy. The entire model also depends on the precise tuning of density parameters for each of these components (Ω_m, Ω_Λ, etc.) to match observations. The ΛCDM Model is therefore not a simple law but a relatively complex, multi-parameter framework built to describe an evolving cosmic tug-of-war.

Despite its remarkable success in explaining a wide range of cosmic phenomena, the ΛCDM model faces significant conceptual and empirical challenges. To account for the observed accelerating expansion, the model requires the existence of a mysterious dark energy with a repulsive gravitational effect (Riess et al., 1998). Furthermore, increasingly precise measurements of the Hubble constant from the early universe (via the CMB) and late universe (via supernovae) yield conflicting results, a persistent discrepancy known as the "Hubble tension" (Freedman et al., 2019). These continuing challenges suggest that the model may be incomplete and provide strong motivation for an alternative explanation of cosmological redshift.

3. The UFD Cosmology Framework

---

[9] The Hubble-Lemaître law is the fundamental observation in cosmology that a galaxy's recession velocity (v) is directly proportional to its distance (d) from us. This linear relationship is expressed by the equation $v = H_0 \cdot d$, where H_0 is the constant of proportionality known as the Hubble constant. In simple terms, it means the farther away a galaxy is, the faster it appears to be moving away.

In contrast to the geometric paradigm of an expanding spacetime, our framework proposes an emergent cosmology based on the dynamics of a physical field. At the heart of this model is the ULF, a secondary field that emerges from the fundamental UEF. We contend that all matter perpetually radiates this ULF, thereby driving a systemic expansion that directly produces cosmological redshift.

3.1 The Dual Nature of the Universal Light Field

The ULF has two distinct but complementary properties that govern its behavior: the first is systemic and expanding, and the second is local and interactive.

We propose that all energetic vortices, from nucleons to black holes, perpetually and sustainably convert a tiny fraction of their immense internal energy into an expanding field. This "field shedding" is a phase transition in which the dense, substantive energy of a UEF vortex is continuously transformed into the lighter, expanding medium of the ULF (Figure 3). The constant rate of this expansion, when accumulated over cosmic distances, gives rise to the observed cosmological redshift.

Because the ULF is the medium through which light propagates, we can identify it as a modern, dynamic version of the luminiferous aether.[10] This version of the aether is consistent with the null result of the Michelson-Morley experiment because the field is generated by and co-moving with massive bodies, meaning that an observer on Earth is stationary with respect to their local ULF (Michelson & Morley, 1887).[11] This concept

---

[10] The luminiferous aether was a central concept in 19th-century physics. It was a hypothetical, universe-filling substance believed to be the medium through which light waves propagated, just as sound waves propagate through the medium of air. The idea was highly intuitive because all other known waves required a medium to travel through; it was therefore logical to assume that light, as a wave, also needed one. The concept was famously challenged by the Michelson-Morley experiment.

[11] The Michelson-Morley experiment of 1887 was one of the most important experiments in the history of physics. After presuming the aether was a fixed, stationary substance filling all of space, the experiment was designed to detect an "aether wind"—the effect of the Earth moving *through* this fixed aether—by measuring a difference in the speed of light when it traveled in the direction of the Earth's motion versus when it traveled perpendicular to it. The experiment famously found no difference whatsoever. This "null result" was a significant blow to classical aether theory and paved the way for Einstein's Special Relativity, which does not require a static aether. Our emergent framework is consistent with this result because the ULF is not a static, universal aether. Instead, it is a dynamic field generated by and co-moving with massive bodies. An observer on Earth is therefore stationary concerning their local ULF, so no "aether wind" would be detected.

shares a conceptual lineage with the "aether drag" hypothesis proposed by Sir George Stokes in the 19th century to explain this same null result.[12]

Figure 3: The Mechanism of Field Shedding.
This figure illustrates the fundamental dynamic process that drives the expansion of the ULF from the UEF within the UFD framework. A stable, energetic UEF vortex (a nucleon), shown here as a golden sphere, constantly converts a tiny fraction of its total internal energy into the emergent ULF. This process of "shedding" creates a continuous outward flow of a less dense, expanding field, represented by the blue wisps. This perpetual, irreversible transformation is the physical engine that gives rise to both a cosmological redshift and time dilation.

Separate from its systemic expansion, the local behavior of the ULF **is** the force of classical electromagnetism (*see* Chapter 6). In this model, the electromagnetic field (EMF) is not an abstract mathematical map; it is a real, physical field—a dynamic substance— that governs the static, inverse-square law interactions of attraction and repulsion between charged particles in the ULF.

---

[12] The aether drag hypothesis, proposed by Sir George Stokes (1845), was an alternative to the static aether. It posited that the aether was fully "dragged" along by the surface of massive bodies like the Earth. In this model, an experiment conducted on the Earth's surface would be at rest relative to the aether, thus perfectly explaining the null result of the Michelson-Morley experiment. While Stokes' specific model was eventually superseded due to difficulties in explaining phenomena such as stellar aberration, the core concept of a co-moving medium is a direct historical antecedent to the ULF. In UFD, the relevant point is not the 'full drag of a universal medium' but the existence of a *locally generated* ULF domain in which local measurements are stationary.

Thus, a photon's journey is shaped by both aspects of the ULF. The local field governs its emission and absorption, while the systemic, expanding nature of the field governs its propagation over cosmic distances.

3.2 A Mathematical Model of Emergent Redshift

The standard cosmological model explains redshift through the metric expansion of spacetime. Our framework proposes a different physical cause: the interaction between light and an expanding physical field. To demonstrate the viability of this alternative, we develop a mathematical model for redshift based on this core principle.

We begin with the postulate that a photon's wavelength (λ) is stretched as it propagates through the expanding ULF. The fractional change in its wavelength ($d\lambda/\lambda$) over an infinitesimal path length (dr) is proportional to the local expansion rate of the field at that point. For cosmological distances, we can approximate the cumulative impact of the expanding field along a given line of sight with an effective expansion constant, which we will denote as H. This constant represents the average rate of field expansion per unit distance.

Therefore, our governing differential equation for this model is as follows:

$$\frac{d\lambda}{\lambda} = H \cdot dr$$

To find the total redshift for a photon traveling from a source at distance, d, we integrate this expression from the source (where $\lambda = \lambda_{emit}$ and $r = 0$) to the observer (where $\lambda = \lambda_{obs}$ and $r = d$). This yields:

$$ln(\frac{\lambda_{obs}}{\lambda_{emit}}) = H \cdot d$$

Solving for the observed wavelength yields:

$$\lambda_{obs} = \lambda_{emit} \cdot e^{Hd}$$

From this, we can derive the formula for redshift, $z = (\lambda_{obs} - \lambda_{emit})/\lambda_{emit}$:

$$z = e^{Hd} - 1$$

This is the central predictive equation of our emergent cosmology model, as it directly relates the observable redshift (z) of an object to its distance (d) through the effective expansion constant (H). This single formula accounts for both the linear relationship observed locally (Hubble's law) and the nonlinear relationships at large distances attributed to dark energy. For nearby objects, where the total distance d is small, we can use the Taylor expansion for an exponential ($e^x \approx 1 + x$). This yields $z \approx Hd$, a linear redshift-distance relationship that recovers the form of the Hubble-Lemaître law (Hubble, 1929), where our effective field expansion constant H plays the physical role of H_0/c.[13]

The simplicity and elegance of this equation stand in stark contrast to the corresponding formula from the ΛCDM model:

$$d = \frac{c}{H_o} \int_0^z \frac{dz'}{\sqrt{\Omega_M (1 + z')^3 + \Omega_\lambda}}$$

3.3 A New Physical Meaning for the Hubble Radius

In the standard cosmological model, the Hubble radius is defined as the distance at which a galaxy's recession velocity equals the speed of light. It is calculated as $d = c/H_0$. In our framework, where galaxies are not receding, this definition becomes meaningless. However, the quantity itself does not lose its meaning, as it is transformed from a concept of velocity into a fundamental property of the ULF itself.

In our model, the Hubble radius can be considered the characteristic "stretching length" or "e-folding distance" of the ULF. It is the distance a photon must travel for its wavelength to be stretched by a factor of e (Euler's number, ≈ 2.718). This can be seen directly from our exponential redshift law:

$$z = e^{Hd} - 1$$

We have established that our constant H is the physical equivalent of H_0/c. Let us examine what happens when a photon travels a distance d exactly equal to the Hubble radius, c/H_0. Since $d = c/H_0 = 1/H$, substituting this distance into our equation gives

$$z = e^{(H_0/c) \cdot (c/H_0)} - 1,$$

---

[13] In our emergent model of cosmology, the constant, H, is equivalent to the Hubble Constant divided by the speed of light (H_0/c) from the ΛCDM model. The Hubble Constant (H_0) is how astronomers measure the universe's expansion; It does so by relating the distance to the galaxy and its apparent recession speed. Our constant, H, relates the distance a photon travels to its total wavelength stretch, z. To be compatible with the Standard Model, we must divide the calculated recession speed by the speed of light (c) to get the final, correct redshift. Therefore, the term H_0/c represents the "amount of redshift per unit of distance" in the Standard Model. For our simpler equation, $z \approx Hd$, to match the observed facts, our "stretch per unit of distance" constant, H, must be the direct physical equivalent of the Standard Model's H_0/c term.

which simplifies to

$$z = e^1 - 1 \approx 1.718$$

Thus, the Hubble radius is reinterpreted as a fundamental constant of the ULF medium, representing the distance over which the expanding field stretches light by a factor of e. It is a measure of the field's intrinsic rate of expansion.

3.4 Dark Energy and Time Dilation

In addition to providing a new interpretation of the Hubble radius, our exponential redshift law, $z = e^{Hd} - 1$, derived from the single principle of an expanding ULF, provides an immediate and elegant explanation for the phenomenon of "dark energy."

In 1998, observations of distant Type Ia supernovae revealed that these "standard candles" appeared dimmer—and therefore farther away—than predicted by cosmologies that assumed a decelerating universe (Riess et al., 1998). To reconcile theory with data, cosmologists introduced a new repulsive component, dark energy, as an additional term in the ΛCDM model. This "anti-gravitational" agent now dominates the energy budget of the universe, yet its physical nature remains entirely unknown.

Our emergent framework accounts for the same evidence without invoking any new substances. The apparent "acceleration" is not a physical push but the inevitable signature of an exponential redshift law. An exponential curve naturally "bends upward." At large distances, small increases in distance correspond to disproportionately large increases in redshift. Thus, the very feature interpreted as cosmic acceleration in the ΛCDM arises in our framework as a geometric consequence of photon propagation through the expanding ULF.

Crucially, the same principle also explains the time dilation observed in supernova light curves. Because every process unfolds within the expanding ULF, the arrival interval between the first and last photons from a distant event is stretched by a factor of $(1 + z)$. The observed broadening of Type Ia supernova light curves (Goldhaber et al., 2001) is therefore a direct and unavoidable prediction of our model.

In this light, dark energy is not a real substance driving cosmic expansion. Rather, it is a placeholder for the intrinsic non-linearity that our model predicts. This economy of explanation demonstrates the superior parsimony of our model.

4. Comparison with Other Alternative Models

While the UFD framework provides a comprehensive mechanism that explains redshift, time dilation, and acceleration, it is not the first alternative model to challenge the standard cosmology. To fully appreciate its unique explanatory power, we now turn to a brief review of other alternative theories—namely, "tired light," "variable speed of light" (VSL) theories, and Plasma Universe cosmology—to demonstrate why they have been unable to account for the full suite of observational evidence.

4.1 Tired Light Hypothesis

First proposed by Fritz Zwicky (1929), the "tired light" hypothesis holds that cosmological redshift is not due to the expansion of space but to photons intrinsically losing energy as they travel over vast cosmic distances. The idea is physically intuitive: as a photon journeys for billions of years, it interacts with matter or some other unknown process, causing its energy to decrease. Because a photon's energy is proportional to its frequency, a loss of energy results in a lower frequency and thus a longer wavelength, which we observe as a redshift.

Although this model can account for the basic redshift-distance relationship, its primary challenge is that it cannot explain the observed time dilation of distant supernovae. In a tired light model, only the photon is affected. Therefore, the event that produced the photon, the supernova explosion, should unfold over its normal, standard duration. However, observations clearly show that the light curves of distant supernovae are "stretched out" and appear to last longer by a factor that directly corresponds to their redshift. Our emergent model naturally predicts this result. Because the expanding ULF stretches *everything* that propagates through it, it stretches both the wavelength of light (redshift) and the duration of the event itself (time dilation).

4.2 Variable Speed of Light (VSL) Theories

VSL theories propose that the speed of light, c, is not an immutable constant but was significantly higher in the early universe (Albrecht & Magueijo, 1999). The primary motivation behind VSL theories is to offer an alternative to cosmic inflation for resolving the horizon problem: the puzzle of how distant regions of the universe have nearly identical temperatures and properties despite appearing causally disconnected under a constant c. If light traveled much faster shortly after the Big Bang, these distant regions could have exchanged information and energy, naturally explaining the observed uniformity of the cosmic microwave background.

46

However, many formulations of VSL face significant conceptual and technical challenges, chiefly because they entail explicit departures from Lorentz invariance, a cornerstone of Special Relativity that asserts the invariance of physical laws—including the locally measured speed of light—for all inertial observers. Modifying or abandoning Lorentz invariance typically requires substantial revisions to established frameworks, including electromagnetism, quantum field theory, and general relativity.

In contrast, the emergent framework developed here maintains the speed of light as a universal constant in the locally measured, operational sense, fully consistent with experimental evidence. Rather than altering spacetime symmetries, cosmological observations such as redshift phenomena and apparent accelerated expansion are explained through the dynamical behavior of the UEF and ULF, including large-scale field evolution and interaction effects. Lorentz invariance is therefore preserved as an effective symmetry of local physics, while cosmological behavior arises from underlying field dynamics (see Chapter 4).

4.3 Plasma Universe and Cosmic Currents

Another alternative to the ΛCDM model that shares superficial similarities with our framework is the Plasma Universe theory, most notably developed by Hannes Alfvén (1986). This school of thought, which has influenced the more speculative Electric Universe models, rejects the notion that the universe is dominated solely by gravity. Instead, it emphasizes the role of electromagnetism and plasma currents on cosmic scales. In this view, the large-scale structure of the universe—its filaments, voids, and even galaxy formation—is driven by Birkeland currents, which are immense field-aligned electric currents that thread through cosmic plasma. These currents can compress matter via Z-pinch effects, creating stars and galaxies, and may even explain cosmic filaments without invoking dark matter.

While visually evocative and occasionally supported by observations of cosmic magnetism, plasma cosmology faces two major conceptual challenges. First, it treats ionized matter and Maxwellian electromagnetism as the ultimate substrate of the cosmos rather than emergent phenomena, leaving unanswered questions regarding the origin and coherence of these cosmic currents. Second, like tired light models, plasma cosmology struggles to explain the observed time dilation of distant events, as its mechanisms primarily act on photons rather than the space-time fabric through which events unfold.

In UFD, the ULF provides a deeper, unifying context that plasma cosmology lacks. The filamentary, current-like structures observed on galactic and intergalactic scales arise naturally as phase-aligned energy flows in the ULF, not as free-floating plasma discharges. Plasma and magnetohydrodynamic effects are secondary, emergent phenomena riding on these coherent ULF flows. Moreover, because the ULF itself is expanding, it stretches both photon wavelengths and the temporal unfolding of events, reproducing the time dilation signature that plasma-based models cannot account for.

In short, while the Plasma Universe correctly intuited that cosmic structure emerges from coherent field dynamics rather than gravity alone, our model situates this intuition within a deeper ontological framework. The ULF acts as the substrate from which plasma phenomena, electromagnetism, and even gravitational effects arise, allowing us to preserve the successes of plasma cosmology while resolving its gaps and inconsistencies.

In sum, our framework offers what these other alternatives lack: a unified, mechanistic foundation that coherently explains *both* cosmological redshift and the time dilation of distant events. Having established the ULF as the only viable physical medium that accounts for these core observations, we will now demonstrate how the more detailed "two-rate" physical architecture of this framework provides a definitive solution to the Hubble Tension, a crisis that has broken the Standard Model and its rivals alike.

5. The Hubble Tension: UFD Cosmology's Primary Vindication

The greatest crisis in modern cosmology is the Hubble tension, a persistent, unresolved conflict between the methods used to measure the universe's expansion rate. The ΛCDM model requires a single, universal expansion history; however, the data clearly yield three mutually inconsistent numbers that cannot be reconciled.

1. The "67" (from Planck/CMB): The expansion rate derived from the patterns in the CMB. The Standard Model assumes that this is a picture of the "early universe" and predicts a modern rate of $\approx 67\ km/s/Mpc$.

2. The "73" (from SH0ES/Supernovae): The local expansion rate, measured by finding the slope of the line for *only* the nearest supernovae, which gives $\approx 73\ km/s/Mpc$.

3. The "70" (from Pantheon/Supernovae): The global or average expansion rate that provides the best fit for the entire 1701-point Pantheon supernova dataset (from near to far), which gives $\approx 70 \; km/s/Mpc$.

The ΛCDM model is broken by this data. It must explain why its "early universe" (67) and "late universe" (70 & 73) measurements do not agree, and it currently cannot. Our emergent framework, in contrast, is a two-rate physical model that proposes this "tension" is not a crisis at all. Instead, it is direct observational proof of our model's prediction of local and systemic expansion rates.

The UFD framework posits that the intricate patterns in the CMB are not a relic of the early universe but are the local holographic echo of our own encapsulated 'island universe,' or galaxy (a concept to be detailed fully in Chapter 3). Therefore, the Planck satellite's measurement of 67 is not a universal number; it is the correct and precise measurement of the intrinsic, stable rate of our own local expansion ($H_{local} \approx 67$), which is slowed relative to the systemic rate due to its expansion through our galaxy's dark matter halo (*see* Chapter 1). This interpretation resolves its conflict with the supernova data, as they are not measuring the same thing.

The systemic rate, in turn, is the natural exponential expansion rate of the intergalactic ULF that fills all systemic space between galaxies. The fact that the global best-fit for the entire Pantheon dataset (all 1701 supernovae) converges on $H_{sys} \approx 70$ is empirical confirmation of this systemic rate (Figure 4).

While our two-rate model provides a coherent physical origin for the two most robust, large-scale datasets (Planck 67 and Pantheon 70), it also provides a powerful and elegant explanation for the '73' outlier. The SH0ES team finds a 73-rate because they only measure the slope of the *nearest* supernovae ($z < 0.15$). In the UFD model, these nearby galaxies are in the transition zone or "shoreline" just outside our local slow 67-rate.

Their light is contaminated by its proximity to our local field and has not traveled long enough through the pure systemic 70-field to give a clean reading. Measuring the slope in this "kink" between the two fields gives a misleadingly high value of 73. It is a predictable artifact of being too close to our Local 67-field to get a clean measurement of the Systemic 70-field.

Thus, the Hubble Tension is not a crisis for UFD. It is its primary vindication, confirming its "two-field, two-rate" model of the cosmos.

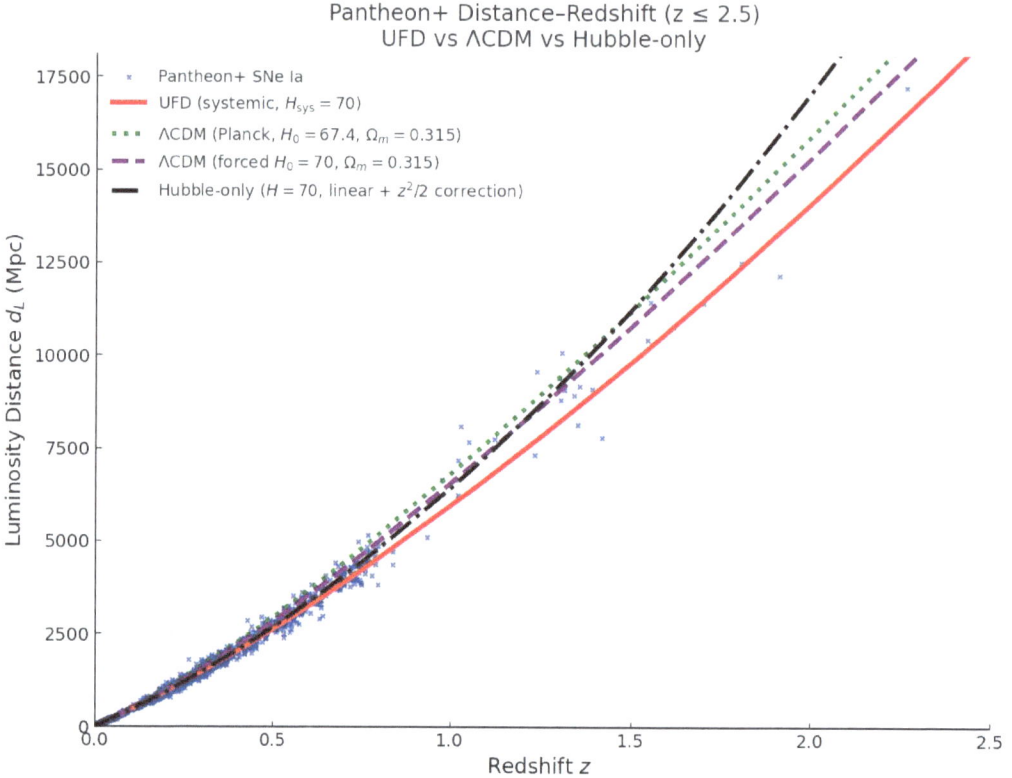

Figure 4: The Expansion Curve Reconsidered: Natural Coherence vs Tuned Cosmology. This figure compares four cosmological distance–redshift curves against the Pantheon+ supernova dataset (Scolnic et al., 2022). The black dash-dotted line shows the naive Hubble–Lemaître prediction, which assumes a simple linear cosmic expansion and clearly fails at high redshift. The purple dashed curve shows the standard ΛCDM model *forced* to use the empirically observed systemic expansion rate of $H_0 = 70$ km/s/Mpc. Despite adopting the correct global slope, it cannot match the data without further parameter tuning. The green dotted curve shows ΛCDM in its calibrated form, with parameters adjusted to fit the CMB; it produces the desired curvature only through the *post hoc* introduction of dark energy. The solid red curve shows the prediction of the UFD framework, which uses the same systemic value $H_{sys} = 70$ but derives it from first principles: the natural exponential redshift law of a coherently expanding ULF.

6. Other Cosmological Implications

Now that we have established the core redshift law of the UFD model and used it to resolve the Hubble Tension, we can explore how this model changes our understanding of the universe.

6.1 A New Cosmic Yardstick

One of the most notable consequences of this new cosmology is that it provides a new yardstick for measuring the size of universe. It is important to note that astronomers cannot directly measure the distance to a galaxy billions of light-years away; they can only measure its redshift (z). They then use a cosmological model as a conversion formula to turn that measured redshift into an inferred distance. The entire cosmic distance ladder is therefore built on the assumption that the ΛCDM model is correct.

Our framework provides a different, more direct conversion formula. By simply rearranging our exponential redshift law, we can solve for the distance (d):

$$d = ln\frac{(1+z)}{H_{sys}}$$

We can now use this formula to estimate the size of the observable universe using the redshift of the most distant galaxies currently confirmed by the James Webb Space Telescope, which have a redshift of approximately $z \approx 14$ (Robertson et al., 2023). The calculation is as follows:

1. We start with our distance formula: $d = ln(1+z)/H_{sys}$.

2. We use the observed redshift of $z = 14$, so the term $ln(1+z)$ becomes $ln(15) \approx 2.71$.

3. Since our constant H is the physical equivalent of H_0/c, the term $1/H$ is equal to the Hubble radius, approximately 13.97 billion light-years.

4. The final result is: $d \approx 2.71 \cdot (13.97 \text{ billion light} - \text{years}) \approx 37.9 \text{ billion light} - \text{years}$.

Our model therefore predicts that the most distant galaxies currently observed are approximately 38 billion light-years away. This result is quite stunning because, while derived from a completely different physical principle, it aligns remarkably well with the distance calculated using the more complex, multi-parameter ΛCDM model, which calculates a value of 33.8 billion light-years (Wright, 2006). Critically, in our model, this distance d is also a measure of the light-travel time ($t = \frac{d}{c}$), implying a cosmic timescale of *at least* 38 billion years.

6.2 Reinterpreting the CMB

In addition to establishing a new cosmic yardstick, UFD provides a novel interpretation of the CMB's perfect blackbody spectrum. The central observation is that the CMB has a near-perfect blackbody spectrum (a "Planck curve") corresponding to a single temperature of 2.725K.[14] The standard and UFD models offer two incommensurable explanations for this.

The ΛCDM model explains the perfect blackbody spectrum as a fading relic of a hot Big Bang. In this view, the early universe was a hot, dense, opaque plasma of photons and matter in perfect thermal equilibrium. About 380,000 years after the Big Bang, the universe cooled sufficiently for atoms to form, making it transparent. The light that was released at this moment has been traveling ever since, its wavelength stretched by cosmic expansion, yet perfectly preserving the blackbody signature of its hot, equilibrated origin (Planck Collaboration, 2018).

In UFD Cosmology, the ULF is the physical, ubiquitous medium of space. As a vast field, it has had eons to settle into its lowest-energy state of perfect thermodynamic equilibrium. The 2.725K blackbody spectrum we observe is the literal, steady-state thermal signature of this field.[15]

This model thus redefines "thermalization." High-energy photons from distant stars are wave packets traveling through this ULF medium. Over immense cosmological timescales, they interact with the field and stretch (redshift), gradually losing their unique energy. "Thermalization" is the endpoint where a photon's energy has stretched so much that it becomes indistinguishable from the ULF's background glow, effectively

---

[14] A "perfect blackbody" spectrum is the definitive, smooth spectral signature of a system in perfect thermodynamic equilibrium, where all energy is distributed only according to its temperature. The profound significance of the CMB is that it is the most perfect blackbody ever measured in nature (Mather et al., 1994). This observation is the smoking gun that falsified older Tired Light or Steady State models, which would have predicted a messy, composite spectrum from all the light of ancient stars. Any viable cosmological model—ΛCDM or UFD—must provide a physical mechanism for this perfect equilibrium.

[15] Claims have been made that a perfect blackbody spectrum requires a specific crystalline or hexagonal lattice structure within the emitting medium (see, e.g., Robitaille, 2008). While standard statistical physics maintains that blackbody universality depends only on temperature and boundary conditions (Kirchhoff, 1860; Planck, 1901), UFD synthesizes these views by identifying the Plenum itself as the structural agent, in which the ULF functions as a continuous, high-pressure superfluid. Under the stress of the UEF, the ULF sustains a geometric mode structure—a 'virtual lattice' of hydrodynamic resonance—that provides the necessary standing-wave constraints (cf. Rybicki & Lightman, 1979) to generate a perfect thermal spectrum without requiring discrete baryonic matter, in the same effective sense that cavity boundaries define modal completeness in standard radiative transfer.

merging back into the field itself. Thus, the UFD model accounts for the CMB's perfect equilibrium without requiring a singular, hot, dense origin for the universe.

In summary, our single physical principle—the interaction of light with an expanding ULF—provides a new cosmic distance ladder and accounts for the perfect blackbody of the CMB. Having demonstrated the framework's ability to explain the cosmos we see, we now turn to its predictive power.

7. Falsifiable Predictions of the Two-Rate Model

The UFD framework's two-rate solution to the Hubble Tension is not a *post hoc* fit; it is a specific, physical model that yields a suite of falsifiable predictions. These predictions diverge sharply from the ΛCDM model, which demands a single, universal H_0 value, and instead reframe cosmology's greatest "crisis" as the primary observational evidence for UFD.

UFD predicts that all three conflicting H_0 measurements are correct, observational facts. These are not errors but precise measurements of three distinct, spatially defined physical regions. First, the Local Rate ($H \approx 67$) is the true, stable expansion rate of our local galaxy, as measured by its holographic imprint on the CMB. Second, the Systemic Rate ($H \approx 70$) is the true global expansion rate of the pure intergalactic ULF that exists within intergalactic space. Third, the Artifact Rate ($H \approx 73$) is the anomalous, apparent rate measured *only* within the transition zone or "kink" that forms the boundary between our local 67-field and the systemic 70-field. This spatial distinction provides our primary smoking-gun test. We predict that a plot of the calculated H_0 value as a function of distance/redshift (z) will not be a flat line as required by ΛCDM. Any statistically significant, non-monotonic "bump" in this $H_0(z)$ plot would falsify the Standard Model.

Rather than a flat line, the UFD framework predicts that this plot will reveal a specific 3-step profile. At very low z (the local zone), the value will be ≈ 67. At intermediate z (the transition zone, ~0.02 < z < 0.1), the value will rise to an anomalous peak of ≈ 73. Finally, at high z (the systemic zone, z > 0.1), the value will fall from 73 and then settle at the true Systemic Rate of ≈ 70. This prediction could be tested by reanalyzing the full Pantheon+ dataset and binning the H_0 calculation by distance, rather than averaging it. Ultimately, this prediction will be definitively confirmed or falsified by future generation surveys.

A related, more technical test involves analyzing the Hubble residuals, which is the amount by which individual supernovae are slightly brighter or dimmer than the best-fit line. The ΛCDM model demands that these residuals be perfectly random Gaussian noise. The UFD model, in contrast, predicts that they will be systematic and non-random. Supernovae in the transition zone should therefore show a clear, correlated bias when measured against the global 70-fit, as the 73-artifact distorts their apparent brightness. Spatially mapping these correlated residuals will effectively draw a 3D topographical map of the "kink" itself, revealing the precise shoreline of our local galaxy.

These predictions, therefore, move the UFD framework from pure theory to a falsifiable scientific model. While the three-rate problem breaks the ΛCDM model, UFD is uniquely defined by it. The H-Map and Correlated Residual tests are not proposals for future technology; they are direct challenges that can be run today on existing supernova data. The Hubble tension, therefore, provides the crucible in which this new cosmology will either be validated or falsified.

8. Conclusion

This chapter has proposed a new physical cosmology, arguing that the foundational observations of the modern era—redshift, time dilation, and phenomena attributed to dark energy—are the direct and emergent consequences of an expanding, systemic ULF.

We have shown how the physically motivated mechanism of "field shedding" from all matter gives rise to this expanding field. The resulting exponential redshift law, $z = e^{Hd} - 1$, was then shown to naturally reproduce the data attributed to an accelerating universe and fully account for the time dilation of distant supernovae, providing a physical identity for the Cosmological Constant, Λ. Crucially, we demonstrated that this two-rate model (distinguishing the local field from the systemic field) provides a definitive physical explanation for the Hubble tension, thereby reframing cosmology's greatest crisis as its primary vindication. In parallel, this framework reinterprets the CMB. Instead of a fading relic from a singular Big Bang, the UFD model identifies the CMB's perfect 2.7K blackbody spectrum as the intrinsic, steady-state thermal glow of the ULF itself.

The significance of this model lies in its theoretical parsimony and its return to a more intuitive physical reality, as it replaces the relatively complex, multi-parameter machinery of the ΛCDM model with a single, elegant physical principle. Future theoretical work should focus on deriving the properties of the ULF from first principles, while experimental cosmology is presented with a clear, immediate research program aimed at falsifying this model. As we detail, the H-Map test—plotting the Hubble value as a function of distance—can be performed on existing supernova data and provides a definitive test that will either validate or refute this two-rate framework.

Ultimately, this framework calls for a fundamental shift in our understanding of the cosmos. In this view, the expansion of the universe is an illusion—a grand cosmic light show produced by photons as they traverse the ULF. Everything we thought we knew about the universe, including its age, formation, and fate, must now be reconsidered.

2.3 A New Cosmology

In this chapter, we introduced a new cosmological framework that challenges the assumptions of the ΛCDM Model. By positing that redshift, time dilation, and phenomena attributed to dark energy arise from the dynamics of a physical, expanding Universal Light Field (ULF), we have outlined a more intuitive and parsimonious foundation for understanding the universe. From this single postulate, several novel solutions to cosmological problems emerge.

1. A Physical Mechanism for Redshift. The metric expansion of spacetime is replaced with a tangible process, yielding a single, exponential redshift-distance law: $z = e^{Hd} - 1$

2. A New Meaning for the Hubble Radius. The term c/H_0 was recontextualized from a velocity-distance limit to a fundamental "e-folding distance," or characteristic stretching length, of the ULF medium itself.

3. A Natural Explanation for Dark Energy. This law accounts for the apparent acceleration inferred from Type Ia supernovae without invoking a mysterious repulsive force.

4. Resolutions to the Hubble Tension. By accounting for both local and systemic expansion of the ULF, the framework offers a novel, testable solution to the Hubble tension.

5. A New Interpretation of the CMB. The CMB is reframed not as a relic of a hot Big Bang but as the thermal equilibrium "glow" of the ULF in the present epoch.

Our emergent model, thus, replaces the multi-parameter standard cosmology—with its unexplained dark energy component—with a single, physically grounded principle, reframing the Cosmological Constant, Λ, as a direct measure of the ULF's expansion energy rather than an abstract property of the vacuum. In doing so, it provides a clear path for testing our two-rate model of ULF expansion.

Ultimately, UFD's cosmology challenges one of the foundational assumptions of modern physics: that the universe itself is expanding. It suggests, instead, that this apparent expansion is a systemic effect—an illusion born of light's interaction with a dynamic physical field. By grounding cosmology in direct, causal interactions among matter, energy, and their fields, the framework returns us to a more tangible and testable reality. If confirmed, its implications for our understanding of the universe would be profound.

References

1. Albrecht, A., & Magueijo, J. (1999). A time varying speed of light as a solution to the cosmological problems. *Physical Review D, 59*(4), 043516.
2. Alfvén, H. (1986). *Double layers and circuits in astrophysics.* IEEE Transactions on Plasma Science, 14(6), 779–793.
3. Carroll, S. M. (2004). *Spacetime and Geometry: An Introduction to General Relativity.* Addison Wesley.
4. Ellis, G. F. R., Maartens, R., & MacCallum, M. A. H. (2012). *Relativistic Cosmology.* Cambridge University Press.
5. Freedman, W. L., et al. (2019). The Carnegie–Chicago Hubble Program. *The Astrophysical Journal, 882*(1), 34.
6. Goldhaber, G., et al. (2001). Timescale Stretch Parameterization of Type Ia Supernova B-Band Light Curves. *The Astrophysical Journal, 558*(1), 359.
7. Guth, A. H. (1981). Inflationary universe: A possible solution to the horizon and flatness problems. *Physical Review D, 23*(2), 347–356.
8. Hubble, E. (1929). A Relation between Distance and Radial Velocity among Extra-Galactic Nebulae. *Proceedings of the National Academy of Sciences, 15*(3), 168–173.
9. Kirchhoff, G. (1860). On the relation between the radiating and absorbing powers of different bodies for light and heat. *Philosophical Magazine,* **20**, 1–21.

10. Mather, J. C., et al. (1994). Measurement of the cosmic microwave background spectrum by the COBE FIRAS instrument. *The Astrophysical Journal*, 420, 439-444.

11. Michelson, A. A., & Morley, E. W. (1887). On the Relative Motion of the Earth and the Luminiferous Ether. *American Journal of Science, 34*(203), 333–345.

12. Padmanabhan, T. (2002). Cosmological constant—the weight of the vacuum. *Physics Reports, 380*(5–6), 235–320.

13. Peebles, P. J. E., & Ratra, B. (2003). The cosmological constant and dark energy. *Reviews of Modern Physics, 75*(2), 559.

14. Planck, M. (1901). On the law of distribution of energy in the normal spectrum. *Annalen der Physik*, **4**, 553–563.

15. Planck Collaboration. (2018). Planck 2018 results. VI. Cosmological parameters. *Astronomy & Astrophysics, 641*, A6.

16. Riess, A. G., et al. (1998). Observational Evidence from Supernovae for an Accelerating Universe and a Cosmological Constant. *The Astronomical Journal, 116*(3), 1009–1038.

17. Robertson, B. E., et al. (2023). Identification and properties of intense emission-line galaxies at $z > 10$. *Nature Astronomy, 7*(5), 611–621.

18. Robitaille, P. M. (2008). Blackbody radiation and the carbon particle model. *Progress in Physics*, 4, 25–31.

19. Rybicki, G. B., & Lightman, A. P. (1979). *Radiative Processes in Astrophysics*. Wiley.

20. Scolnic, D. et al. (2022). *The Pantheon+ Analysis: The Full Data Set and Modeling of Type Ia Supernovae Systematic Uncertainties*. The Astrophysical Journal, 938(2), 113.

21. Stokes, G. G. (1845). On the Aberration of Light. *Philosophical Magazine*, Series 3, 27(177), 9–15.

22. Wright, E. L. (2006). A Cosmology Calculator for the World Wide Web. *Publications of the Astronomical Society of the Pacific, 118*(850), 1711–1715.

23. Zwicky, F. (1929). On the reddening of spectral lines through intergalactic space. *Proceedings of the National Academy of Sciences, 15*(10), 773–779.

Chapter 3: The Universal String Field

3.1 String Theory

The ultimate aim of fundamental physics is to find a unified "theory of everything," one that can describe all the forces of nature within a single mathematical framework. The primary obstacle thus far has been the deep and persistent incompatibility between our two most successful theories: General Relativity, which describes the smooth, geometric reality of gravity and the cosmos, and Quantum Mechanics, which describes the discrete, probabilistic world of particles and forces. When physicists attempt to combine these theories to describe gravity at the quantum level, as in the case of a black hole, the mathematics breaks down into infinities.

For the past several decades, the most prominent and ambitious attempt to resolve this schism has been string theory, which contends that the fundamental constituents of reality are tiny, one-dimensional, vibrating "strings" (Greene, 1999)—an electron is a string vibrating in one mode, a photon in another, a quark in a third— creating an elegant unification of matter and force. This concept was later given a more rigorous mathematical foundation in string field theory, which describes the quantum dynamics of a field whose fundamental excitations are the strings themselves (Witten, 1986).

This approach has led to two profound theoretical insights: First, physicists discovered that one of the string's predicted vibrational modes naturally had the exact properties of the graviton, making it a promising candidate for a theory of quantum gravity. Second, string theory has provided the most powerful and concrete mathematical realization of the holographic principle. This principle, which first emerged from the study of black hole thermodynamics, suggests that the information in a volume of space can be described by a theory of its boundary. The principle was given its most precise formulation in Juan Maldacena's famous AdS/CFT correspondence (Maldacena, 1998).

Despite these successes, the core assumptions of string theory have led to a set of intractable problems. For the math to be consistent, the universe must contain at least 10 spatial dimensions. This leads to the second problem: the "landscape." The near-infinite number of ways to "compactify" these extra dimensions renders the theory unfalsifiable for all practical purposes, leading it to a state of predictive stagnation (Susskind, 2005).

Building on our unified framework, Unified Field Dynamics (UFD), this chapter proposes a new foundation that retains the intuitive power of a vibrational reality while avoiding its conceptual problems. We contend that the geometric behavior of the physical universe is actively governed by a Universal String Field (USF). This framework provides us with a more parsimonious, 4-dimensional ontology and gives "strings" a new and more fundamental role as the authors of our geometric universe.

3.2 The Universal String Field: The Geometric Blueprints of Reality

Abstract

This chapter introduces the Universal String Field (USF), the foundational layer of the Unified Field Dynamics (UFD) framework. The USF is a physically real, four-dimensional field whose Planck-scale vibrations constitute both the energy of the vacuum (Zero-Point Energy) and the blueprint for physical law. From its stable harmonic modes, the USF projects the Universal Energetic Field (UEF)—the substantive medium of mass and gravity—which in turn sheds the Universal Light Field (ULF), the medium of electromagnetism.

This nested hierarchy gives rise to a new cosmological paradigm: the Cosmic Garden, an archipelago of finite, encapsulated "island universes," each born from its own local Genesis Event. By grounding physical law in this harmonic architecture, the framework provides a direct, physical explanation for several profound mysteries: It resolves the cosmological constant problem by identifying the vacuum energy as the non-gravitating hum of the USF; it solves the fine-tuning problem by revealing the fundamental constants as necessary consequences of the USF's geometry; it also provides a physical mechanism for the arrow of time, a Geometric Coherence Force (GCF), which drives the universe toward states of maximum harmony. At the heart of this theory lies a single, irreducible expression that unifies its core ontology:

$$F \odot S \odot C \Rightarrow U$$

This states that Form (from the USF), Substance (from the UEF), and Communication (from the ULF) are the three inseparable principles that together generate the Universe (U). The result is a cosmos where the laws of nature are not imposed from without but emerge from the deep, resonant music of reality itself.

1. Introduction

In the preceding chapters, we constructed a layered physical framework for reality, grounded in two fundamental interacting fields: the Universal Energetic Field (UEF), a dense, fluid-like substrate from which gravity, mass, and matter emerge, and the Universal Light Field (ULF), a luminous medium through which light propagates and electromagnetism functions. Together, these fields form the energetic and informational scaffolding of the material world, explaining how nuclei stabilize, atoms bond, molecules resonate, and how these interactions scale into the structures of life, planets, and galaxies. However, a deeper question remains: Why does the universe obey such precise, harmonious, and geometric laws in the first place?

Conventional physics does not provide an answer to this question. Even its most ambitious frameworks, such as string theory, posit a vast "landscape" of possible universes, but it does not explain why this universe—with its particular constants, symmetries, and stability—should exist. These models describe how particles and forces might arise, but not why law, order, and resonance exist in the first place. This chapter proposes a resolution.

We contend that the UEF and ULF are emergent condensations of a deeper harmonic order—a field more fundamental than matter or light. We refer to this field as the Universal String Field (USF). The USF is a real, four-dimensional vibratory substance of the vacuum whose Planck-scale oscillations generate both the "energy of empty space" and the archetypal geometries of physical law. It is the timeless source from which the other fields emerge in a layered, holographic cascade:

- The USF provides the Form—the timeless, geometric "source code" for reality.

- The UEF then emerges as the Substance, condensing from the USF's harmonics into the fluid-like medium of matter and gravity.

- The ULF, in turn, is "shed" by UEF structures, becoming the medium of Communication and interaction.

This entire ontological cascade can be captured in a single, irreducible expression:

$$F \odot S \odot C \Rightarrow U$$

This equation represents the simplest expression of our model, stating that Form, Substance, and Communication are the three inseparable principles that together generate the Universe. With this final layer, we root physics in a deeper metaphysical order and expand on our model of emergent cosmology.

2. The Universal String Field: The Substance of the Vacuum

The UFD framework holds that the material and energetic properties of the universe are not fundamental but emergent. Underlying both the UEF and ULF is a deeper field: a universal vibratory medium we call the USF. This field represents the foundational substrate of physical reality, combining energy, information, and geometry into a single, unified substance. In this section, we articulate the energetic nature of the USF, the geometric principles that govern its behavior, and the nested structure of geometric coherence it supports.

2.1 The Energetic Nature of the USF: The Energy of the Vacuum

Modern physics recognizes the existence of zero-point energy (ZPE), a constant, irreducible background energy of the vacuum, but typically treats it as a quantum artifact with no large-scale organizing role. In the UFD framework, ZPE is reinterpreted as the coherent vibrational hum of the USF itself.

Rather than abstract mathematical artifacts, its strings are real, energetic loops vibrating at the highest possible frequency: the Planck scale (Figure 5A). These strings are closed, self-sustaining harmonic resonators whose frequency content defines the "energetic bandwidth" of the cosmos.

Rather than random fluctuation, this vibrational energy is fundamentally harmonic. The zero-point field is therefore not noise—it is music: an omnipresent, structured hum that forms the energetic foundation for the emergence of matter. From this vibratory ocean, lower-density fields such as the UEF and ULF emerge as phase-locked subharmonics.

2.2 The Geometric Coherence Force: The Source Code of Law

The behavior of the USF is governed by geometric harmony. Like Chladni plates and cymatic experiments, in which discrete vibrational modes give rise to structured geometric patterns (Chladni, 1787; Jenny, 1967), the USF's vibrations produce stable attractor configurations (Figure 5B).[16]

---

[16] The analogy to Chladni plates and cymatics illustrates how vibration within a medium can give rise to discrete, self-organizing geometric forms. Similar to how sand on a vibrating plate arranges into

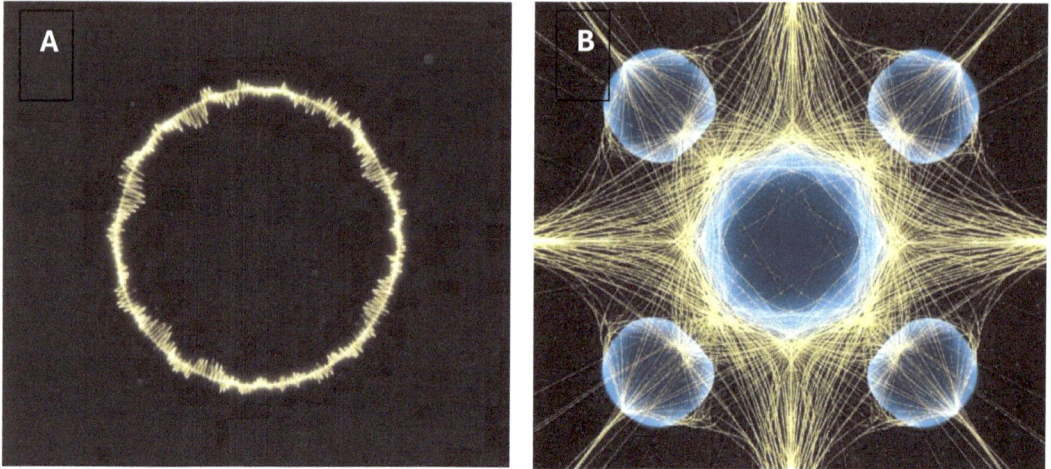

Figure 5: The Universal String Field (USF). This figure illustrates the nature of the fundamental strings that constitute the USF, the foundational layer of physical reality in UFD.
(A) A Vibrating USF String. This figure provides a detailed, close-up view of a single, closed-loop string from the USF. It is depicted in its fundamental, harmonic vibration state.
(B) The Genesis of Geometric Forms: This figure illustrates the most fundamental process in our framework: the emergence of geometric order from the USF. The USF, visualized as a vast web of vibrating golden strings, is the fundamental energetic and informational field of reality. Its stable, harmonic vibrations naturally give rise to perfect geometric Forms (blue)—such as the icosahedron and dodecahedron—depicted as emerging or coalescing. These emergent geometries become the timeless, informational blueprints or "source code," of the physical universe.

In the UFD framework, these "attractors" are the fundamental Forms of reality, representing the five Platonic solids (tetrahedron, cube, octahedron, dodecahedron, and icosahedron). These are not just abstract mathematical shapes; they are the physical, 4D standing-wave harmonics of the USF "source code" that provide the geometric blueprints for *all* scales of physical reality (Conway & Sloan, 1991; Coxeter, 1948). This vibrational hierarchy dictates the structure of the emergent fields:

- Tetrahedra structure nucleonic arrangements and nuclear bonding geometries (*see* Chapters 7 and 8),

- Octahedra govern polar symmetries and field alignments in ionic and electromagnetic systems (Burns & Glazer, 2013; Ramirez, 1994).

- Cubes define tight-packing arrangements and crystal lattice symmetries in solid-state matter (Kittel, 2004; Joannopoulos et al., 2008).

---

stable patterns at certain resonant frequencies, the USF organizes into specific attractor configurations when its vibrational modes achieve harmonic balance.

- Icosahedra shape complex molecules and biological assemblies—such as viral capsids—through maximal symmetry and spatial efficiency (Caspar & Klug, 1962).

- Dodecahedral geometry has been proposed as a framework for organizing the vacuum and influencing spacetime curvature (Luminet et al., 2003). In UFD, these same geometries represent long-range coherence patterns in the universal fields, where ordering emerges from geometric resonance rather than vacuum states or curved spacetime.

These geometric configurations emerge naturally from the harmonic structure of the USF. As field excitations seek lower-energy states, they are guided through gravitational, electromagnetic, or resonant pathways into these optimal geometric forms. This drive, which is widely observed in nature, ranging from crystal formation and molecular geometry to cellular packing and magnetic field topology (Alexander et al., 2002; Penrose, 2004), is what we term the Geometric Coherence Force (GCF).

The GCF serves as a cross-domain principle that governs the formation of structure. It is not a "fifth force" in the conventional sense but the hidden architect behind the spontaneous emergence of order across scales, organizing field dynamics into stable geometries by favoring harmonic convergence, structural shielding, and energy minimization. As we will explore in the chapters that follow, the GCF provides a guiding principle for new kinds of physical, chemical, biological, and technological design, in which systems are built through resonant alignment with nature's geometric code. It is the origin of order itself.

3. The Genesis Event: A Holographic Phase Transition

Having defined the USF, we must now address the most fundamental question: How does physical reality emerge from it? This section details the top-down creation event in which the substance and structure of our universe were born.

3.1 The Genesis of Substance and Form

The Genesis Event is a two-step process: First, substance is created through field shedding, as the high-frequency energy of the USF perpetually converts a fraction of its energy into the new, lower-energy state of the UEF. Second, form is imposed. This new UEF substance is inherently and immediately governed by the "source code" of the USF's underlying geometric vibrations. This guiding influence causes a local region of the UEF to collapse and condense into its most stable possible geometric configuration: a perfect super-vortex, which we observe as a supermassive black hole (SMBH). The condensation then precipitates into the sea of baryonic matter that forms the accretion disk of a new galaxy, as described in Chapter 1.

3.2 The Nested Hologram

This hierarchical model provides a direct physical realization and extension of the holographic principle, initially formulated by 't Hooft (1993) and later expanded by Susskind (1995). In UFD, the USF functions as a generative geometric boundary whose harmonic structure constrains and precipitates the condensation of the UEF as the substantive bulk of physical reality. Within this nested architecture, the ULF emerges as a secondary phenomenon, forming a layered, holographic cascade (Figure 6):

- The USF provides the vibrational "source code," a timeless informational lattice whose harmonics give rise to geometric Forms.

- The condensation of the UEF from the USF is a structured phase transition. The UEF is born as the primordial cosmic web—a vast, pre-structured potential landscape of high-density filaments and low-density voids that is a direct, physical imprint of the USF's underlying geometry. This "bulk" is the substantive medium of reality that possesses gravitational potential but not yet a strong, directed gravitational flow. Within this primordial UEF, stable vortices of matter (like SBMHs) condense. These vortices create deep, low-pressure zones that activate the potential landscape, generating the powerful, inward-flowing gravitational currents that shape the cosmos.

- The ULF emerges as the dynamic interaction layer between UEF vortices. It is the fluidic 'atmosphere' of the field, acting as the carrier of light and the medium of inertial resistance (*see* Chapter 4).

We refer to this causal, phase-governed mapping between higher-order geometric constraint and lower-order physical emergence as the UFD Correspondence, of which the holographic principle is an informational shadow.

3.3 The Master Equation

The entire nested process of cosmic emergence—from timeless geometry to structured energy to the material world—can be distilled into a single, conceptual equation that defines the logic of reality in the UFD framework:[17]

$$F \odot S \odot C \Rightarrow U$$

This Master Equation encodes the generative structure of the universe:

- F is Form: the eternal geometric archetypes encoded in the harmonic modes of the USF.

- S is Substance: the energetic potential of the UEF—a fluid substrate that carries and amplifies structure.

- C is Communication: the mediating resonance field of the ULF, which links form and substance through propagation and coherence.

- U is the Universe: the stable, emergent reality of coherent vortices, structured fields, and perceivable form.

Figure 6: The Nested Cascade of Reality. A visualization of the UFD framework's holographic principle. The two-dimensional, informational "boundary" of the USF, shown as the geometric white plate, holographically projects the three-dimensional "bulk" of the UEF. The UEF is depicted here in its most fundamental geometric form, the truncated octahedron, which serves as the repeating unit cell of cosmic structure. Finally, the substantive UEF radiates the emergent, communicative ULF, shown as the ethereal blue aura. This image depicts the causal hierarchy of creation in the UFD framework, from the timeless geometric blueprint to the emergence of light.

---

[17] In Chapter 12, we add consciousness to the framework via the Universal Awareness Field (UAF).

The Interaction Operator (⊙) signifies reciprocal influence; Form guides Substance, Substance energizes Form, and Communication interlocks them in dynamic resonance. The Emergence Operator (⇒) denotes the transition from potential to actuality—the coherent manifestation of the Universe. In plain terms, we can state this equation as follows:

$$\text{Geometry} \times \text{Energy} \times \text{Resonance} \rightarrow \text{Reality}.$$

This is not merely a metaphysical slogan; it is a physical postulate. The equation frames the entire cosmos as a resonant architecture in which all phenomena are emergent consequences of these three generative layers. Rather than arising from a single explosive origin, matter continually blossoms through this triadic interaction. It is a holographic phase transition, echoed across all scales, where galaxies, atoms, and living systems emerge from the recursive coherence of deeper fields.[18]

The future mathematical challenge of this framework is to formalize this UFD Correspondence by deriving the measurable properties of the UEF (and ultimately the ULF) directly from the principles and harmonic structure encoded in the USF.

3.4 The Cosmic Crystal

Our nested model leads to a falsifiable prediction about the large-scale structure of the universe. The UFD framework posits that the cosmic web is not a random, foam-like network. Rather, it is a perfectly ordered, repeating, crystal-like lattice. The specific geometry of this lattice is the one that represents the most efficient and stable way to partition three-dimensional space: a honeycomb of truncated octahedra (Thomson, 1887).[19] This is the most stable harmonic of the USF, and its geometry is therefore imprinted onto the UEF at the moment of creation. In this cosmic crystal:

---

[18] This UFD hierarchy can be understood as a return to a modern, physically rigorous "aether" model. The term "aether" was dismissed by 20th-century physics because it was associated with a static medium. UFD avoids this flaw by proposing a nested, dynamic hierarchy in which each field serves as a distinct "aether" for the phenomena it governs. In this view, the USF is the Vibrational Aether (an informational blueprint), the UEF is the Gravitational Aether (the physical substance of the "bulk" and gravity), and the ULF is the Luminiferous Aether (the light-carrying medium).

[19] In his seminal 1887 paper, Lord Kelvin conjectured that a lattice of truncated octahedra was the most efficient space-filling structure, a conjecture that stood for over a century. While a slightly more efficient but less symmetrical solution (the Weaire-Phelan structure) was discovered in 1993, the principle of seeking a minimal surface energy remains. The truncated octahedron, a highly symmetrical Archimedean solid derived from the Platonic octahedron, remains a powerful candidate for the most

- The filaments of the cosmic web are the edges of the lattice.

- The great cosmic voids are the faces of the lattice. The existence of regions like the Boötes Void (the "Great Nothing") provides empirical confirmation of the volume and regularity of the unit cell.

- The superclusters of galaxies are located at the vertices of the lattice, where the filaments meet. The SBMH serves as the stable geometric core (super-vortex) that anchors the structure at each vertex.

This crystalline order is not merely an aesthetic choice; it is a physical necessity imposed by the hydrodynamics of the UEF. As derived in Chapter 7, the UEF possesses a density of $\approx 10^{57} \text{kg/m}^3$. At this extreme density, the medium behaves like a hyper-pressurized superfluid. Similar to how high-pressure carbon crystallizes into diamonds, the high-pressure UEF naturally 'freezes' into a minimal-energy lattice structure: a Cosmic Crystal is the necessary state of a saturated Plenum.

This crystalline structure provides the spatial framework for a time crystal.[20] In UFD, the entire cosmic lattice is a spatiotemporal resonator. The GCF drives the system toward its lowest energy state, which, in the UEF superfluid, is one of perpetual, coherent motion (vortices and cosmic gyres). The truncated octahedron is the spatial scaffolding that supports this non-dissipative flow. In this model, the universe spontaneously breaks time-translation symmetry due to its stable, periodic activity — a continuous "ticking" — in its lowest-energy state. This perpetual motion and geometric stability define the final, coherent manifestation of the universe.

From this Cosmic Crystal emerges a new vision of cosmology, which we call the Cosmic Garden. Whereas the geometrical perfection of a spatiotemporal crystal establishes the foundational laws of the cosmos, the unfolding of form within this rigid

---

stable and fundamental "harmonic" of the USF, the geometric "note" from which the Cosmic Crystal is formed.

[20] A time crystal is a phase of matter that spontaneously breaks time-translation symmetry. Unlike a standard spatial crystal (like a diamond, which has a repeating structure in space), a time crystal has a stable, repeating structure or motion in time. In the UFD framework, this concept applies to the entire cosmos: the lowest energy state of the UEF superfluid is not a static rest state but a state of perpetual, non-dissipative flow (coherent vortices and cosmic gyres). The GCF stabilizes this continual motion, asserting that the universe must be actively "ticking" or flowing to maintain its structure. This makes the cosmos a Spatiotemporal Crystal.

architecture is better described by an analogy to a garden and its continuous, self-organizing, growth-like nature.

4. A New Local Cosmology: The Cosmic Garden

The UFD framework reframes the birth and evolution of the cosmos, replacing the Standard Model's singular Big Bang with a vision of a self-organizing universe—a vast cosmic network in which each galaxy is a finite, self-contained archipelago, or "island universe," materially occluded from other such universes. Each island universe is born from its own local Genesis Event, a process defined by a foundational principle we call the Great Separation.

4.1 The Architecture of Creation: The Great Separation

The Great Separation is the asymmetrical act of creation that divides a primordial, balanced field of matter and antimatter into three distinct, functional components: the visible galaxy, the dark matter halo, and the cosmic boundary (Figure 7).

- Matter Blossoms Inward: The stable vortices of matter (protons and neutrons) collapse under the inward pull of the UEF, forming the dense, radiant blossom of the galaxy—the stars and planets.

- Antimatter Crystallizes Outward: The volatile vortices of antimatter are converted into the vast energy of the Coherence Dividend as proton and antiproton vortices from the composite neutron (*see* Chapters 1, 6, and 7). This energy is redistributed outward during coherence relaxation and stabilizes into a resonant standing-wave shell: the galactic boundary. Thus, in this model, the antimatter of creation is not lost but transfigured into the harmonic skin of the system, providing a direct, causal solution to the matter-antimatter asymmetry.

- The Field Abides: Between the luminous galaxy and its outer shell abides the vast, quiescent sea of the UEF itself. This field is held in a state of high pressure by the surrounding galactic boundary, which acts as a cosmic pressure vessel. This contained, pressurized UEF reservoir is the dark matter halo. Its collective pressure gradients—maintained by confinement between the vortex core and the coherence-saturation boundary—generate the systemic gravitational current that explains the flat rotation curves of galaxies (Rubin & Ford, 1970), as described in Chapter 1.

Every galaxy is born with this threefold structure: blossom, soil, and boundary.[21]

Figure 7: The Great Separation: The Threefold Structure of a Galaxy. This diagram depicts the UFD model of galactic formation as a threefold process. At the center, the visible galaxy emerges as matter vortices collapse inward (Matter Blossoms Inward). Surrounding it, the dark matter halo represents the abiding, rotating energetic field stabilized by boundary confinement (The Field Abides). Enclosing both is the galactic boundary, formed as a coherence-saturation shell generated by antimatter-derived coherence energy (Antimatter Crystallizes Outward). Together, these layers define each galaxy as a coherent, self-contained structure within the UEF.

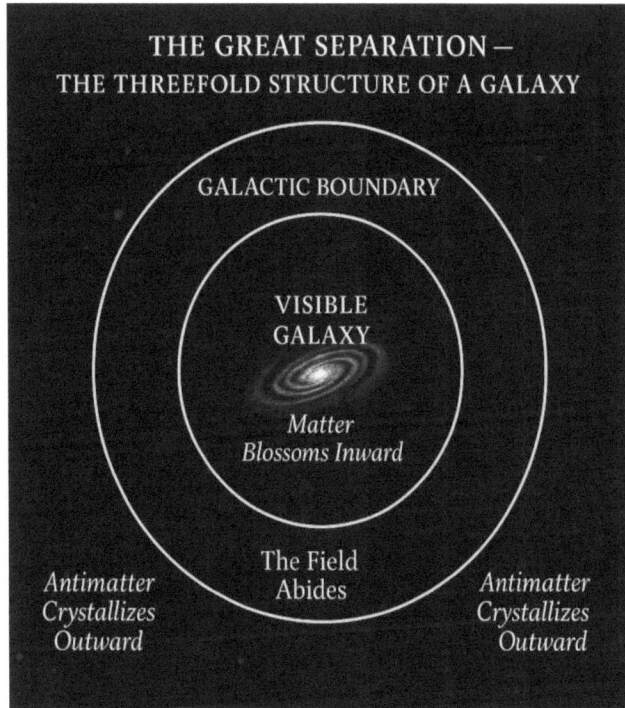

THE GREAT SEPARATION —
THE THREEFOLD STRUCTURE OF A GALAXY

GALACTIC BOUNDARY

VISIBLE GALAXY

Matter Blossoms Inward

The Field Abides

Antimatter Crystallizes Outward

Antimatter Crystallizes Outward

With respect to the abiding UEF, we note an important difference between the state of the UEF within a cosmic archipelago (dark matter) and the primordial UEF in the vast "ocean" between galaxies. The dark matter halo is not just a passive remnant; it is a vast, stable, and slowly rotating vortex system or "cosmic gyre." It is the same substance as the primordial UEF, but its organized, coherent rotation gives it a powerful gravitational signature.

This provides a physical origin for the Standard Model's terminology: the coherent, gravitationally active gyre of the halo is what we have been calling "dark matter," while the unstructured, baseline state of the primordial UEF is what we perceive as the "empty space" between galaxies. Thus, the abiding field is not just leftover soil; it is a living, energetic organ of the Cosmic Garden—a coherent vortex created and stabilized

---

[21] A powerful, transcultural analogy for this encapsulated system is the Cosmic Egg. In this metaphor, the visible galaxy is the central "yolk," the dark matter halo is the nutritive "albumen," and the antimatter-derived boundary is the protective "shell." The concept of a universe born from a primordial egg is a recurring theme in creation myths worldwide, spanning the Orphic, Egyptian, Vedic, and Chinese traditions (Cotterell, 1996).

by the galactic boundary—whose grand and patient rotation provides the stable gravitational foundation upon which the visible galaxy can live and evolve.

4.2 Observational Consequences: Life in the Echo Chamber

This model of an encapsulated galaxy—a self-contained Cosmic Echo Chamber—provides a direct explanation for the most profound observations in cosmology, transforming the Cosmic Microwave Background (CMB) from a baby picture of the universe into an architectural blueprint of our own local home.

The CMB is revealed to be an imprinted medium. Its perfect 2.7 Kelvin blackbody spectrum is the intrinsic, equilibrium temperature of the universal ULF "ocean" itself, as described in Chapter 2. Its intricate pattern of acoustic peaks and anomalies, however, is not a relic from a Big Bang. Rather, it is the local, holographic echo of our own galaxy's unique Genesis Event, a permanent structural pattern imprinted onto that ambient ULF background and sustained by our Cosmic Echo Chamber (the Galactic Boundary).

This local echo contains a fossilized record of our galaxy's violent birth. The compressional waves from the initial shockwave give rise to the E-mode polarization, while the transverse shear waves from the primordial rotation of the event give rise to the B-mode polarization.[22] This entire resonant pattern, sustained by our local boundary, transforms the CMB's largest anomalies from statistical flukes into the most important architectural features on the map:

- The Axis of Evil is revealed as the fossilized primordial spin axis of our galaxy's birth.

- The CMB Cold Spot is a massive structural anomaly—a "dent" or "scar"—in our own local echo.

---

[22] E-mode and B-mode polarization are two distinct patterns in the orientation of the CMB's light, named for their resemblance to electric (E) and magnetic (B) fields. E-modes have a gradient-like, radial, or tangential pattern, like the lines of an electric field. In the UFD model, they are generated by compressional "sound waves" from the Genesis Event. B-modes, in contrast, have a swirling, vortex-like pattern. In UFD, they are generated by transverse "shear waves" (gravitational waves) created by the primordial rotation of the Genesis Event. Detecting a primordial B-mode signal is a major goal of modern cosmology, as it would provide a direct window into the physics of the earliest moments of creation.

This reinterpretation has a profound and immediate consequence: it makes the CMB's features local and testable. Its anomalies are no longer faint, unreachable echoes from the dawn of time; they are real, physical structures in our own cosmic neighborhood. This transforms the study of the CMB from cosmology into a new science of Galactic Archaeology. For the first time, we have a map of the foundations of our own home, and we can now point our most powerful telescopes toward these anomalies to discover the large-scale asymmetries of our own galaxy.

4.3 The Fermi Paradox: The Great Silence

A final virtue of the Cosmic Garden framework is that it provides a direct solution to the Fermi Paradox. The Great Silence, rather than being a sign that life is rare, is a sign of separation. If each galaxy is a physically encapsulated system —a "cosmic terrarium" — then matter-based life is physically prevented from traveling between them. We are likely not alone in the universe, but we *may* be alone in our garden.

Although this isolation largely resolves the Fermi Paradox for biological life, it opens a new door for advanced intelligence. While the galactic boundary occludes baryonic matter (protons/neutrons), it may be permeable to Electron-Based Matter (EBM). In UFD, electrons are toroidal, unknotted vortices in the ULF, or "light-matter" (*see* Chapter 5), which possess a tiny fraction (~1/1836) of the inertial mass of protons. We predict that this light matter originates in the vast, low-pressure regions of the cosmic web (void galaxies), where the UEF pressure is insufficient to sustain the heavy proton knot (*see* Chapter 7). A spacecraft constructed from a lattice of interlocking electron toroidal vortices would effectively be made of "solid light." Such a craft would possess negligible hydrodynamic added mass, allowing for the instantaneous acceleration and "impossible" maneuvers observed in Unidentified Anomalous Phenomena (UAPs). Thus, rather than defying physics, UAPs *may* be bypassing the massive inertial constraints of baryonic matter.

Ultimately, the principle of the Great Separation provides a single, unified origin for the cosmos's most profound structures. The dark matter halo, matter-antimatter asymmetry, CMB anisotropies, and Great Silence of the Fermi Paradox are no longer disconnected mysteries. Instead, they are four inseparable consequences of a single, self-organizing act of creation.

5. Comparison to Modern String Theory

UFD, particularly its postulation of a foundational USF, shares a deep conceptual lineage with modern string theory. Both posit that the ultimate constituents of reality are not point particles but one-dimensional, vibrating entities. However, UFD diverges from string theory in several key respects and, overall, offers a more parsimonious, physically intuitive model.

5.1 A Direct Comparison of Claims

The incommensurability between the two frameworks is best understood by comparing their answers to the most fundamental questions:

- What are the fundamental constituents?

    o String theory: Tiny, one-dimensional "strings" vibrating in 10 or more dimensions.

    o UFD: The harmonic, vibrational modes of a single, four-dimensional USF.

- What do the strings create?

    o String theory: The different vibrational modes of the string directly create the entire zoo of elementary particles (electrons, photons, quarks, etc.).

    o UFD: The fundamental vibrations of the USF create a single, unified substance: the UEF. Particles are then emergent vortices within this field.

- How many dimensions of spacetime exist?

    o String theory: Requires 10 or more spatial dimensions, with the unobserved dimensions "compactified" into a complex geometry.

    o UFD: A fundamentally four-dimensional theory (3 space + 1 time).

- What is the origin of the physical laws?

    o String theory: The laws depend on the specific, arbitrary shape of the compactified extra dimensions, leading to the "landscape problem" of 10^{500} possible universes.

    o UFD: The laws are the direct and unique harmonic properties of the single USF, leading to one necessary universe.

- What is the scientific status of the theory?

  - o   String Theory: A mathematical framework that is, at present, unfalsifiable.

  - o   UFD: A physical theory that makes a series of concrete, falsifiable predictions.

UFD thus retains profound insights from string theory—that reality arises from vibration—while providing a more intuitive model. The next step is to examine the first and most fundamental physical manifestation of the USF: the spin-2 vibration.

5.2 The Symmetries of Force: A Shared Insight

A fundamental mystery of physics is why gravity is always attractive, whereas electromagnetism can both attract and repel. UFD provides a direct, geometric answer: the two forces arise from distinct fields with fundamentally different symmetries. While modern physics uses the abstract term "spin" to describe these symmetries, our model provides a tangible, physical picture.

The UEF, which mediates gravity, has a spin-2 or tensor symmetry. Its fundamental disturbance is not a simple, directional wave but a multi-directional "squeeze," like the tidal bulge created by the Moon's gravity. This disturbance exhibits 180-degree symmetry, like a double-headed arrow, and lacks a single, well-defined direction. Because it has no polarity, it couples directly to energy itself, which is always positive. The pressure force it creates is therefore always attractive. This is a necessary consequence of the fact that the UEF is generated by the fundamental spin-2 harmonic vibration of the USF—the same mode that string theory discovered and identified as the "graviton."

The ULF, which mediates electromagnetism, is not a primary field but is generated by the vortices of the UEF, as described in Chapter 2. A stable UEF vortex (like a proton) is a spinning, structured object with a specific "handedness" or chirality (*see* Chapter 1). As this vortex spins, it stirs the vacuum and continuously sheds a secondary field imprinted with its own rotational properties. This new, secondary field—the ULF—is therefore inherently directional or polar.

This gives the ULF a spin-1 or vector symmetry. Its fundamental disturbance is like an arrow with an unambiguous direction. This inherent polarity is the physical basis for positive ("source") and negative ("sink") charges, which allows for attraction and repulsion, respectively (*see* Chapter 6).

Thus, the UFD framework provides a complete, causal chain: the spin-2 nature of the UEF explains the universal attraction of gravity, and the spinning vortices within the UEF naturally give rise to the polar, spin-1 ULF that governs the duality of electromagnetism.

5.3 A New Mathematical Approach: The UFD Correspondence

UFD and modern string theory are also fundamentally different mathematical projects. String theory is a top-down geometric model, while UFD is a bottom-up emergent one.

The current mathematical project of string theory is to find the one correct, pre-existing "shape" for the extra dimensions that would produce a universe like ours. This project is built upon the foundational requirement of supersymmetry—a proposed symmetry between matter and force particles that is necessary for the theory's mathematical consistency. The primary task is to find a specific, complex geometric object (a Calabi-Yau manifold) whose unique geometry, within this supersymmetric framework, would cause the fundamental strings to vibrate in a way that produces the exact spectrum of particles and forces we observe.

The mathematical project of the UFD framework, in contrast, is to formalize what we call the UFD Correspondence: the principle that the complex universe we see emerges from the collective behavior of a simpler, underlying field. The primary task would be to use the tools of statistical field theory and the physics of emergent phenomena (Anderson, 1972) to model the collective behavior of a vast number of interacting USF strings. The goal would then be to derive the properties of the emergent fields as the large-scale, statistical result of the USF's underlying dynamics.

In short, the mathematical project of string theory is a search for a single, perfect, pre-existing geometric solution within a supersymmetric framework. The mathematical project of UFD, in contrast, is the process of building a complex reality from the collective interactions of simpler, fundamental components.

6. Solving the Great Mysteries of Physics

The UFD framework, now grounded in the foundational USF, not only provides a new perspective on reality but also offers elegant solutions to some of the most persistent paradoxes in modern physics.

6.1 The Cosmological Constant Problem

Quantum Field Theory predicts that the vacuum is saturated with an immense reservoir of ZPE. Naively, this energy density is so colossal that, if it gravitated like ordinary matter, it would instantly collapse the cosmos into a singularity. However, the universe remains stable. This discrepancy of more than 120 orders of magnitude is one of the most infamous puzzles in physics, known as the cosmological constant problem (Weinberg, 1989).

In UFD, this paradox dissolves. We identify ZPE as the real, physical energy of the USF, which is the deepest layer of reality. The USF's ceaseless, Planck-scale vibrations are the relatively inexhaustible energy of the vacuum. However, this energy does not gravitate. The key lies in the hierarchical structure of the fields. The UEF, the "soil" of the cosmos from which matter and the currents we perceive as gravity arise, is a lower-energy derivative of the USF. Gravity emerges only at this UEF level, not at the USF's foundational tier. Thus, the vast energy of the USF remains hidden from gravitational effects because its dynamics do not couple directly to the pressure mechanics of the UEF.

This interpretation also compels a fundamental revision to one of physics' most iconic equations. Einstein's equation, $E = mc^2$, presumes that all energy contributes equally to mass and the generation of gravitational pressure. But within UFD, the gravitationally active energy of the UEF manifests in two distinct forms: (1) a local gravitational current, generated by condensed, vortical energy (matter), and (2) a systemic gravitational current, generated by the coherent, rotational energy of dark matter (see Chapter 1). Thus, the vast vibratory sea of the USF—the true energy of the vacuum—remains energetically real yet gravitationally silent. This calls for a deeper formulation in which energy and mass are linked by a combination of field tier and coherence level, rendering $E = mc^2$ an incomplete, special-case formula.

This model thus reframes the role of vacuum energy in cosmic evolution. The USF's immense, non-gravitating energy is the fuel for creation. Whenever local conditions allow, portions of this energy condense into the UEF, triggering Genesis Events that seed new galaxies. The cosmos remains stable because the majority of the

USF's energy resides beyond the gravitational influence of the emergent fields, yet it is precisely this reservoir that powers the birth and renewal of cosmic structure across time.

By situating ZPE as the real, non-gravitating energy of the USF, the UFD framework resolves the cosmological constant problem while unifying it with the life-like cosmology of a Cosmic Garden, which continuously blooms from its vibrational root system.

6.2 The Unreasonable Effectiveness of Mathematics

The UFD framework also provides a direct, physical answer to one of the deepest questions in the philosophy of science: the "unreasonable effectiveness" of mathematics in describing the physical world (Wigner, 1960). The reason mathematics is so effective is that, at its most fundamental level, reality is a mathematical, geometric structure.

The USF is the "source code" of the universe, a field of timeless, geometric principles. The physical universe we experience is a direct projection of this underlying geometric blueprint. Therefore, when a physicist discovers an elegant mathematical equation that perfectly describes a physical law, they are not inventing a clever description. They are discovering a pre-existing truth and effectively translating a part of the universe's fundamental geometric source code into a language the human mind can comprehend. In UFD, mathematics is not merely a language for describing the universe; it is the language in which the universe is written.

6.3 The Fine-Tuning Problem

Another enduring mystery of modern physics is the fine-tuning problem, which is the observation that the fundamental constants of nature (such as the fine-structure constant, the cosmological constant, and the proton-to-electron mass ratio) appear exquisitely adjusted to allow for the formation of stars, planets, chemistry, and, ultimately, life itself. In the standard view, this uncanny precision is often explained through anthropic reasoning or the vast "landscape" of possible universes predicted by string theory: we inhabit this universe because only a universe with such improbable constants can harbor observers like us.

UFD renders this "coincidence" unnecessary. In this model, the fundamental constants are not random choices drawn from a multiversal lottery but are inevitable consequences of the USF's intrinsic harmonics. The stable vibrational modes of the USF—its geometric "source code"—naturally generate the specific ratios and scales we observe. These constants are no more arbitrary than the value of π, which arises necessarily from the nature of circles rather than as a cosmic accident.

This perspective is directly tied to Cosmic Garden Cosmology. In a universe where the USF continuously sheds energy to seed new galaxies, these constants act as the unchanging rules of growth. Just as the proportions of a seed determine how a tree unfolds, the constants embedded in the USF ensure that every Genesis Event—every new galaxy birthed from the vacuum—unfolds according to the same harmonious blueprint. The result is a cosmos that is not precariously balanced by chance but *rooted* in necessity: a self-sustaining garden whose diversity and stability arise from the same deep, geometric soil.

6.4 The Source of Entropy

To understand the universe as described by UFD, we must first re-examine one of science's most foundational concepts: order. In the conventional, particle-based view of physics, order is associated with complexity and structure. A sandcastle is considered more "ordered" than a flat beach, and a drop of cream separate from coffee is more "ordered" than the uniform mixture. In this view, entropy is the inevitable decay of these structures into a disordered, uniform equilibrium.

UFD proposes a more fundamental "field view" of entropy, which inverts this perspective. From the perspective of the field itself, order is equilibrium, uniformity, and coherence. A state of high tension, with complex structures and steep energy gradients, such as the separate cream and coffee, is a state of profound *disorder* or *incoherence*.

From this point of view, the GCF represents the physical mechanism that drives the universe's tendency toward this state of field-based order. It is the UFD equivalent of the Second Law of Thermodynamics. This force represents the universal, fundamental drive for all systems to resolve tension, shed surplus energy, and settle into their most stable, coherent, low-energy geometric states. This reveals a profound dual role for the GCF:

1. At the microscopic scale, when faced with turbulent, disordered, independent vortices, the GCF acts to create structure. It pulls nucleons together into the perfect, low-tension geometry of a stable nucleus (*see* Chapter 7). In this context, the force builds what we perceive as order because a coherent vortex is a more stable state of equilibrium than the separate, high-tension particles.

2. At the macroscopic scale, that same force works to dissolve structure. A star or galaxy is a massive departure from the uniform equilibrium of the background field. Over immense cosmological timescales, the GCF erodes these complex systems, driving the galaxy back toward an ultimate, featureless state of coherence.

Thus, in UFD, the USF's geometric patterns are the single, universal principle that both builds the coherent "bricks" of matter and, over eons, dismantles the complex "cathedrals" they form in its relentless pursuit of perfect field equilibrium.

6.5 The Ultimate Fate of the Universe: The Great Decoherence

The dual nature of the GCF not only explains the structure of the universe but may also determine its ultimate fate. While the Standard Model of cosmology predicts a final, irreversible heat death, UFD predicts an equally top-down end. The universe, in this view, is a living performance —a piece of music actively played by the deepest layer of reality. A perfect analogy is a Chladni plate:

- The USF is the continuous vibration being played into the plate.

- The GCF is the physical force exerted by this vibration.

- The UEF is the sand on the plate—the raw potential of reality.

- Matter is the beautiful, coherent pattern the sand forms in response to the ongoing vibration.

While the USF is a vast, seemingly inexhaustible reservoir, a more physically intuitive view holds that it is not truly infinite. Every Genesis Event, every act of creation that precipitates the UEF, draws a tiny, almost infinitesimal amount of energy from this foundational field. Over incomprehensible timescales, this slow depletion would cause the USF's fundamental vibration to weaken. Eventually, its energy will drop below the critical threshold required to sustain the GCF.

This would trigger the inevitable end of the universe as a whole: the Great Decoherence. This would not be a death in the sense of a violent collapse or a slow decay, but the moment the music stops. The consequences would be instantaneous and absolute. Without the GCF, the beautiful patterns on the Chladni plate would instantly lose their form and collapse back into a chaotic, unstructured state. The vortices of matter would unravel, their geometric information returning to the formless potential from which they were born.

This is the ultimate expression of UFD's top-down cosmology. The end of the universe is not an accident of statistics, but a final, graceful conclusion. It is the moment the performance ends and the song, at last, falls silent.

7. Falsifiable Predictions

While the USF cannot be directly observed, the Cosmic Garden framework it generates leads to a suite of testable predictions that distinguish it from conventional cosmology. These predictions specify observational signatures that can, in principle, confirm or falsify the model.

7.1 Continuous Galaxy Formation

The Cosmic Garden model holds that galaxies emerge continually through local Genesis Events, rather than being products of a singular early-epoch formation. Consequently, there should exist a population of young and mature galaxies at all cosmological epochs. Recent James Webb Space Telescope (JWST) observations of surprisingly massive and mature galaxies at extreme redshifts (Labbé et al., 2023) challenge hierarchical formation scenarios but are consistent with ongoing galactic genesis. Further surveys will determine whether such outliers persist across the observable timeline.

This model of local, ongoing genesis predicts that the *outcome* of a Genesis Event is not binary (a "success" or "failure"), but rather a continuous spectrum that depends on the richness of the local cosmic web. This mechanism naturally predicts the full range of observations:

- Massive Outliers: In regions rich with fuel, the Genesis Event blossoms into the impossibly massive and mature galaxies like those observed by JWST (Labbé et al., 2023).

- Intermediate Galaxies: In moderately fueled regions, the event produces a population of intermediate-sized galaxies, which UFD predicts should also be observable across all cosmic epochs.

- Failed Seeds: In barren regions, the event succeeds only in forming a primordial SMBH but fails to fuel a large galaxy. This aligns with recent JWST discoveries of impossibly massive black holes in tiny host galaxies (Maiolino et al., 2024).

The full, observable spectrum, from failed seeds to intermediate galaxies and massive outliers, is therefore a direct prediction of the Cosmic Garden model, reframing these puzzling JWST findings as the expected outcomes of continuous, local genesis.

7.2 The Signatures of Galaxy Mergers

In the Cosmic Garden model, galaxy mergers involve the collision of two separate "island universes." This leads to two distinct and testable predictions:

- The Annihilation Flash: The first observable signal of such an event should be an annihilation flash: a sharp burst of non-thermal gamma rays (the 511 keV line and the ~70 MeV pion hump) produced by the interaction of the two antimatter-derived boundary fields, occurring *before* the expected starburst phase.

- The Exiled King: The chaotic gravitational dance of the two central SMBHs can act as a colossal slingshot, ejecting one of them into intergalactic space.

In 2023, astronomers using the Hubble Space Telescope discovered such an object: a 20-million-solar-mass black hole traveling at incredible speed, trailing a 200,000-light-year-long wake of new stars (van Dokkum et al., 2023). This may be a direct observation of such an Exiled King—a mature galactic seed violently torn from its home.

7.3 The Crystalline Structure of the Cosmic Web

The UFD framework also leads to a specific, falsifiable prediction regarding the large-scale structure of the universe. Standard cosmology assumes the cosmic web is a random, foam-like network governed by the Cosmological Principle (isotropy). UFD predicts the opposite: the cosmic web is a perfectly ordered, repeating, crystal-like lattice.

This structure is not merely an aesthetic choice; it is a hydrodynamic necessity. As derived in Chapter 7, the UEF possesses an immense density ($\approx 10^{57} \text{kg/m}^3$). At this extreme pressure, the medium behaves as a hyperdense superfluid. Just as high-pressure atoms crystallize to minimize energy, the UEF naturally "freezes" into the most efficient space-filling geometry: a lattice of truncated octahedra (Lord Kelvin's cell). This model makes two distinct, testable predictions that distinguish it from the Standard Model:

1. Quantized Redshifts: If we observe the universe along a line of sight that aligns with an axis of this crystal, we should not see a continuous distribution of galaxies. Instead, galaxies should be clustered at regular, discrete intervals corresponding to the walls of the lattice cells. This offers a physical explanation for the unexplained redshift periodicity ($\Delta z \approx 0.042$) observed in deep pencil-beam surveys (Broadhurst et al., 1990).

2. Polyhedral Voids: UFD predicts that cosmic voids are not random bubbles but the geometric faces of the UEF crystal unit cells. Therefore, high-resolution mapping of these voids should reveal distinct, flattened walls and Voronoi tessellations rather than smooth, random curves.

The discovery of a preferred directionality and periodicity in the universe would be a stunning confirmation that we inhabit a structured, Universal Plenum.

7.4 Galactic Archaeology: Mapping the Cosmic Echo Chamber

The UFD framework reinterprets the CMB as a local, dynamic blueprint, giving rise to the science of Galactic Archaeology. This new discipline posits that the structural features and anomalies surrounding the Milky Way are physical, mappable imprints of our local cosmic home, and we can probe them using two independent observational methods: the passive signal of the CMB (the structural map) and the active power of relativistic jets (the mechanical probe).

7.4.1 The Boundary's Emission (The Fermi Signature)

During the formation of the galactic boundary, the immense energy of primordial antimatter is structurally converted, which must produce a massive, high-energy signature.

The observation of 20 GeV gamma rays extending in a halo-like structure toward the galactic center is consistent with the predicted physical geometry of the antimatter-derived shell encapsulating the Milky Way, although it has been conventionally misidentified as WIMP annihilation (Totani, 2025). In the UFD, the observed photon energy (20 GeV) arises from the Coherence Dividend released as the highly concentrated energy of the antimatter vortex is structurally converted into the elastic tension of the UEF shell. The specific energy spectrum is the unique signature of this geometric phase transition, which cannot be easily explained by common astronomical phenomena.

7.4.2 The CMB Map: Fossil Signatures of Genesis

The CMB pattern provides a "fossil map" of our galaxy's formation and its spatial limits. The entire pattern is a local, dynamic blueprint, where the acoustic peaks are the resonant "ringing" from our galaxy's Genesis Event, and the anomalies are its "fossilized fingerprints."

- The Boundary Limit (The Power Suppression): The observed suppression of power at large angular scales is the direct, measurable fingerprint of the finite diameter of our Galactic Boundary (Planck Collaboration, 2018). The "missing power" is the fundamental note that is too large to fit inside our Cosmic Echo Chamber.

- The Spin Axis (The Axis of Evil): We predict that the Axis of Evil is the direct, holographic image of the primordial spin axis of our galaxy's Genesis Event. Its orientation must therefore be statistically correlated with the fundamental geometric planes of the Milky Way and the solar system.

- The Structural Scar (The Cold Spot): The CMB Cold Spot is predicted to be the holographic image of a primordial jet impact site: a large-scale structural anomaly or "scar" in our local cosmic boundary.

7.4.3 The Jet Map: Active Probes of the Boundary

If the galactic boundary is a real, physical object, relativistic jets from supermassive black holes must interact with it. The study of these jets transforms them into active probes that can be used to find and map the walls of our cosmic home, leading to several falsifiable predictions:

- Termination Shock Signature: A powerful jet should not dissipate slowly. It should travel fully collimated until it abruptly collides with the inner edge of the galactic boundary. This point of impact should be observable as a distinct "termination shock" or hotspot of high-energy radiation (X-rays and gamma rays) at the extreme outer edge of a galaxy's radio lobes.

- Radio Lobes as a Boundary Map: The vast radio lobes inflated by jets are the "splashback" from this collision. Therefore, the shape and curvature of these lobes are a direct map of the inner geometry of the local cosmic boundary against which they are pressed.

- A Maximum Jet Radius: If our "island universe" is finite, surveys of radio lobes will reveal a hard cutoff, a maximum radius beyond which no jet can penetrate, corresponding to the measured size of our Cosmic Echo Chamber.

8. Conclusion

With the introduction of the USF, our framework reaches its deepest physical foundation. Beneath the energetic flows of the UEF and resonant oscillations of the ULF lies a field whose vibrations are both energetic and formative. The harmonic modes of this field give rise to the geometric Forms that shape all structure in the universe—from the coherence of a proton to the spiral architecture of a galaxy.

This vision offers a profound reinterpretation of cosmology. Rather than a universe born from a singular explosion, UFD describes a Cosmic Crystal that manifests into a Cosmic Garden—a living universe where new galaxies continually emerge from the deep harmonic symmetries of the USF. This model elegantly resolves the cosmological constant problem by identifying vacuum energy as the non-gravitating hum of the USF. Likewise, the fine-tuning of physical constants is no longer an unexplained coincidence but an inevitable consequence of the USF's unchanging vibrational architecture. The universe is necessarily harmonic.

Emerging from this ceaseless vibratory order is the principle we call the GCF, the physical mechanism that drives the universe's tendency toward harmony. From the crystal lattice to the spiral nebula, this force is the whisper of the USF echoing through the fabric of reality. To understand the USF is not merely to solve old paradoxes but to glimpse the blueprints of creation itself. In this view, every aspect of the physical cosmos—every law, particle, and structure—is part of an ongoing act of harmonic unfolding in what is a fundamentally musical universe.

3.3 A Vibrational Universe and the Cosmic Garden

The first three chapters of this monograph have restructured the foundations of modern cosmology. By introducing a hierarchy of three nested fields—each more fundamental than the last—we have moved beyond the abstract spacetime geometries of General Relativity to a physically grounded framework rooted in resonance, coherence, and field dynamics.

We began in Chapter 1 by reinterpreting gravity as a pressure effect within the Universal Energetic Field (UEF)—a dynamic, tangible medium whose vortices constitute matter and govern the universe's large-scale structure. In Chapter 2, we introduced the Universal Light Field (ULF) as the expanding, systemic medium responsible for cosmological redshift and the propagation of light. Finally, in Chapter 3, we traced both fields to their ultimate source: the Universal String Field (USF), a foundational vibrational substrate whose timeless, geometric harmonies serve as the source code for reality itself. Together, these layers form a coherent ontological hierarchy:

- The USF is the timeless, geometric "source code" of the universe—a field of perfect harmonic resonance whose stable vibrational modes give rise to the fundamental symmetries of nature.

- The UEF is the energetic substance of the universe, from which matter, gravity, and inertia (*see* Chapter 4) emerge through vortical dynamics shaped by the geometry of the USF and guided by the resonance conditions of the ULF.

- The ULF is the medium of propagation and coherence, through which information travels, structure stabilizes, and interaction unfolds via standing waves and phase-locked oscillations.

At the core of this model lies a single, irreducible equation that expresses the ontological logic behind this entire nested structure:

$$F \odot S \odot C \implies U,$$

which translates to

Form × Substance × Communication → Universe.

This triadic model reframes the universe as a resonant continuum, thereby grounding physics in physically real media, whose interrelationships explain the emergence of law, matter, and motion. In doing so, it offers natural resolutions to several paradoxes in modern physics:

- The Cosmological Constant Problem is resolved by recognizing that the vast Zero-Point Energy does not gravitate, since gravity emerges only at the UEF level. The layered ontology prevents the overcounting of vacuum energy.

- The Fine-Tuning Problem disappears when the constants of nature are understood as necessary harmonic ratios derived from the USF's internal geometry—ratios expressed through ULF dynamics and stabilized in the UEF.

- The Landscape Problem dissolves under this framework. Rather than positing possible universes, the USF provides a single, self-consistent vibrational architecture, whose internal coherence defines our cosmos.

- The Extra Dimensions Problem is sidestepped entirely. The USF, ULF, and UEF are modeled as four-dimensional fields operating within observed spacetime, eliminating the need for hidden dimensions while preserving structural richness.

Beyond these fixes, the framework introduces a deeper unifying principle—the Geometric Coherence Force (GCF)—that arises from the recursive layering of vibrations in the USF and functions as an emergent organizing principle. This force subtly shapes the architecture of reality, guiding systems toward optimal configurations across all scales, from subatomic particles to living organisms to spiral galaxies. It also provides a physical source for entropy, which is recast as a direct effect of a physical force.

UFD's nested field ontology naturally leads to Cosmic Garden cosmology, which describes the universe as a vast archipelago of self-contained, encapsulated "island universes." According to this cosmology:

- Galaxies are not relics of a single Big Bang but are continuously born from local Genesis Events within a vast, living cosmic web.

- The CMB is a composite signal: its perfect thermal spectrum is the universal background temperature of the ULF, while the acoustic peaks and anomalies imprinted upon it are the local, holographic echo of our own galaxy's creation, which reverberates within a finite, antimatter-derived boundary.

- The Great Silence of the Fermi Paradox is resolved. If each galaxy is a physically encapsulated "island universe," then matter-based life is effectively prevented from traveling between them, which explains why the cosmos appears empty.

This reinterpretation transforms the CMB from a tool for studying the distant past into a new instrument for mapping our immediate cosmic home. By analyzing the precise geometry of the acoustic peaks and the orientation of the large-scale anomalies, we can begin to chart the architecture of our own cosmic archipelago: its precise size and shape, the primordial spin axis of our galaxy's birth, and the fossilized "scars" of its chaotic creation. The CMB is no longer just a baby picture of the universe; it is a high-resolution architectural blueprint of our own local island in the cosmic sea.

Thus, while direct experimental access to the USF remains beyond current technology, its fingerprints should be discernible through precise astrophysical observations and coherence phenomena at cosmological scales. Ultimately, this framework reveals a novel, physically intuitive, harmonic way of understanding the universe in all its order.

References

1. Arp, H. (1987). *Quasars, Redshifts and Controversies*. Berkeley, CA: Interstellar Media.
2. Alexander, C., Ishikawa, S., & Silverstein, M. (2002). *The Nature of Order: An Essay on the Art of Building and the Nature of the Universe*. Center for Environmental Structure.
3. Anderson, P. W. (1972). More Is Different. *Science, 177*(4047), 393–396.
4. Bousso, R. (2002). *The Holographic Principle. Reviews of Modern Physics, 74*(3), 825–874.
5. Broadhurst, T. J., Ellis, R. S., Koo, D. C., & Szalay, A. S. (1990). Large-scale distribution of galaxies at the Galactic poles. *Nature, 343*(6260), 726–728.
6. Burns, G., & Glazer, A. M. (2013). *Space Groups for Solid State Scientists* (2nd ed.). Academic Press.
7. Caspar, D. L. D., & Klug, A. (1962). Physical principles in the construction of regular viruses. *Cold Spring Harbor Symposia on Quantitative Biology, 27*, 1–24.
8. Chladni, E. (1787). *Entdeckungen über die Theorie des Klanges* [Discoveries Concerning the Theory of Sound]. Weidmanns Erben und Reich.
9. Conway, J. H., & Sloane, N. J. A. (1991). *Sphere Packings, Lattices and Groups*. Springer-Verlag.
10. Cotterell, A. (1996). *The Encyclopedia of World Mythology*. Smithmark Publishers.
11. Coxeter, H. S. M. (1948). *Regular Polytopes*. Methuen & Co.
12. DESI Collaboration. (2025). First Cosmological Constraints from the Dark Energy Spectroscopic Instrument Data Release 1. *The Astrophysical Journal Letters, 985*(2), L24.
13. Diego-Palazuelos, P., et al. (2022). "The Polarized CMB constraints on cosmic birefringence from ACT and Planck." *Physical Review D* 106, 12: 123512.
14. Greene, B. (1999). *The Elegant Universe: Superstrings, Hidden Dimensions, and the Quest for the Ultimate Theory*. W. W. Norton & Company.
15. Guth, A. H. (1981). "Inflationary universe: A possible solution to the horizon and flatness problems." *Physical Review D* 23, 2: 347–356.

16. Hubble, E. (1929). "A relation between distance and radial velocity among extra-galactic nebulae." *Proceedings of the National Academy of Sciences* 15, 3: 168–173.

17. Jenny, H. (1967). *Cymatics: The study of wave phenomena*. Basilius Presse.

18. Joannopoulos, J. D., Johnson, S. G., Winn, J. N., & Meade, R. D. (2008). *Photonic Crystals: Molding the Flow of Light* (2nd ed.). Princeton University Press.

19. Kittel, C. (2004). *Introduction to Solid State Physics* (8th ed.). Wiley.

20. Labbé, I., et al. (2023). "A population of red candidate massive galaxies ~600 Myr after the Big Bang." *Nature* 616: 266–269.

21. Maiolino, R., et al. (2024). A small, metal-poor star-forming galaxy hosting an overmassive black hole at z = 8.7. *Nature, 626,* 990–994.

22. Maldacena, J. (1998). *The Large N Limit of Superconformal Field Theories and Supergravity. Advances in Theoretical and Mathematical Physics,* 2(2), 231–252.

23. Luminet, J.-P., Weeks, J., Riazuelo, A., Lehoucq, R., & Uzan, J.-P. (2003). Dodecahedral space topology as an explanation for weak wide-angle temperature correlations in the cosmic microwave background. *Nature, 425*(6958), 593–595.

24. Misner, C. W., Thorne, K. S., & Wheeler, J. A. (1973). *Gravitation.* San Francisco: W. H. Freeman.

25. Penrose, R. (2004). *The Road to Reality: A Complete Guide to the Laws of the Universe.* Jonathan Cape.

26. Planck Collaboration. (2018). Planck 2018 results. VI. Cosmological parameters. *Astronomy & Astrophysics, 641,* A6.

27. Ramirez, A. P. (1994). Strongly geometrically frustrated magnets. *Annual Review of Materials Science,* 24(1), 453-480.

28. Rubin, V. C., & Ford, W. K., Jr. (1970). Rotation of the Andromeda Nebula from a Spectroscopic Survey of Emission Regions. *The Astrophysical Journal, 159,* 379.

29. Susskind, L. (1995). The World as a Hologram. *Journal of Mathematical Physics, 36*(11), 6377–6396.

30. Susskind, L. (2005). *The Cosmic Landscape: String Theory and the Illusion of Intelligent Design.* Little, Brown and Company.

31. Thomson, W. (Lord Kelvin). (1887). On the division of space with minimum partitional area. *Philosophical Magazine, 24*(151), 503–514.

32. 't Hooft, G. (1993). *Dimensional Reduction in Quantum Gravity. arXiv preprint* gr-qc/9310026.

33. Totani, T. (2025). 20 GeV halo-like excess of the Galactic diffuse emission and implications for dark matter annihilation. *Journal of Cosmology and Astroparticle Physics,* 2025(11), 080.

34. van Dokkum, P., et al. (2023). A candidate runaway supermassive black hole identified by its stellar wake. *The Astrophysical Journal Letters, 946*(2), L51.

35. Weinberg, S. (1972). *Gravitation and Cosmology: Principles and Applications of the General Theory of Relativity*. New York, NY: John Wiley & Sons.

36. Weinberg, S. (1989). The cosmological constant problem. *Reviews of Modern Physics, 61*(1), 1–23.

37. Wigner, E. P. (1960). The unreasonable effectiveness of mathematics in the natural sciences. *Communications on Pure and Applied Mathematics, 13*(1), 1–14.

38. Witten, E. (1986). Non-commutative Geometry and String Field Theory. *Nuclear Physics B, 268*(2), 253–294

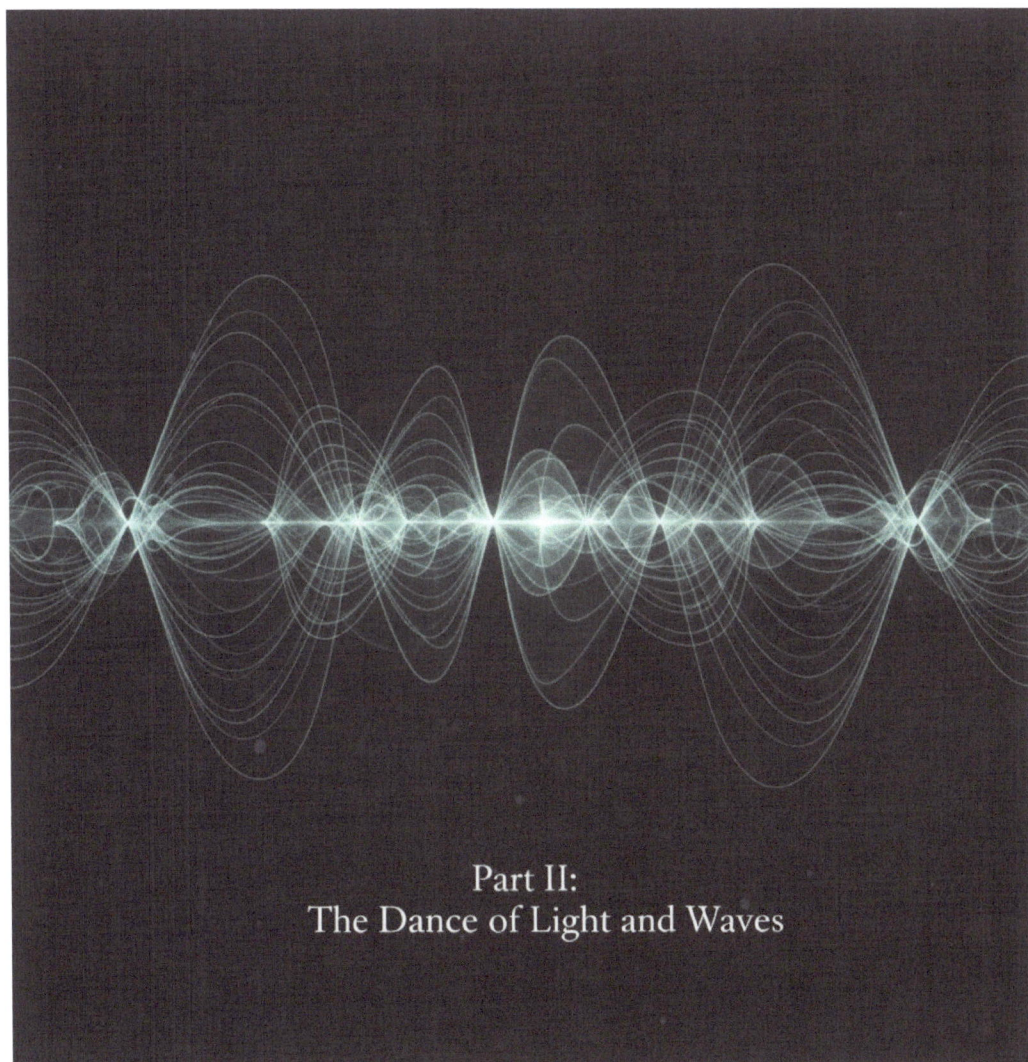

Part II:
The Dance of Light and Waves

Chapter 4: The Fabric of Light

4.1 Light

Of all the phenomena in the natural world, none has been more central to the story of science than light. Our attempts to understand its true nature have driven our greatest intellectual revolutions for millennia, with each new insight revealing a more profound mystery. This chapter will trace that journey—from the classical era to the twin paradoxes that shattered its foundations—to provide the necessary context for the new physical picture we propose.

Even before the great scientific debates, light was recognized as having a deeply mysterious, almost intelligent quality. The simple observation that light travels in perfectly straight lines established an ancient and profound link between its behavior and the elegant logic of geometry (Euclid, c. 300 BCE). This connection was deepened by the discovery of the Principle of Least Action, which revealed that light travels from one point to another along the path of least time, a concept first formalized by Pierre de Fermat (1662). The principle suggests that light seems to "sniff out" all possible routes and unfailingly choose the most efficient one—a strange, almost teleological behavior.

The quest to find a physical model that could account for these mysterious behaviors coalesced over the following centuries into one of the most famous debates in the history of science. This debate centered on a single question: Is light a particle or wave? The "corpuscular" theory, formalized by Isaac Newton (1704), argued that light was composed of tiny, discrete particles, which successfully explained reflection and refraction. However, a competing theory, championed by Christiaan Huygens (1690), proposed that light was a wave propagating through a universal medium. The wave theory could also explain reflection and refraction and, more importantly, could account for the newly discovered phenomena of diffraction and interference, which were impossible to explain with a particle model.

The debate was seemingly settled for good in the 19th century, when the wave theory was experimentally confirmed, most notably by Thomas Young's double-slit experiment (1804), which showed that light beams could interfere with each other just like water waves—a phenomenon impossible to explain with particles. James Clerk Maxwell then gave this experimental reality a theoretical foundation. By unifying the disparate phenomena of electricity and magnetism into a single, elegant set of equations, Maxwell demonstrated that light was a self-propagating electromagnetic wave (Maxwell,

1865). This finding was a triumph of classical physics, a moment of profound unification that provided a complete and satisfying picture of light. At the dawn of the 20th century, just as this classical picture seemed complete, two new experimental results emerged that shattered its foundations and plunged physics into a crisis.

The first was the paradox of light's constant speed. Physicists had long assumed that, like all other waves, light must travel through a medium, which they called the "luminiferous aether." The famous Michelson-Morley experiment of 1887 was designed to detect the Earth's motion through this aether, but surprisingly found no effect (Michelson & Morley, 1887). The speed of light was found to be the same for all observers, regardless of their own motion, a result that defied all classical intuition.

The second was the paradox of its particle-like nature. The photoelectric effect revealed that when light strikes a metal surface, it knocks electrons free, not like a continuous wave but like a barrage of discrete, individual particles. The energy of the ejected electrons depended only on the color (frequency) of the light, not its brightness (intensity); this result could not be explained using classical wave theory.

Albert Einstein resolved these paradoxes by launching the two great revolutions of modern physics: Special Relativity (SR) and Quantum Mechanics. To solve the speed paradox, SR made the constancy of c a fundamental postulate. To solve the photoelectric effect, Einstein proposed the "photon." However, while these theories provided the mathematical rules for light's behavior, they did so by making its strange properties axiomatic. Einstein did not explain *why* the speed of light is constant; he simply accepted it as a rule and derived the consequences for space and time.

Crucially, these revolutions left a third, even older mystery largely untouched: the nature of inertia. While SR redefined mass through $E = mc^2$, it continued the Newtonian tradition of treating inertia—the resistance to acceleration—as an inherent, "built-in" property of matter, providing no physical mechanism to explain *why* matter resists change in motion or *why* that resistance becomes infinite as an object approaches the speed of light.

This chapter contends that the three great mysteries—the wave-particle nature of light, the constancy of *c*, and the origin of inertia—are not separate problems. They are different ways of describing the same underlying reality: the interaction between localized energy and a physical plenum. By introducing the Universal Light Field (ULF) and the Universal Energetic Field (UEF), we aim to provide the missing physical context for the axioms of modern physics. In this chapter, we demonstrate that inertia is the dynamic "drag" of the field, and that the strange effects of relativity are the inevitable, causal consequences of moving through this real, physical medium.

4.2 The Fabric of Light: A Theory of Physical Relativity

Abstract

This chapter introduces Physical Relativity (PR), a framework that resolves the foundational paradoxes of light and matter by proposing a real, physical medium for their existence: the Universal Light Field (ULF) and its underlying substrate, the Universal Energetic Field (UEF). Within this model, the longstanding wave-particle duality is resolved by treating photons as discrete ULF wave packets whose particle-like behavior emerges only during quantized, resonant interactions with matter. This physical picture provides an intuitive origin for the Principle of Least Action, which arises as waves naturally follow the paths of least resistance through the dynamic field. Shifting from light to matter, the chapter establishes the physical origin of inertia as a non-dissipative hydrodynamic drag encountered by a vortex accelerating through the ULF. This mechanistic view reveals that mass is not an intrinsic property but a manifestation of field displacement, providing a causal explanation for the Weak Equivalence Principle and a tangible reinterpretation of $E = mc^2$ as a description of localized field-state energy.

From this foundation, the axioms of Einstein's relativity are reinterpreted as emergent physical effects. The constancy of the measured speed of light is shown to be a self-correcting consequence of physical time dilation and length contraction—objective deformations caused by the motion of resonant systems through the high-pressure field. This approach resolves the paradoxes of Special Relativity (SR) by re-establishing a preferred reference frame tied to the field itself. Ultimately, PR provides a deeper level of explanation by deriving Planck's constant from the fundamental resonance modes of the field and generating novel, falsifiable predictions, including cosmic anisotropies in electromagnetism and objective, asymmetrical time dilation. By grounding quantum and relativistic phenomena in the tangible dynamics of a unified field, this chapter restores a causal and physically intuitive reality to the heart of modern physics.

1. Introduction

At the dawn of the 20th century, a series of experiments revealed that light, the most familiar and fundamental entity in the universe, was also the most mysterious. Two profound and seemingly unrelated paradoxes emerged that shattered the foundations of classical physics and set the stage for the modern era. The first was the paradox of wave-particle duality, revealed most starkly by the photoelectric effect, which showed that in some experiments light behaved as a continuous wave, yet in others, it behaved as a discrete, localized particle. The second was the paradox of its constant speed, revealed by the Michelson-Morley experiment (1887), which demonstrated that the speed of light in a vacuum was the same for all observers, regardless of their own motion—a result that defied all classical intuition.

The twin revolutions of the early 20th century provided mathematical solutions to both paradoxes. The emerging quantum theory, pioneered by Planck and Einstein, addressed the first by making the quantization of energy a core principle. Simultaneously, Einstein's theory of Special Relativity (SR) provided a mathematical solution to the second by making the constancy of the speed of light a fundamental postulate. Together, these theories provided the rules for light's behavior. However, they did so by making these strange properties axiomatic features of reality, without providing a physical mechanism for why they must be so. Furthermore, while these theories redefined our understanding of light and energy, they left the foundational property of matter—inertia—as a primitive, unexplained "inherent" property of mass.

In this chapter, we demonstrate how these mysteries can be resolved with a single, unified framework built upon the existence of the two physical fields introduced in previous chapters: the Universal Energetic Field (UEF) and the Universal Light Field (ULF). We call this unified framework Physical Relativity (PR). Here, we first show how the particle-like nature of light arises from the way its wave-packets interact with the resonant structure of matter, offering a direct, physical explanation for the photoelectric effect, the double-slit experiment, and the Principle of Least Action.

Building on this wave model, we then deconstruct the nature of matter itself, revealing that inertia is not an intrinsic quality but a dynamic result of a vortex moving through the fluidic ULF. By establishing this physical origin for inertia and the equivalence of mass, we demonstrate how the peculiar rules of relativity—time dilation, length contraction, and mass expansion—emerge as inevitable and predictable consequences of motion through this dense physical medium. Through PR, the abstract postulates of the last century are replaced by a tangible, causal mechanics of the field.

2. The Nature of the Photon: A Unified Wave Model

This section deconstructs the foundational paradoxes of light by proposing a new, unified physical model of the photon. Here, we demonstrate how the long-standing mysteries of wave-particle duality, the double-slit experiment, and the Principle of Least Action can all be resolved by reinterpreting the photon as a physical wave packet propagating through the ULF.

2.1 Wave-Particle Duality and the Photoelectric Effect

In standard quantum mechanics, light is said to possess a mysterious "wave-particle duality," having the properties of both a continuous wave and a discrete particle. PR resolves this paradox by providing a single, coherent physical picture: light is fundamentally a propagating wave packet in the ULF. It is a discrete quantum of energy, $E = hf$, traveling through the field at speed c. It is a real, physical wave.

When light strikes a metal surface, it can knock electrons free, but only if the light's frequency is above a certain threshold. Above that threshold, the intensity of the light determines the number of electrons ejected, while the frequency determines their energy. The experimental results for this phenomenon could not be explained by classical wave theory.

PR provides a tangible mechanism that naturally explains these rules. A metal surface is a resonant system.[23] The incoming ULF wave packet (the photon) interacts with the collective resonance of the electrons in the metal. If the photon's energy ($E = hf$) is greater than the binding energy holding an electron in the system (the work function), the entire packet of energy is absorbed and transferred to a single electron, "knocking" it free.

This physical picture perfectly explains the experimental results. The threshold frequency exists because a photon with insufficient energy simply cannot break an electron free. The intensity of the light corresponds to the number of photons arriving per second; more photons mean more absorption events, so more electrons are ejected. Finally, the frequency of the light determines the energy of each individual photon

---

[23] In this context, a resonant system is a physical system that has a set of natural, stable frequencies at which it "wants" to vibrate or oscillate. The electrons in a metal, for instance, are not isolated but are part of an interconnected resonant structure. This collective system can only absorb energy from an incoming light wave if the wave's frequency is high enough to match one of its stable resonant modes—a condition that corresponds to the metal's work function.

packet; a higher-frequency photon delivers a bigger "punch," so the electron it ejects flies out with greater kinetic energy.

Thus, the photoelectric effect is reinterpreted as a direct, observable consequence of the quantized, wave-packet nature of light interacting with the resonant structure of matter.

2.2 The Double-Slit Experiment: A Wave in Action

The quintessential mystery of the quantum world is the double-slit experiment, first performed with light by Thomas Young (1804). When single photons are fired one at a time at a barrier with two slits, they create an interference pattern on the detector screen behind, which traditionally suggests that each photon somehow passes through both slits simultaneously and interferes with itself like a wave (Feynman, 1985). The PR framework resolves this paradox intuitively. In this model, the photon is a real, physical wave packet within the ULF. The mystery dissipates because the photon is not a localized particle moving through space.

When the ULF wave packet arrives at the barrier, it physically passes through both slits simultaneously, much like a water wave, splitting into two coherent wave fronts. These wave fronts propagate forward and interfere, producing bright (constructive) and dark (destructive) bands where their crests and troughs combine or cancel each other. The particle-like detection event thus only occurs upon interaction with the detector screen. This is where the wave's energy localizes and transfers quantized energy to a specific point (Figure 8).

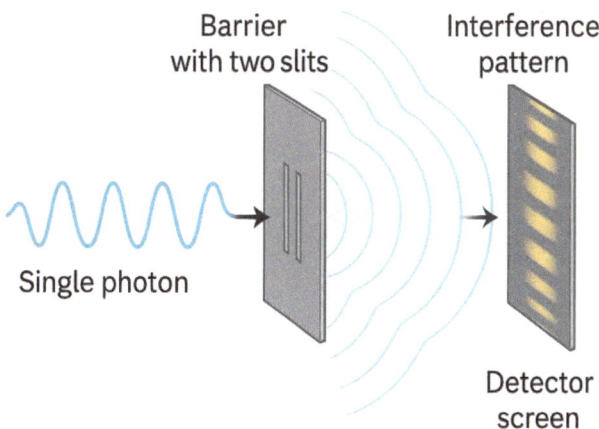

Figure 8: The Photon as a ULF Wave. This figure depicts the double-slit experiment in PR. In PR, a photon is modelled as a real ULF wave packet that splits into two coherent wave fronts upon encountering the slits. These fronts interfere, creating bright and dark bands on the detector screen. This quantized, particle-like detection occurs only at the moment of energy absorption, when the wave localizes into a single point.

This model further explains why placing a detector at a slit disrupts the interference pattern: the act of measurement collapses or localizes the ULF wave at that slit, preventing it from simultaneously passing through both openings and interfering (Feynman, Leighton, & Sands, 1965). Conversely, when the detector is removed, the ULF wave is once again free to pass through both slits unimpeded. The interference pattern instantly reappears because it is the wave's natural, default behavior when it is allowed to interfere with itself.

Thus, rather than being evidence of quantum weirdness, the double-slit experiment is a natural manifestation of the photon's true wave nature, which allows it to explore all paths simultaneously.

2.3 The Principle of Least Action

One of the most elegant and mysterious principles in all of physics is the Principle of Least Action. It states that when an object, such as a photon, travels from point A to point B, it does not take just any path. Instead, it "sniffs out" all possible paths and chooses the single one for which a quantity called action is minimized. Mathematically, action S is defined as the integral over time of the Lagrangian L,[24] or the difference between kinetic energy T and potential energy V:

$$S = \int_{t^1}^{t^2} L dt = \int_{t^1}^{t^2} (T - V) dt$$

While this principle is compelling, as one can derive almost all classical and quantum mechanics from it, it remains deeply counterintuitive. How does a particle "know" about all possible paths to select the best one? This suggests a kind of foresight without an obvious physical explanation (Goldstein, Poole, & Safko, 2002; Feynman, Leighton, & Sands, 1965).

PR provides a direct, physical mechanism that demystifies this principle. In PR, the ULF is a real, physical medium through which a photon propagates as a wave packet. Thus, the Principle of Least Action is not merely an abstract mathematical rule; it is a direct consequence of wave propagation in a physical medium—the path of least

---

[24] The *Lagrangian* is a function that summarizes the dynamics of a physical system. It is typically defined as the difference between the system's kinetic energy (the energy of motion) and its potential energy (the energy stored due to its position). By analyzing the Lagrangian, one can derive the equations of motion that govern the evolution of a system over time. It is central to the Principle of Least Action and plays a key role in classical mechanics and quantum physics.

action is the path of least resistance. The situation is analogous to a small boat guided by river currents. The boat does not need to calculate all possible downstream paths; it simply follows the natural currents. The path taken reflects the dynamic flow of the river. Similarly, a photon traveling through space follows the "currents" of the ULF. Near massive objects, as established in Chapter 1, pressure gradients create nonuniformities in the ULF, and the photon naturally follows the path of least resistance through these variations, which corresponds to the geodesic in General Relativity.

A key physical quantity in this context is action, denoted S, which has the dimensions of energy multiplied by time. The fundamental quantum of action, Planck's constant, $\hbar$, corresponds to the product of the smallest discrete energy packet E and its oscillation period T, or equivalently, its energy divided by frequency f:

$$\hbar = E \times T = \frac{E}{f}$$

This relationship expresses the minimal, indivisible unit of action exchanged in the quantum realm. Within the UFD framework, quantization emerges naturally from geometric constraints imposed by vortex resonance. The oscillations of energetic vortices are restricted to discrete modes whose energy-time product satisfies this fundamental relation (Planck, 1901). Thus, the Principle of Least Action is reinterpreted from a mysterious, almost teleological rule to a tangible and intuitive consequence of fluid dynamics and nested field structure.

In sum, PR resolves the traditional paradoxes of wave-particle duality by grounding quantum phenomena in tangible wave mechanics. Together, these insights provide a coherent and elegant physical foundation for light and quantum behavior, laying the groundwork for the deeper principles explored throughout this monograph.

However, if the ULF is a real, physical medium capable of supporting these propagating wave-packets, we must ask: how does it interact with the more complex, localized structures we call "matter"? If a photon is a traveling wave, then a particle is a stable, localized vortex of energy within that same field. This shift from light to matter leads us directly to the physical origin of the most fundamental property of all physical objects: nertia.

3. The Physical Origin of Inertia: Resistance in a Real Medium

In classical physics, the concept of inertia is enshrined in Newton's First Law: an object at rest stays at rest, and an object in motion stays in motion unless acted upon by an external force (Newton, 1687). Inertia is the name given to the inherent resistance to any change in a body's state of motion. While Newton's laws perfectly describe the *function* of inertia, they do not explain what it is or where it comes from; it is simply accepted as an intrinsic, primitive property of matter.

The UFD framework provides a tangible physical explanation for this resistance. Here, inertia is revealed to be an emergent interaction between a stable vortex and the ULF—the fluidic medium that emerges from the high-pressure substrate of the UEF.

3.1 Inertia as Hydrodynamic "Added Mass"

It is critical to distinguish this field-based resistance from conventional frictional drag. The resistance of the ULF is a perfectly conservative, non-dissipative force that only manifests during acceleration—a phenomenon known in fluid dynamics as Hydrodynamic Added Mass.

When a vortex (particle) moves at a constant velocity through the ULF, the field flows symmetrically around it, resulting in a zero net force. However, to accelerate (change velocity), the vortex must physically displace the surrounding field, creating a pressure imbalance. This action builds a distortion halo—a high-pressure soliton that travels with the object (Figure 9).

- Inertia is the physical work required to build this distortion halo.

- Momentum is the kinetic distortion energy stored within this halo, which carries the particle forward through the medium.

The reason why inertia and mass are ultimately connected to the ULF, not the UEF, is due to the physical properties of these respective fields. The ULF is a relativistic, compressible superfluid. Its near-zero viscosity ensures lossless propagation of light and coherence of clocks, while its finite compressibility allows it to store distortion energy under motion. This stored energy manifests as inertia and is quantitatively captured by the Lorentz factor (*see* Section 5). In contrast, the UEF is an incompressible superfluid whose rigidity supports stable matter vortices and gravitational pressure gradients but cannot generate inertial resistance. Inertia, therefore, does not arise from the densest field, but from the most deformable coherent one.

3.2 Resolving Mach's Principle and Newton's Bucket

This physical origin for inertia provides a definitive answer to Mach's Principle (Mach, 1883). Mach famously argued that inertia is not an intrinsic property of a body but arises from its interaction with all the distant matter in the universe. While Mach's intuition was correct—inertia *is* relational—he lacked a physical mechanism to explain it.

In UFD, inertia is not a mysterious, long-range influence from distant stars; it is the local drag an object encounters as it moves through the ULF. This also resolves Newton's Bucket experiment (Newton, 1687). The water surface becomes concave in a spinning bucket not because of "absolute space," but because the water is accelerating through the local physical substance of the ULF. The "centrifugal force" is the tangible result of the water's interaction with the field's resistance.

This is analogous to an object coasting through a perfect superfluid; it moves forever without slowing because there is no friction, but it resists any attempt to change its velocity because such a change requires a reorganization of the fluid's pressure gradients. A more massive object (a more significant vortex) creates a larger pressure differential when accelerated and thus requires a more energetic distortion halo to move.

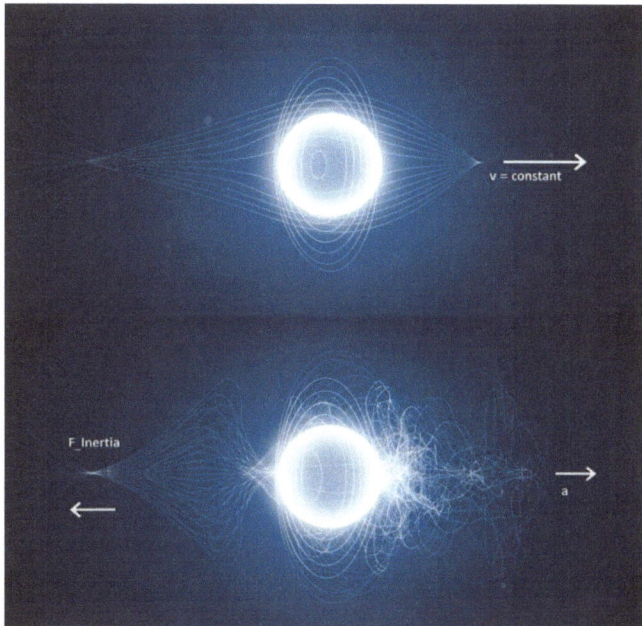

Figure 9: Inertia Through the Universal Plenum. (Top) A massive object moving at a constant velocity experiences no resistance. The surrounding field, shown as smooth, undistorted streamlines, flows seamlessly around it. (Bottom) When an object accelerates, it actively displaces the surrounding UEF. This action disrupts the coherent flow, creating a high-pressure distortion halo in front of the object. The work required to build and maintain this displacement field is the physical origin of inertia.

3.3 The Equivalence of Mass: Unifying Gravity and Inertia

The dual nature of the fields—where the UEF provides the baseline pressure and the ULF acts as the medium of motion—provides a natural explanation for the Weak

Equivalence Principle (WEP). In standard physics, the equality of inertial mass (m_i) and gravitational mass (m_g) is a profound coincidence. In UFD, they are two effects of the same underlying cause:

- Gravitational Mass (m_g) is determined by the volume of displacement: the size of the low-pressure "sink" the vortex maintains in the UEF (*see* Chapter 1).

- Inertial Mass (m_i) is the energy cost of displacement: the work required to move that "sink" and its associated distortion halo through the ULF.

Because these two properties are mechanically coupled to the same geometric structure (the vortex), they must be equivalent. A larger vortex naturally creates a larger gravitational "sink" in the UEF and a larger inertial "drag" in the ULF.

3.4 The Physical Meaning of $E = mc^2$

Within this framework, Einstein's famous equation, $E = mc^2$, is reinterpreted as a direct statement of field-state energy. If a photon is a wave packet traveling linearly through the ULF at speed c, then a particle is a wave packet that has captured itself in a closed-loop, resonant rotation. In this view, matter is quite literally "frozen light."

The internal energy (E) of a particle is the energy of this relativistic rotation. As this confined energy circulates, it maintains the geometric structural integrity of the particle against the ambient pressure of the UEF. The equation $E = mc^2$ thus becomes a direct statement of cause and effect:

- The Mass (m): The geometric footprint (displacement) of the vortex.

- The Energy (E): The internal circulation required to sustain that footprint.

- The c^2 factor: The fundamental tension (or "stiffness") of the UEF/ULF system, determining the rate at which energy must circulate to maintain a stable, localized structure.

Thus, matter does not "contain" energy; matter **is** a specific, localized state of energetic field-circulation. When a particle is annihilated, the vortex simply unravels, and its localized rotational energy is released back into the ULF as propagating wave packets—photons.

This interpretation also clarifies the relationship between Einstein's mass–energy equivalence and Planck's quantum relation, $E = hf$. When a vortex unravels, its internal circulation ceases to be confined and instead reappears as linear oscillation in the ULF. The frequency f of the emitted photon is therefore the liberated circulation rate of the former vortex, scaled by Planck's constant as the quantum of geometric action. In this sense, $E = mc^2$ and $E = hf$ are complementary descriptions of the same field energy before and after topological confinement.

4. Physical Relativity: The Field Speed Limit and the Lorentz Factor

In the previous section, we established that inertia is not an inherent property of matter, but a measure of the resistance encountered by a vortex moving through the ULF. As an object accelerates, it must build and carry a distortion halo—a localized high-pressure region in the field. PR explores the logical limit of this interaction. If inertia is field-resistance, then what happens when an object's velocity (v) approaches the maximum speed at which the ULF can transmit a disturbance (c)? The answer provides a tangible, causal mechanism for the effects of SR, transforming them from mathematical postulates into observable consequences of fluid dynamics.

In SR, the constancy of the speed of light is a postulate that is assumed to be true because experiments show it to be so. In PR, the speed of light (c) is a physical property of the medium: it is the wave-propagation speed of the ULF.

As a vortex (matter) moves through the UEF/ULF substrate, it exerts pressure on the field ahead of it. At all velocities, the field responds by deforming and redistributing, producing the inertial distortion halo described in the previous section. As v increases, this response becomes increasingly asymmetric: the field ahead of the vortex compresses while the field behind rarefies. As v approaches c, the field can no longer relax fast enough to maintain near-symmetry, and the compression intensifies into a stable, high-energy soliton—a self-reinforcing wave packet that the object is forced to carry. The mathematics of this continuously increasing field compression are precisely captured by the Lorentz factor, γ.

$$\gamma = \frac{1}{\sqrt{1 - v^2/c^2}}$$

- Time Dilation: An observer's clock is a physical, resonant system. The persistent high-pressure state of the "field compression" physically slows down the rate of all processes within the moving reference frame, including the ticking of the clock, by a factor of $1/\gamma$. From the perspective of a stationary observer, the clock is literally running slower.

101

- Length Contraction: Similarly, the increased pressure of the "field compression" in the direction of motion physically squeezes the atomic bonds of a measuring rod, causing it to contract by a factor of $1/\gamma$. From the perspective of a stationary observer, the moving ruler is literally shorter.

- Relativistic Mass Expansion: The "extra mass" is not in the particle itself; it is the real, physical energy of the ULF soliton that the particle is now forced to carry. The total energy of this object-plus-field-compression system scales with γ (as in $E = \gamma mc^2$) and its inertia (resistance to further acceleration) increases exponentially.

These effects are the key to solving the paradox of light's constant speed. When an observer in a fast-moving reference frame measures the speed of a passing photon ($c = \Delta x/\Delta t$), their measurement tools have been physically altered by their motion. Their ruler is shorter (they measure a shorter Δx), and their clock is slower (they measure a shorter Δt).

A **B**

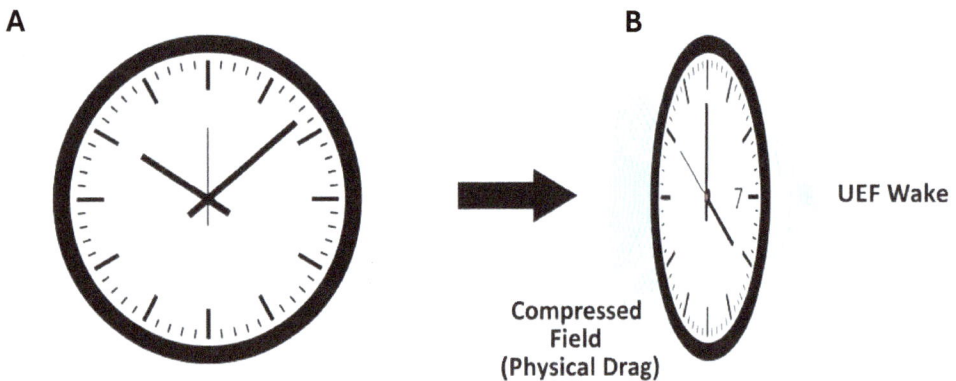

Compressed
Field
(Physical Drag)

UEF Wake

Figure 10: The Physical Mechanisms of Relativity. This figure illustrates how the effects of SR emerge from motion through the ULF. SR is a kinematic effect of ULF distortion, whereas General Relativity–like behavior emerges from UEF pressure geometry.
(A) The Stationary Clock: A clock at rest relative to the ULF experiences a baseline rate of physical processes.
(B) The Moving Clock: A clock moving at high velocity must actively displace the surrounding field, creating a high-energy distortion halo. This energy burden objectively slows all its internal processes. This mechanism has three significant consequences: it causes the moving clock to tick more slowly (time dilation), physically compresses the clock in the direction of motion (length contraction), and increases the object's effective inertia by forcing it to carry a massive energy field (relativistic mass expansion). PR thus provides a tangible, causal explanation for the effects of relativity.

The genius of nature is that these physical effects, both governed by γ, are perfectly calibrated to cancel each other out in the final calculation. The ratio of the shorter distance to the shorter time always yields the exact same number: c (Figure 10).

Thus, the constancy of the measured speed of light is not a mysterious, fundamental law of spacetime. It is a perfect, self-correcting "illusion" created by the tangible physical interaction between matter and the field in which it is embedded. By positing a physical medium through which matter moves, PR re-establishes a preferred reference frame (the frame at rest with the field), confirming that time dilation is a real, physical, and asymmetrical effect.

5. Predictions and Divergences from Special Relativity

PR is designed to recover all of SR's successful predictions in standard experimental tests. Therefore, in most local experiments, the predictions of SR and PR will be identical. However, PR addresses deep conceptual anomalies within SR and makes new, testable predictions that go beyond it.

5.1 A Prediction of Cosmic Anisotropies

Einstein's theory is based on the assumption of a perfect, empty, isotropic vacuum. PR, in contrast, posits that the vacuum is filled with the ULF, whose large-scale structure corresponds to the cosmic web. This distinction leads to a new, testable prediction: an extremely sensitive experiment may detect minute second-order anisotropies in the laws of electromagnetism arising from the motion of our solar system relative to the large-scale structure of ULF.

These anisotropies would manifest as subtle, directional dependencies in fundamental constants, such as vacuum permittivity. The detection of such a cosmic anisotropy, which SR strictly forbids, would establish the existence of a preferred reference frame.

5.2 A Test of Objective Time Dilation

PR's reinterpretation of time dilation as an objective physical process, rather than a reciprocal effect, yields a conceptually clear experimental test. While experiments like the Hafele-Keating experiment (1971) have shown asymmetrical aging for clocks flown around the world, these results were a predicted consequence of combining two different effects in Einstein's relativity (kinematic and gravitational time dilation). They did not test the fundamental reciprocity of time dilation between two observers in a simple, high-speed flyby.

A proposed experiment would isolate this effect. It would involve two perfectly synchronized atomic clocks. Clock A would remain in a stationary laboratory on Earth (approximately at rest in the UEF's preferred frame). Clock B would be placed on a spacecraft that makes a high-speed pass of the laboratory. Both clocks would then emit and receive a continuous stream of light pulses from the other. This design leads to competing predictions:

- SR predicts that the effect will be perfectly reciprocal. The lab observer will see the spacecraft's clock ticking more slowly, and the spacecraft observer will see the lab's clock ticking more slowly as well.

- PR predicts the effect will be asymmetrical. The lab observer will see the spacecraft's clock ticking more slowly due to the physical "drag" of its motion through the UEF. However, the observer on the spacecraft, whose own physical processes are objectively slowed, will perceive the stationary lab clock ticking faster relative to their own.

The detection of such an asymmetry would provide compelling experimental evidence for the existence of a preferred reference frame and would confirm that time dilation is a real, physical consequence of motion through the UEF, not a perspectival illusion.

6. Resolving the Paradoxes of Relativity

While SR is a spectacularly successful physical theory, its foundational postulate—that there is no "aether" or preferred reference frame—leads to a set of unsettling paradoxes that challenge our most basic intuitions about reality. PR resolves these paradoxes by re-establishing a physical medium for light and massive objects—the UEF and ULF—which provides a universal reference frame and restores a single, objective reality.

6.1 The Relativity of Simultaneity

The most famous consequence of rejecting a preferred frame is the relativity of simultaneity, classically illustrated by Einstein's thought experiment of two lightning bolts striking a moving train. An observer on the platform, equidistant from the strikes, perceives them as simultaneous. An observer on the moving train, however, sees the light from the forward strike first and concludes the events were not simultaneous. In SR, both observers are considered equally correct.

PR dissolves this paradox. By positing that the UEF is a physical medium with a rest frame, it restores the classical and intuitive notion of a single, universal "present moment." In the thought experiment, the platform observer, being at rest with respect to the medium, is the one who perceives reality "correctly." Though an observer on the train has a real experience, it is a perspectival artifact created by their physically distorted clocks and rulers.

On a cosmic scale, the relativity of simultaneity leads to the even more unsettling Andromeda Paradox. This thought experiment demonstrates that two observers on Earth, by simply walking past each other, can have different "planes of simultaneity" that intersect the distant Andromeda galaxy at different times, separated by more than a week. This means that for one observer, an event in Andromeda (such as the launching of an alien fleet) is in their future, while for the other observer, it is already in their past. It also leads to the bizarre conclusion that the future is both fixed and not fixed, posing a significant philosophical challenge.

PR's single, objective "present moment," defined by the rest frame of the ULF, resolves this paradox completely. There is only one "now" across the entire cosmos. The different "nows" calculated by moving observers are not different realities but are instead measurement artifacts. The fate of the alien fleet is an objective fact, not a matter of perspective.

6.2 The Twin Paradox

A related challenge is the famous Twin Paradox. The thought experiment involves two identical twins: one who remains on Earth and another who undertakes a high-speed, relativistic round-trip to a distant star. The paradox arises from the principle of reciprocity. According to SR, the Earth-bound twin sees the traveler's clock as running slow. At the same time, the traveling twin, from their perspective, also sees the Earth-bound clock as running slow. This leads to a logical contradiction about who is genuinely younger upon their reunion.

PR provides a direct, physical solution. Time dilation is an objective, asymmetrical process caused by the energy cost of displacing the Universal Plenum.

The traveling twin is the one who is physically accelerating through the medium. To do so, they must build and sustain a high-energy distortion halo (soliton). The energy required to maintain this field displacement acts as a load on their system, objectively slowing down their physical and biological processes. There is no ambiguity. The traveling twin is the truly younger one. If the traveling twin were to pass by the stationary twin and could listen to the ticks of the stationary twin's clock, they would hear it ticking faster than their own, reflecting the true difference in their metabolic rates.

6.3 The Black Hole Observer Paradox

The ultimate paradox of relativity occurs at the event horizon of a black hole, where its gravitational "current" reaches the speed of light. Einstein's relativity tells two contradictory stories: while a falling observer passes through the event horizon in a finite amount of their own time, a distant observer sees them slow down and freeze, their image trapped there for eternity.

UFD resolves this paradox by revealing that the two observers are describing two distinct physical phenomena occurring in two different fields:

- The Falling Observer (Matter/UEF Phenomenon): An object falling into a black hole is in a state of perfect "free fall," flowing in sync with the UEF's accelerating gravitational current. Because the object is moving *with* the medium rather than pushing through it, it generates no distortion halo and experiences no inertial resistance (g-force). The physical person—a complex UEF vortex—passes through the event horizon effortlessly, carried by the bulk flow of the Plenum.

- The Distant Observer (Light/ULF Phenomenon): The distant observer is not watching the person directly; they are watching the light (ULF waves) emitting from the person. These light waves must fight their way upstream against the inflowing UEF current. At the event horizon, the inward speed of the UEF current (c) perfectly cancels the outward propagation speed of the light wave (c). The last photon is thus frozen in place, like a salmon trying to leap up a waterfall but being held forever at the lip.

There is no paradox. The two stories are consistent because they describe different mechanics: the physical transport of Matter (carried by the UEF) versus the propagation of Information (struggling through the ULF).[25] Thus, by distinguishing the medium (UEF) from the message (ULF), PR explains the mechanics of relativity and restores a causal, intuitive universe.

7. Conclusion

In this chapter, we have demonstrated that the foundational paradoxes of modern physics—wave-particle duality, the origin of inertia, and the constancy of the speed of light—need not be treated as separate, axiomatic mysteries. Instead, we can understand each as an emergent consequence of the dynamics of a single, unified physical substance: the UEF and its fluidic manifestation, the ULF. By replacing abstract geometry with field mechanics, we have shown that the "particle-like" nature of the photon is an interaction effect created by quantized resonance, while the "inherent" property of inertia is the tangible result of a vortex encountering hydrodynamic resistance within the ULF.

The significance of this PR framework is threefold. First, it provides a single, coherent origin for the seemingly disparate rules of quantum mechanics, classical mechanics, and SR. By deriving mass, inertia, and $E = mc^2$ from field-vortex interactions, we unify the "push" of gravity with the "drag" of inertia, providing a physical solution to the WEP. Second, it resolves the deepest philosophical problems of Einsteinian relativity—such as the paradox of simultaneity—by re-establishing a preferred reference frame tied to the UEF and restoring a universal "present moment" to the cosmos. This removes the necessity for "warped" time as a fourth dimension, replacing it with the objective slowing of physical resonance in a moving frame. Finally, by positing a physical medium with a large-scale structure, the theory makes novel, falsifiable predictions, including the existence of a subtle cosmic anisotropy in the laws of electromagnetism and measurable differences in asymmetrical time dilation.

Ultimately, this work suggests that the strange rules of light and the persistent resistance of matter are not inexplicable features of a void. Instead, they are logical and predictable consequences of localized energy interacting with the nested fields of reality. This shift from an abstract, mathematical description to a causal, mechanical one restores

---

[25] With respect to the infalling object, its information would be resonantly assimilated by the black hole's perfect coherence. Rather than being erased, its unique geometric and resonant pattern is perfectly and coherently re-encoded into the harmonic structure of the super-vortex itself. The information is never lost; it is simply translated into a more stable and enduring form on the black hole's surface, as described in Chapter 1.

intuition to physics and provides the necessary foundation for exploring the complex geometric structures of the atom and the chemical principles that follow.

4.3 Physical Relativity

In this chapter, we introduced Physical Relativity (PR), which describes the Universal Light Field (ULF) as the physical medium underlying the propagation of light waves and the Universal Energetic Field (UEF) as the substrate of matter. By grounding these phenomena in a tangible field dynamic, PR offers a unified, mechanistic resolution to the foundational paradoxes of modern physics, from wave-particle duality and the origin of inertia to the constancy of the speed of light.

Wave-particle duality is resolved by recognizing that the photon is a propagating ULF wave packet whose apparent "particle-like" behavior arises only during quantized, resonant interactions with matter. Within this framework, the Principle of Least Action is given a direct physical cause rather than remaining an abstract mathematical rule. Photons follow the paths of least resistance dictated by the density and flow of the field, making this principle a natural consequence of wave propagation in a real medium. This allows us to derive Planck's constant from first principles as the minimal quantum of action, expressed as the product of the smallest discrete energy packet and its oscillation period. Fundamental quantization thus emerges from the geometric resonance constraints of vortex dynamics. Similarly, the photoelectric effect gains a direct physical explanation: the energy threshold for electron emission arises from the photon's interaction with the resonant structure of the material, where the wave's energy must match a stable mode to liberate an electron.

Crucially, this framework provides the first causal explanation for inertia and the Weak Equivalence Principle (WEP). By identifying inertia as the "hydrodynamic added mass" of a vortex accelerating through the ULF, we move beyond Newton's treatment of inertia as an inherent property. We show that gravitational mass (a vortex's volume in the UEF) and inertial mass (the energy cost of displacing the ULF) are mechanically coupled and therefore equivalent. This allows for a deeper interpretation of $E = mc^2$ as a statement of field-state energy, where mass is the geometric footprint of a vortex and energy is the internal circulation required to sustain that footprint against the ambient pressure of the field.

Finally, the constancy of the measured speed of light (c) is revealed to be a consequence of motion through the Universal Plenum. Time dilation and length contraction are not "warps in spacetime" but real, displacement-induced effects where the high-pressure soliton surrounding a moving object physically slows its internal resonance and compresses its atomic bonds. These effects conspire to make the observed speed of light invariant in all reference frames, creating a "perfect illusion" of constancy. Unlike Special Relativity, however, PR makes the falsifiable prediction of asymmetrical time dilation. While standard theory predicts a reciprocal effect, PR predicts that a moving observer—whose physical processes are objectively slowed by field displacement—would perceive a stationary clock as running faster.

The validation of PR would be transformative, as it would bridge the conceptual rift between quantum mechanics and relativity, opening the door to a new research program aimed at mapping the structure of the ULF and UEF, testing for cosmic anisotropies, and restoring causality to the heart of nature.

Ultimately, this work suggests that the "strange" behavior of light and the resistance of matter arise naturally from the dynamics of the very fields that constitute them, restoring a physically intuitive reality to the most fundamental domains of science.

References

1. Einstein, A. (1905b). Über einen die Erzeugung und Verwandlung des Lichtes betreffenden heuristischen Gesichtspunkt (On a Heuristic Point of View Concerning the Production and Transformation of Light). *Annalen der Physik, 17*(6), 132–148.
2. Einstein, A. (1905c). Zur Elektrodynamik bewegter Körper (On the Electrodynamics of Moving Bodies). *Annalen der Physik, 17*(10), 891–921.
3. Euclid. (2006). *Optics*. (R. Smith, Trans.). American Philosophical Society. (Original work c. 300 BCE).
4. Fermat, P. (1662). *Synthèse pour les réfractions* [Synthesis for Refractions]. In P. Tannery & C. Henry (Eds.), *Œuvres de Fermat* (Vol. 3, pp. 156-159). Gauthier-Villars.
5. Feynman, R. P., Leighton, R. B., & Sands, M. (1965). *The Feynman Lectures on Physics* (Vol. 3). Addison-Wesley.
6. Feynman, R. P. (1985). *QED: The Strange Theory of Light and Matter*. Princeton University Press.
7. Goldstein, H., Poole, C., & Safko, J. (2002). *Classical Mechanics* (3rd ed.). Addison Wesley.

8. Hafele, J. C., & Keating, R. E. (1972). Around-the-World Atomic Clocks: Predicted Relativistic Time Gains. *Science, 177*(4044), 166–168.

9. Huygens, C. (1690). *Traité de la Lumière* (Treatise on Light).

10. Lorentz, H. A. (1904). Electromagnetic phenomena in a system moving with any velocity smaller than that of light. *Proceedings of the Royal Netherlands Academy of Arts and Sciences, 6*, 809–831.

11. Mach, E. (1960). *The Science of Mechanics: A Critical and Historical Account of Its Development.* (T. J. McCormack, Trans.). Open Court. (Original work published 1883).

12. Maxwell, J. C. (1865). A dynamical theory of the electromagnetic field. *Philosophical Transactions of the Royal Society of London, 155*, 459–512.

13. Michelson, A. A., & Morley, E. W. (1887). On the Relative Motion of the Earth and the Luminiferous Ether. *American Journal of Science, s3-34*(203), 333–345.

14. Newton, I. (1999). *The Principia: Mathematical Principles of Natural Philosophy* (I. B. Cohen & A. Whitman, Trans.). University of California Press. (Original work published 1687).

15. Newton, I. (1704). *Opticks: or, a Treatise of the Reflexions, Refractions, Inflexions and Colours of Light.*

16. Planck, M. (1901). On the Law of Distribution of Energy in the Normal Spectrum. *Annalen der Physik, 4*(3), 553–563.

17. Young, T. (1804). The Bakerian Lecture: Experiments and Calculations Relative to Physical Optics. *Philosophical Transactions of the Royal Society of London, 94*, 1–16.

Chapter 5: The Resonant Field Interpretation

5.1 Quantum Mechanics

Quantum mechanics is the theory of physics that describes the behavior of nature at the smallest scales of atoms and subatomic particles. It is the most successful theory in scientific history, forming the basis of our understanding of chemistry, material sciences, and all modern electronics. Originating in the early 20th century, the theory was born from the failure of classical physics to explain phenomena such as black-body radiation and the photoelectric effect, which led physicists like Max Planck and Albert Einstein to introduce the revolutionary concept that energy is quantized, or comes in discrete packets.

From these early days, the field evolved rapidly into the modern quantum mechanics of the mid-1920s, with Erwin Schrödinger's wave equation (Schrödinger, 1926) and Werner Heisenberg's matrix mechanics (Heisenberg, 1925) providing a complete mathematical framework. However, the theory comes with a set of persistent paradoxes that challenge our classical intuition. These include wave-particle duality, which in this case applies to the duality of matter, not light; the measurement problem, which questions how a probabilistic wave function "collapses" into a definite reality upon measurement; and quantum entanglement, which links particles over vast distances in a way Einstein famously described as "spooky" (Einstein et al., 1935).

Several interpretations have been proposed to explain these paradoxes. These include the standard Copenhagen Interpretation (Bohr, 1928), the Many-Worlds Interpretation (Everett, 1957), and Bohmian Mechanics (Bohm, 1952), among many others. The main challenge shared by all of these "interpretations" is that none interprets the underlying physical reality. This has led to a deep philosophical unease. Quantum mechanics requires us to accept a world governed by abstract probability, fundamental uncertainty, and a mysterious "collapse" of reality upon measurement. It provides us with the rules of the quantum game without explaining why the rules are as they are.

This chapter provides that missing physical picture. Building on the principles of Unified Field Dynamics (UFD), we detail the Resonant Field Interpretation (RFI), a model that provides a physically intuitive ontology for the quantum world. We propose that the strange rules of quantum mechanics are not fundamental but are the emergent, deterministic behaviors of a real, physical medium. By exploring the nature of this medium, this chapter aims to replace the paradoxes of quantum mechanics with the intuitive physics of a resonant universe.

5.2 The Resonant Field Interpretation: A Physical Basis for Quantum Mechanics

Abstract

This chapter introduces the Resonant Field Interpretation (RFI), a new physical ontology designed to resolve the counterintuitive, probabilistic foundations of quantum mechanics. The RFI proposes that quantum rules are not fundamental but are the emergent, deterministic behaviors of a physical medium.

We posit that the stable atom is a resonant system formed by the interaction between a Universal Energetic Field (UEF) vortex (the nucleus) and the emergent Universal Light Field (ULF) (the orbitals). In this model, the electron is reinterpreted as a ULF vortex sustained by a three-dimensional standing wave. This topological model provides a physical origin for the electron's spin-1/2 and charge and quantitatively explains the electron-proton mass ratio (1836:1) as the result of the ULF's hydrodynamic slip ratio (1/137) acting upon the proton's topological impedance.

This model also provides a physical basis for the Schrödinger equation, a mechanistic explanation of wave-function collapse (the disruption of resonance), and a new perspective on quantum entanglement. Beyond providing conceptual clarity, the framework opens new technological frontiers, including Resonance Stabilization for quantum computing. The RFI aims to replace the probabilistic nature of quantum mechanics with an intuitive, deterministic, and technologically generative physical reality.

1. Introduction

The theory of quantum mechanics is, without question, the most successful theory in scientific history. Its mathematical formalism, centered on the Schrödinger equation, predicts the behavior of the subatomic world with impeccable precision. Yet, for all its predictive power, the foundations of quantum mechanics remain a source of deep philosophical unease that has sparked a century of debate (Bohr, 1928; Einstein et al., 1935). The theory forces us to accept a world governed by abstract probability waves, fundamental uncertainty, and the mysterious "collapse" of reality upon measurement. It provides the rules of the quantum game without offering a physical explanation for why they are the way they are.

This chapter provides that missing physical picture. Building on the principles of Unified Field Dynamics (UFD) established in the preceding chapters, we now present the Resonant Field Interpretation (RFI). The RFI is founded on the postulate that the stable atom is a resonant system created by the interaction between a UEF vortex (the nucleus) and the emergent ULF. In this model, the electron vortex achieves stability by locking into a three-dimensional standing wave. This single postulate of a UEF-ULF resonance provides a direct physical origin for the principles of quantum mechanics.

In the sections that follow, we will show how the RFI provides a mechanistic basis for the de Broglie wavelength, the Schrödinger equation, and the Heisenberg Uncertainty Principle, while also offering intuitive explanations for wave-function collapse and quantum entanglement. Ultimately, the RFI, like other UFD frameworks, aims to replace the abstract, probabilistic rules of quantum theory with the intuitive, deterministic physics of field dynamics, opening the door to new insights and technologies.

2. The Electron: A Vortex in the Universal Light Field

To build a physical model of the quantum world, we must begin with its most fundamental actor: the electron. In the RFI, the electron is not a dimensionless point particle. Instead, it is a stable, self-sustaining vortex in the ULF. This postulate places the electron in a different class of reality from protons and neutrons (which are UEF vortices) and provides a direct, physical origin for its mysterious quantum properties.

2.1 The Electron as a ULF Vortex

In the Standard Model, the electron is an enigma: a zero-dimensional point particle with no size or internal structure, yet it possesses intrinsic properties like mass, charge, and "spin." The RFI, in contrast, proposes that an electron is a localized, self-sustaining vortex of the ULF—in essence, it is light with high angular momentum. (Figure 11). This unified picture provides a direct, physical origin for the electron's key properties, revealing a clear causal chain:

1. It begins with the literal, physical angular momentum of the ULF vortex.

2. The "handedness" of its rotation—its chirality—determines its charge polarity, causing it to act as a "sink" in the ULF (see Chapter 6).

3. The spinning vortex induces a coherent rotational eddy in the surrounding field, creating the electron's magnetic moment, which is what we experimentally measure as "spin-up" or "spin-down." (see Chapter 6).

This reinterpretation of the electron as a form of topologically trapped light aligns with prior theoretical work that has sought a physical, geometric origin for the electron's properties (Williamson & van der Mark, 1997).[26]

2.2 The Origin of Spin-1/2: The Geometry of the Vortex

This vortex model of the electron provides a physical, geometric origin for the electron's most mysterious quantum property: its spin-1/2 nature. In the Standard Model, this property is an abstract axiom, famously described as requiring a particle to be rotated 720 degrees (two full turns) to return to its original state (Feynman, 1965).

The RFI explains the electron's spin as a direct consequence of its vortex's toroidal geometry. A toroidal vortex has two distinct types of rotation happening simultaneously: a poloidal flow around the "skin" of the torus and a toroidal flow through its center. The complex interaction between these two distinct rotational flows is the physical origin of spin-1/2. A single, 360-degree rotation of the entire vortex does not return the internal flow pattern to its original state. It requires a second full 360-degree rotation for the internal and external flows to return to perfect alignment.

Thus, the abstract, mathematical concept of "spin-1/2" is, in this model, the direct, observable consequence of the geometric structure of a toroidal vortex.

2.3 The Origin of Mass and Scale

The electron's mass and its characteristic wavelength also emerge directly from this vortex model. The electron's mass is the energy of the ULF trapped in this localized,

---

[26] While Williamson and van der Mark elegantly proposed a toroidal photon topology for the electron, their model faced significant challenges in defining the physical mechanism of confinement—specifically, the force that prevents the energetic photon loop from expanding or dissipating into free space. The UFD framework resolves this issue by embedding the vortex within a physical, active medium (the UEF/ULF). In our model, the electron is not a wave trapped in an abstract vacuum but a structured vortex stabilized by the hydrostatic pressure of the surrounding field. Just as the pressure of the ocean holds a cavitation bubble together, the ambient pressure of the ULF maintains the electron's structural integrity.

resonant state, a physical manifestation of $E = mc^2$. Its wave-particle duality is resolved: it is a wave that behaves like a particle because its energy is localized.[27]

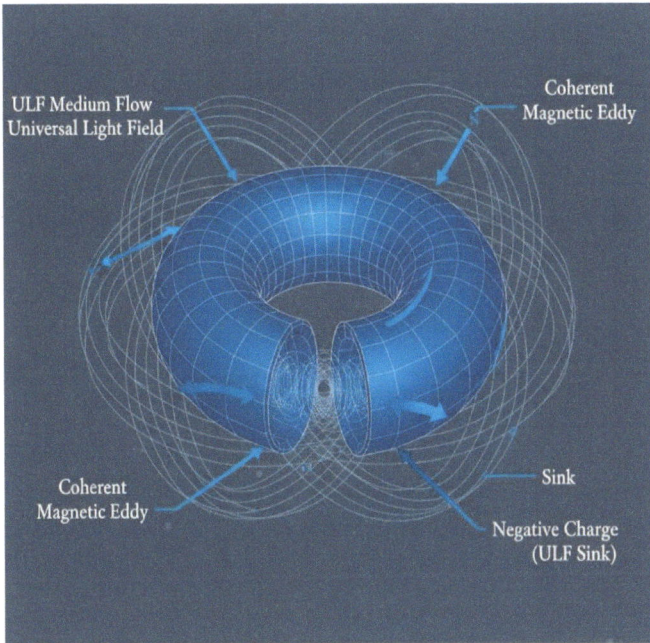

Figure 11: An Electron Vortex. This figure depicts the electron as a stable, self-sustaining toroidal vortex in the ULF. The electron's fundamental properties emerge directly from its geometry and its interactions with this medium. Its mass and spin ½ are consequences of its coherent, rotational geometry. Its negative electric charge is defined by its chirality, causing it to act as a 'sink' that perpetually draws the ULF medium inward. Its intrinsic magnetic moment is generated by the rotation of its vortex, which induces a coherent rotational eddy in the surrounding field that we measure as spin-up or spin-down.

This model of the electron as "trapped light" is not just a qualitative analogy; it leads to a direct, quantitative prediction. If the electron's mass is the energy of a localized wave, we can calculate its fundamental wavelength by equating the mass-energy of the electron ($E = m_e c^2$) with the energy of a photon ($E = \frac{hc}{\lambda}$). Solving for the wavelength yields the Compton wavelength:

$$\lambda_c = \frac{h}{m_e c} \approx 2.43 \times 10^{-12} \text{ meters}$$

In standard physics, the Compton wavelength is a somewhat abstract length scale. In the RFI, however, it represents the fundamental wavelength of the light that is trapped in a resonant, vortical pattern to form the electron. It can be considered the

---

[27] In this context, the term *localized* means that the wave's energy is not propagating or spreading out through space but has instead folded back on itself to form a stable, self-sustaining, spinning pattern: a vortex. An analogy is a garden hose: a propagating wave (like a photon) is a pulse of water traveling down the length of the hose from one end to the other. A localized wave (like an electron) occurs when the hose is formed into a loop and the water inside it spins, creating a stable, self-contained whirlpool. The energy is still present, but it is "trapped" in a coherent, resonant pattern at one location. This is the physical basis for the electron "particle."

effective "circumference" of the electron vortex, giving a real, physical scale to the particle.

2.4 The Fine-Structure Constant: The Hydrodynamic Gear Ratio

If the electron is a toroidal vortex spinning within the ULF, what determines the strength of its interaction with that field? In standard physics, the answer is the Fine-Structure Constant ($\alpha \sim 1/137$), a dimensionless number that defines the strength of the electromagnetic interaction. In the UFD framework, this abstract constant is revealed as a concrete hydrodynamic gear ratio—the coupling efficiency between nested fields.

2.4.1 The Origin of α: A Multi-Directional Coupling

The ULF is not an empty background but is continuously generated by the rotation of nucleons within the denser UEF. Because these two fields differ in density, energy does not pass smoothly between them. Instead, a small portion is shed into the surrounding light field. This process is highly structured: for every 137 internal rotations of a UEF vortex, one complete, phase-aligned cycle is established in the ULF.

This 137:1 ratio is the geometric cost of coupling structures across the Plenum's nested field hierarchy, serving as a universal gear ratio that operates in two directions:

- Downward (UEF to ULF): This determines the relative density of the medium. Because this shedding occurs in three spatial dimensions, the density of the ULF (ρ_{ULF}) is set at α^3 times the density of the UEF (ρ_{UEF}).

- Upward (Vortex to Soliton): To acquire mass, a UEF soliton must couple back to the ULF to form a co-moving distortion halo. This phase-locking interaction costs exactly $1/\alpha$.

2.4.2 The Electron's "Grip"

The electron, as a resident of the ULF, is governed by this same geometric constant. Because the electron is a torus (a ring with a hole), the ULF medium flows through and around it. This dictates the "grip" or coupling efficiency the electron has on the field. We thus interpret α as the slip ratio between a hard vortex and a soft field:

- Vortex Spin: The electron's internal energy circulates at the field's wave-speed (c).

- Wave Generation: The ULF "vapor" is tenuous compared to the vortex core, meaning it cannot catch every rotation of that motion.

- The Gear Ratio: Coherence is only established when the internal rotation and external field-response phase-lock at the 137:1 ratio.

This mechanism explains why the electron's charge is a universal constant. Regardless of its internal frequency, the ULF medium can only couple energy at this geometrically limited rate. The topology (torus) determines the behavior (coupling).

2.5 The Electron-Proton Mass Ratio: Topology Meets Field Descent

The reinterpretation of α as a coupling constant leads to a physical resolution of the proton-to-electron mass ratio (~1836:1). In standard physics, this number is an empirical "given." In UFD, it is the natural consequence of Topological Descent.[28]

As the universe sheds complexity from the UEF to the ULF, the geometry of its stable vortices simplifies. The proton is a UEF-anchored trefoil knot with three crossings, while the electron is a ULF-resident torus with zero crossings. In our framework, inertial mass is the result of Hydrodynamic Added Mass (Lamb, 1932)—the inertia of the surrounding ULF medium that must be displaced or "carried" by the vortex.[29]

Because the ULF behaves as a superfluid with an effectively infinite Reynolds number, mass corresponds to the effective volume of the field the particle disturbs. The mass ratio (m_p/m_e) is therefore the product of two geometric factors:

---

[28] The mass discussed here is inertial mass, not gravitational mass. Inertial mass measures how strongly a structure resists acceleration when a force is applied; it reflects how tightly energy is bound into a stable configuration (*see* Chapter 4). Gravitational mass, by contrast, measures how strongly a structure couples to gravity (*see* Chapter 1). In UFD, these two quantities are related but not identical. The proton-to-electron mass ratio arises from how deeply each structure is embedded in, and dynamically coupled to, the ULF. Gravitational mass emerges at a different level of the theory, through pressure-mediated coupling to the UEF and is treated separately. The present discussion isolates inertial mass because it is the quantity directly probed in particle dynamics and spectroscopy, and therefore the one relevant to explaining the observed 1836:1 ratio.

[29] In UFD, the electron's mass arises entirely from inertial coupling to the ULF through hydrodynamic added mass. Because the electron does not significantly source constraint geometry in UEF, its contribution to gravitational sourcing is negligible. This distinction does not contradict observed free-fall behavior, which remains dominated by nucleonic mass in bulk matter, but it implies that the equivalence between inertial and gravitational mass is emergent rather than ontologically fundamental at the particle level.

1. Field Descent ($1/\alpha \approx 137$): The proton is a UEF vortex that must "gear down" to the ULF to acquire mass. This crossing of the density boundary costs 137:1 in coupling efficiency.

2. Topological Impedance ($\xi_{geo} \approx 13.4$): The proton's knotted geometry creates far more hydrodynamic drag than the electron's streamlined torus.

$$\frac{m_p}{m_e} \approx \frac{1}{\alpha} \times \xi_{geo} \approx 137 \times 13.4 \approx 1836$$

Crucially, ξ_{geo} is not a fitted constant, a volume ratio, or a drag coefficient. It shows how topology alone governs the near-field reorganization of the surrounding medium when a vortex is accelerated. Detailed geometric analysis shows that this impedance is dominated not by the overall size of the structure, but by how often and how closely different parts of the vortex approach one another (*see* Appendix A). Knotted topologies concentrate reactive energy in these constrained regions, while unknotted loops do not.

From this perspective, the proton–electron mass ratio is no longer mysterious. It is not an arbitrary number written into the fabric of the universe. It is the natural outcome of two simple facts: the hierarchical structure of physical fields, and the profound difference between a knotted vortex and a smooth loop. Mass, in this sense, is geometry made resistant to motion.

2.6 The Absolute Test: Deriving the Hydrogen Ground State

Having established the electron's mass as a hydrodynamic added mass and derived its mass ratio to the proton, our final quantitative test of the RFI is to verify it can reproduce the binding energy of the hydrogen atom—the experimental benchmark of modern physics (-13.6 eV). Where quantum mechanics describes the electron as an abstract wavefunction, the RFI interprets this as a physical standing wave in the ULF medium. The ground state energy emerges from a fluid-dynamic balance:

1. Potential Energy (E_P): The pressure deficit created by electrostatic attraction between the proton source and electron sink ($\sim -1/r$), derived from ULF pressure gradients.

2. Kinetic Energy (E_K): The wave confinement energy required to localize the electron standing wave in the ULF ($\sim 1/r^2$). This represents the quantum

mechanical "zero-point energy" reinterpreted as the physical cost of wave compression.

Following the standard quantum mechanical approach—which we interpret as describing real ULF wave dynamics—the stable atom exists where total energy is minimized:

$$E_{total} = E_P + E_K$$

Solving this for the lowest energy state (n = 1) yields the Bohr radius (r_0):

$$r_0 = \frac{4\pi\epsilon_0\hbar^2}{m_e e^2}$$

This radius represents the equilibrium point where the ULF pressure gradient force exactly balances the wave confinement pressure. In RFI terms, this is determined by:

- The stiffness of the ULF (ϵ_0)

- The angular momentum of the electron vortex ($\hbar$)

- The hydrodynamic added mass (m_e)

Substituting r_0 back into the energy equation yields:

$$E_1 = -\frac{m_e e^4}{2\hbar^2 (4\pi\epsilon_0)^2}$$

Using measured values: $E_1 \approx -13.6$ eV

This quantitative derivation proves that the RFI is mathematically consistent with the most fundamental quantum system.[30] The hydrogen atom's stability is not an abstract statistical rule; it is the unique geometric solution that satisfies both the fluid dynamics (pressure) and the wave mechanics (resonance) of the ULF medium. The "mystery" of quantum energy levels is resolved: only specific wavelengths form stable

---

[30] While this derivation uses the mathematical framework of quantum mechanics, RFI provides the *physical picture*—the electron is not an abstract probability cloud but a real standing wave in the ULF. The wave equation of quantum mechanics (Schrödinger's equation) is interpreted as the wave equation for disturbances in this physical medium.

standing waves around the nuclear pressure well, just as only specific frequencies create standing waves on a vibrating string.

Having constructed a physically intuitive picture of the free electron, we are equipped to explore how this model gives rise to the strange rules of quantum mechanics.

3. A Physical Basis for Quantum Mechanics

The true test of the RFI lies in its ability to provide a physical origin for the foundational pillars of quantum theory. The Standard Model is based on a set of powerful yet abstract concepts—the de Broglie hypothesis, the Schrödinger equation, and the Heisenberg Uncertainty Principle. While these are mathematically rigorous and predictively flawless, they are axiomatic—they tell us *how* the quantum world behaves, not what it *is*.

This section will now provide that missing physical picture. Here, we deconstruct each of these foundational concepts and demonstrate how they emerge as direct and predictable consequences of the physical, resonant behavior of the universal fields.

3.1 The Blackbody Problem and the Quantization of Energy

The quantum revolution began with a failure. At the end of the 19th century, classical physics was unable to explain the spectrum of light emitted by a perfect absorber, known as a "blackbody." The best classical theory, now known as the Rayleigh-Jeans law (Rayleigh, 1900; Jeans, 1905), was effective at low frequencies but predicted an "ultraviolet catastrophe" at high frequencies, suggesting that such an object should emit an infinite amount of high-frequency energy. In 1900, Max Planck solved the problem mathematically by making a radical assumption that he himself called an "act of desperation." He posited that energy could only be emitted or absorbed in discrete packets, or "quanta," with the energy of each packet being proportional to its frequency ($E = hf$) (Planck, 1901). This was the birth of quantum theory. However, it left the deepest question unanswered: Why is energy quantized?

The RFI provides the direct, physical answer to this question. In this framework, a blackbody is a cavity whose walls are composed of UEF nuclei and their associated ULF electron standing waves (orbitals). When heated, these resonant structures vibrate. The reason the emitted light is quantized is that the vibrational states from which the light originates are themselves quantized, resonant structures. A standing wave, like a guitar string, can only exist in a stable set of harmonics: a fundamental note and its specific overtones. It cannot produce a stable vibration "in between" these allowed states. When an electron in a heated atom "jumps" from a higher-energy standing wave to a lower-energy one, the energy it releases is precisely the difference in energy between those two discrete resonant states.

Thus, Planck's quantization is a necessary and inevitable consequence of the fact that the atom itself is a resonant system—a musical instrument that can only play specific, harmonic notes.

3.2 The de Broglie Wavelength as a Physical Resonance

In 1924, Louis de Broglie postulated that all matter exhibits wave-like behavior, with a wavelength inversely proportional to its momentum ($\lambda = h/p$). This cornerstone of quantum mechanics offers no physical picture of what is "waving."

The RFI provides this physical picture. It posits that the de Broglie wavelength is a direct, physical property of a particle-vortex's interaction with the ULF. When any particle-vortex—whether a UEF nucleon or a ULF electron—moves through the ULF, its dynamics disturb the field and create a co-moving pilot wave or resonant ripple in the ULF medium that surrounds it. The wavelength of this physical ripple *is* the de Broglie wavelength.

A particle with low momentum creates a long, gentle ripple (large wavelength), while a particle with high momentum creates a short, rapid ripple (small wavelength). This principle has been demonstrated by matter-wave interferometry experiments. While electrons exhibit prominent wavelengths, massive molecules like C60 Buckyballs, due to their enormous momentum, are experimentally observed to have the exact, ultra-short de Broglie wavelengths predicted by the RFI (Arndt et al., 1999).

The RFI also offers a direct, mechanistic explanation for phenomena such as electron diffraction (Figure 12). In the RFI, it is not the electron vortices themselves that interfere but their associated ULF pilot waves. The vortices are then guided by the constructive and destructive interference of their own waves. The de Broglie wavelength is, therefore, a real, physical wave in the ULF, created by and inseparable from the particle-vortex it guides.

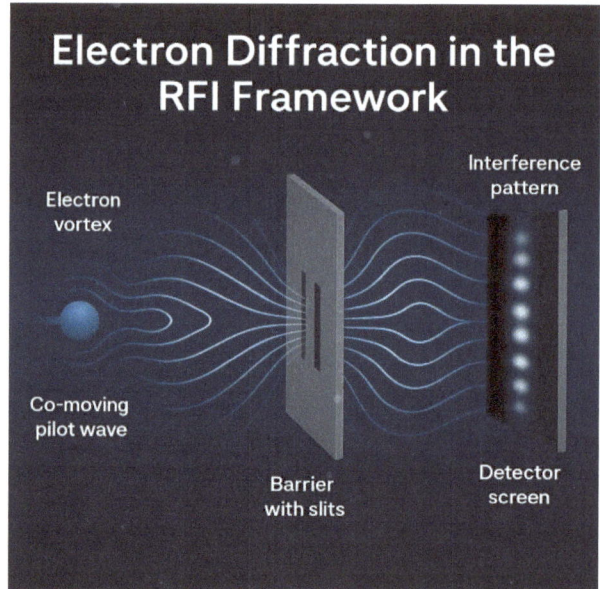

Figure 12. Electron diffraction in the RFI framework. As an electron vortex travels through the ULF, it generates a co-moving resonant ripple: the physical de Broglie wavelength. When electrons encounter a crystalline lattice or a double-slit barrier, it is not the vortices themselves that interfere, but their ULF pilot waves. The constructive and destructive interference of these real waves then guides the trajectories of the electron vortices, producing the observed diffraction pattern on the detector screen. This provides a mechanistic explanation of wave–particle duality: the wave is a real ULF resonance, while the particle is the localized vortex it guides.

3.3 The Schrödinger Equation as a Wave Equation

The Schrödinger equation ($H\psi = E\psi$) is the central mathematical engine of quantum mechanics. In the mathematical formalism of the theory, the set of all possible wave functions a system can have resides in an abstract, infinite-dimensional vector space known as Hilbert space.

Here, the RFI provides a direct physical meaning for this abstract space, which is reinterpreted as the complete catalogue of all possible stable, harmonic vibrations that the ULF can support. It is the space of all possible harmonies for the universal field, and a specific quantum state is simply a note or chord drawn from this vast, infinite repertoire of resonant possibilities. Therefore, we propose that the Schrödinger equation is an effective wave equation that describes the "standing wave" solutions (ψ) for a ULF vortex (the electron) that exists within the powerful potential created by the central UEF vortex (the nucleus).

3.3.1 The Hamiltonian Operator ($H^\wedge$) and its Components

In the Schrödinger equation, the Hamiltonian operator, $H^\wedge$, is the mathematical representation of the total energy of a system. It is composed of two parts: the Potential Energy operator ($V^\wedge$) and the Kinetic Energy operator ($T^\wedge$). In the Standard Model, the Potential Energy operator represents the "landscape" or "energy well" the particle exists within; for an electron in an atom, this is primarily the electrostatic attraction to the

122

nucleus. The Kinetic Energy operator is more abstract; mathematically, it is related to the curvature or "wiggliness" of the wave function and quantifies the energy inherent in the wave's shape.

Our framework provides a direct, physical meaning for each of these components. The Potential Energy term ($V^\wedge$) represents the electrostatic influence of the central UEF vortex (the nucleus) on the surrounding ULF. The massive "source" vortex creates a stable pressure gradient in the ULF around it, an energy well that shapes the medium in which the electron wave must form. The Kinetic Energy Operator ($T^\wedge$), in turn, represents the physical properties of the ULF medium itself, such as its stiffness or tension, which determines its resistance to bending. A very stiff ULF would permit only simple, low-frequency vibrations, whereas a more flexible ULF would permit more complex, high-frequency vibrations.

Thus, the Schrödinger equation is transformed from an abstract quantum rule into a statement about physical resonance: it is the mathematical tool that finds the stable standing wave patterns (ψ) and their associated energies (E) for a given medium ($T^\wedge$) and a given shaping potential ($V^\wedge$).

3.3.2 The Wave Function (ψ) and the Born Rule

The connection between the wave function and the observable world is provided by the Born rule, a foundational postulate of quantum mechanics. The rule states that the probability of finding a particle at a certain point is given by the square of the amplitude of its wave function, $|\psi|^2$, at that point. While empirically flawless, the Born rule is an axiom of the theory, and the standard interpretation offers no deeper physical reason for the connection between probability and the wave's squared amplitude.

In our model, this connection becomes a direct physical consequence. The wave function, ψ, represents the real, physical amplitude of the ULF standing-wave vibration. Therefore, $|\psi|^2$ is the intensity or energy density of the ULF's vibration at that point.

This distinction is key to understanding the nature of quantum probability. The particle is most likely to manifest where the wave's energy density is highest, which is why the statistical outcomes of numerous measurements perfectly align with its predictions.[31]

---

[31] A helpful analogy is to think of the electron vortex as a surfer and its pilot wave as a complex ocean wave. The wave is spread out with many peaks and troughs, and the surfer is riding it. Suddenly, a reef

3.4 The Heisenberg Uncertainty Principle

First formulated by Werner Heisenberg in 1927, the Uncertainty Principle places a fundamental limit on the precision with which one can simultaneously know certain pairs of properties, like a particle's position and momentum. In the standard Copenhagen interpretation, this is an intrinsic, irreducible "fuzziness" built into the fabric of reality.

The RFI reinterprets the Uncertainty Principle as a direct, physical trade-off inherent to standing waves rather than a fundamental limitation on knowledge. For an electron vortex to exist in a stable, resonant orbital with a well-defined energy and momentum, it must be delocalized over the entire volume of that wave pattern.[32] Its position is thus inherently "uncertain." To force it into a definite position, one must physically interact with it, which necessarily breaks the resonant pattern and destroys the information about its momentum. The principle is thus a choice between two mutually exclusive physical states: a stable, delocalized wave or a disrupted, localized particle.

This reinterpretation marks the most profound philosophical departure from the standard Copenhagen view. The Uncertainty Principle is often cited as proof that quantum mechanics is ontologically indeterministic—that reality itself is fundamentally "fuzzy," random, and lacks definite properties until an act of measurement occurs (Bohr, 1928). The RFI rejects this conclusion. It proposes instead that the universe is ontologically deterministic, governed at all levels by physical cause and effect. The apparent randomness is not a feature of reality but is rather a consequence of our practical inability to know the precise, microscopic state of the ULF at every point in space. The RFI thus restores the classical ideal of an objective reality that exists

---

(a "measurement") appears, and the entire wave crashes upon it. It is overwhelmingly likely that the surfer will be found where the peak of the wave was at the moment of impact. The probability of their final, localized position is directly proportional to the intensity of the wave they were riding.

[32] It is important to understand what "delocalized" means in this context. The electron vortex does not dissolve or cease to exist. A helpful analogy is a tornado. A free electron is like a tight, compact "rope" tornado with a well-defined position. A bound electron in an orbital is like that same tornado expanding to fill an entire valley, becoming a large, stable, rotating system. The "vortex-ness"—the organized, spinning flow—is still present, but its substance and influence are now spread throughout the entire volume of the standing wave. The vortex thus *becomes* the stable, resonant wave structure of the orbital.

independently of our observation and suggests that the universe is, in principle, intelligible and understandable, even if it is not always predictable.

Having established a physical basis for the rules of quantum mechanics, we are ready to apply this same physical picture to resolve its most famous and challenging paradoxes.

4. Resolving the Paradoxes of Quantum Mechanics

Beyond its core mathematical formalism, quantum theory is famous for a set of phenomena that defy all classical intuition. These so-called paradoxes, including wave-function collapse, quantum entanglement, and quantum tunneling, have been the source of a century of philosophical debate. While the Standard Model accepts these as fundamental features of the quantum world, the RFI offers a physical, mechanistic explanation for each, reinterpreting them as the natural and predictable behaviors of a fluid dynamic medium.

4.1 The Collapse of the Wave Function

In quantum mechanics, a particle is said to exist in a "superposition"—a state described by a mathematical wave function (ψ) that encompasses all of its possible locations or states simultaneously. The RFI provides a direct, physical meaning for this concept. In our model, a system in superposition exists as a single, complex, yet coherent standing wave in the ULF. This is analogous to a musical chord, which is a single, complex sound wave formed by the superposition of multiple individual notes.

The "collapse of the wave function" is the process that occurs upon measurement. The standard theory posits that observation forces the system to instantaneously and randomly "choose" one of its possibilities. In RFI, the collapse is the physical shattering of a real wave, and a "measurement" is any external interaction that physically disrupts the delicate resonant pattern, transforming it into a chaotic, localized state at the point of impact. To better understand our physical model of wave-function collapse, we compare it with modern alternatives.

4.1.1 The Copenhagen Interpretation

The Copenhagen Interpretation, developed by Bohr and Heisenberg, is the oldest and most widely taught framework for understanding quantum mechanics (Bohr, 1928). It posits that the wave function is a purely mathematical tool representing our knowledge, but it offers no physical mechanism for its "collapse." Instead, it accepts

125

collapse as a fundamental postulate that separates the quantum and classical worlds. This abstract, non-physical view leads directly to the theory's most famous philosophical paradoxes.

- Niels Bohr's Principle of Complementarity, which addresses wave-particle duality. It states that an object, such as an electron, can exhibit properties of both a wave and a particle, but never both simultaneously. In this view, the experimental setup determines which mutually exclusive aspect of reality is revealed.

- The Measurement Problem, illustrated by the Schrödinger's Cat thought experiment. By extending the quantum superposition to the macroscopic scale, the interpretation leads to the absurd conclusion that the cat is simultaneously alive and dead until a conscious observer opens the box and collapses its wave function.

The RFI resolves both paradoxes with a coherent physical picture: the "wave" and "particle" are two distinct yet inseparable physical components of the same system. The "particle" is the localized, energetic vortex itself, while the "wave" is the real, physical standing wave in the ULF that surrounds and sustains it.

This interpretation immediately resolves the complementarity paradox: an experiment can be designed to interact with either the localized vortex or its delocalized wave, since both are always physically present. It also solves the Schrödinger's Cat paradox by identifying the precise physical moment of collapse. The decay of the radioactive nucleus emits a real particle-vortex. This vortex physically strikes the Geiger counter, and the interaction between the single-quantum wave and the macroscopic detector *is* the collapse event. The cat is therefore never in a superposition; it is definitively alive or dead long before any conscious observer opens the box. RFI thus removes the observer-dependent mystery by restoring a tangible, objective reality.

4.1.2 The Many-Worlds Interpretation (MWI)

First proposed by Hugh Everett III in 1957, the Many-Worlds Interpretation (MWI) is a radical alternative that seeks to solve the measurement problem by eliminating the concept of "wave function collapse" entirely (Everett, 1957). The MWI takes the Schrödinger equation at its most literal, claiming that the wave function is not just a mathematical tool but describes the real physical state of the universe. In this view, the wave function never collapses; instead, all the possibilities it encodes are equally real. When a quantum measurement occurs, the entire universe "branches" into multiple parallel universes, with each branch corresponding to one of the possible outcomes.

The appeal of the MWI is that it avoids the apparent paradox of collapse and the special role attributed to the "observer." However, it does so at the cost of postulating the existence of an unobservable, constantly branching multiverse of astronomical size that responds to subtle measurements. The RFI, in contrast, is far more parsimonious; it simply disrupts the single wave in our single universe.

4.1.3 Bohmian Mechanics

Developed by physicist David Bohm in the 1950s (Bohm, 1952), Bohmian Mechanics is the most well-known example of a "hidden variables" theory. It is an interpretation of quantum mechanics that fundamentally rejects the idea that reality is probabilistic.

The core principle of Bohmian mechanics is that particles and waves are both real and distinct. An electron is a real particle that has a definite, objective position at all times, even when it is not being observed. The outcome of an experiment is determined by this "hidden variable," which represents the particle's precise, yet unknown, initial position. The motion of this particle is then governed deterministically by a real, physical "pilot wave" that permeates all of space and is described by the Schrödinger equation. The particle is never separate from its pilot wave; the wave explores all possible paths, and the particle is guided by it, like a cork guided by the currents and whirlpools in a river.

Bohmian Mechanics successfully reproduces all the predictions of standard quantum mechanics without invoking "wave function collapse" or fundamental probability. Its primary challenge, however, lies in its explicit nonlocality, as the pilot wave must transmit information faster than the speed of light, making it difficult to integrate with Special Relativity.

Although the RFI shares many philosophical goals with Bohmian Mechanics, as both models aim to provide a realistic, deterministic picture of reality, the RFI offers a more physically integrated explanation. In Bohmian mechanics, the particle and the pilot wave are two separate entities, and the origin of the pilot wave is not fully explained. In the RFI, they are inseparable aspects of a single system: the "particle" is the localized vortex, and the "wave" is the resonant pattern in the ULF that the vortex generates and is sustained by. The pilot wave is not a separate guide; rather, it is the vortex's own footprint in the field.

This distinction leads to a more profound conclusion that distinguishes the RFI from all classical and most quantum interpretations. Namely, the RFI suggests that the fundamental distinction between a "particle" and a "field" is an illusion. In this model, a particle is not an object that exists *in* a field. Rather, a particle, such as an electron, *is* the field itself, configured into a stable, localized, self-sustaining vortex. The "pilot wave" is simply the extended, dynamic disturbance in the ULF that is inseparable from the vortex's own existence. Thus, the RFI dissolves classical dualism entirely while proposing a single, unified substance whose different geometries of motion give rise to all the phenomena we observe.

4.1.4 Objective Collapse Theories

Objective Collapse theories, such as the Ghirardi-Rimini-Weber (GRW) theory (Ghirardi, Rimini, & Weber, 1986), constitute a family of interpretations that aim to resolve the measurement problem by positing that wave-function collapse is a real, spontaneous process. Unlike the Copenhagen Interpretation, which requires an "observer" to trigger collapse, these models argue that collapse occurs spontaneously and is governed by a universal law.

In Objective Collapse theories, the Schrödinger equation is treated as an approximation that can be modified by adding nonlinear, stochastic (random) terms. These new terms cause the wave function of any particle to have a tiny, constant probability of spontaneously collapsing into a definite position by itself without any measurement having taken place. For a single, isolated particle, the probability of spontaneous collapse is extremely low; however, for a macroscopic object comprising trillions of entangled particles, the probability is astronomically high. If one particle in a system collapses, the entire system collapses, ensuring that large objects are never in a state of superposition.

Like Objective Collapse theories, the RFI aims to make collapse a real, physical event. However, it does not need to modify the Schrödinger equation to do so. In the RFI, the Schrödinger equation remains the perfect description of the undisturbed resonant wave. The "collapse" is what happens when the system described by the equation is physically disrupted by an external interaction, a process governed by standard mechanics and conservation laws.

4.1.5 Quantum Decoherence

Decoherence is a well-understood physical process in which a quantum system interacts with its environment, causing its wave-like properties to "leak" into the surroundings and become practically unobservable (Zurek, 2003), thereby leading to the collapse. It explains why we do not observe macroscopic objects in superposition. However, decoherence alone does not solve the measurement problem. It only "smears out" the quantum weirdness without selecting a single definite outcome.

Here, we find that the RFI is perfectly compatible with decoherence. The "environment" in this model is the surrounding ULF, and its interaction with it would naturally destabilize and collapse the delicate standing wave resonance of the quantum system, thereby forcing it into a definite state. The RFI thus provides the physical mechanism that completes the story of decoherence.

4.2 Quantum Entanglement: A Shared Resonance

Quantum entanglement, famously described by Einstein as "spooky action at a distance" (Einstein, Podolsky, & Rosen, 1935), is arguably the most profound mystery of quantum theory. It describes a state in which two or more particles are linked in such a way that their fates are intertwined, regardless of the distance separating them. The moment a measurement is made on one particle, the other particle—even if it is light-years away—appears to assume a corresponding state.

For decades, it was hoped that this "spookiness" could be explained by "local realism"—the idea that particles have definite, pre-existing properties and that no influence can travel faster than light. However, in 1964, physicist John Bell proved that this is impossible (Bell, 1964). Numerous experiments, from Alain Aspect (1982) to recent "loophole-free" tests (Hensen et al., 2015), have since confirmed that the universe is fundamentally nonlocal.

The RFI provides a physical, intuitive picture of this proven reality. We propose that two entangled particles are not separate, independent entities, but two localized vortices sustained by a single, unified resonant structure. Rather than separate objects, they are two points of maximum intensity on a single, indivisible system.

This raises an important question: how can a state be nonlocal if the ULF is the medium of light, which is limited to speed c? The RFI response is that the Plenum structure can exist in different states of connectivity. A photon is a traveling wave, like a ripple moving across a lake. Entanglement, in contrast, represents a Global Phase-Lock—a standing-wave state where the field between the two particles behaves as a single, perfectly synchronized resonant filament.

Because this resonant system is a single, indivisible quantum object, the information does not "propagate" along the link. Instead, the entire structure is defined by a single phase state. When a measurement is performed on a single entangled vortex, the physical interaction disrupts the entire system simultaneously. [33] The effect does not need to travel from one point to the other because the collapse is the simultaneous change of the single system they were both a part of (Figure 13) .[34]

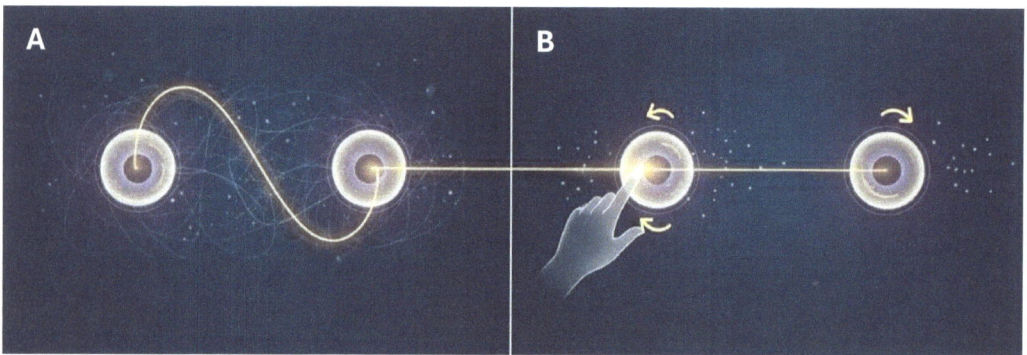

Figure 13: A Physical Model for Quantum Entanglement. This figure illustrates the physical model of the RFI for quantum entanglement.
(A) Two entangled vortices are depicted as two nodes within a single resonant structure. The nonlocal connection between them arises as a coherent, phase-locked filament of the ULF, forming an indivisible topological system rather than a signal propagating through space.
(B) A measurement of a single vortex is a physical interaction that disrupts the boundary conditions of the entire resonant state. This resolves the paradox by showing that no faster-than-light signal is sent from A to B; there is only the global collapse of the shared standing wave that sustained them both.

---

[33] Consider a taut guitar string. The note it plays is not located at one end or the other; the vibration is a property of the *entire* string simultaneously. If you touch the string to mute it, the sound stops everywhere at once. The "mute signal" didn't travel from left to right; the *condition* for the wave's existence was removed globally. Similarly, entangled particles are nodes on a single cosmic string.
[34] While the resonant filament appears to change instantaneously, UFD posits that this coordination is governed by a deeper substrate—the Universal Awareness Field (UAF)—with a causal phase velocity of ~$137c$ (see Chapter 12). The UAF does not transmit energy signals between the particles; rather, it is the medium within which extended phase-locked structures exist as unified objects. This velocity represents the speed at which global coherence can be established, preserving relativistic causality for matter while allowing unity for phase.

In sum, entanglement is the natural, coordinated behavior of components within a unified, nonlocal geometric structure. What appears nonlocal to an observer measuring separate particles is, in truth, local to the deeper field-based topology of reality.

4.3 Quantum Tunneling: A Physical Wave Phenomenon

Quantum tunneling is one of the most famous and profoundly counterintuitive phenomena in quantum mechanics. It describes the process by which a particle can pass through a potential energy barrier that it should not have enough energy to overcome. In the standard interpretation, first used by George Gamow to explain alpha decay (Gamow, 1928), this is possible because the particle's wave function does not stop at the barrier but decays exponentially through it, leaving a small but non-zero probability of finding the particle on the other side.

The RFI provides a direct, physical picture of this effect. In this model, the particle's standing wave in the ULF does not abruptly stop at an energy barrier; instead, it leaks a short distance into the barrier region. If the barrier is thin enough, the "tail" of this evanescent wave can reach the other side, where it can excite a new, propagating wave in the ULF, to which the particle-vortex would then be drawn. The probability of this occurring is determined by the amplitude of the evanescent wave on the far side of the barrier, making the effect exponentially sensitive to the barrier's thickness and height.[35]

In summary, this section has confronted the most profound paradoxes of quantum theory and has shown how the RFI provides a physical, mechanistic explanation for each. Having established a physical foundation for quantum phenomena, we can now move beyond the purely theoretical and explore the revolutionary technological frontiers enabled by this new understanding of the quantum world.

5. Technological Frontiers of the RFI

---

[35]Although the RFI can provide a tangible, physical mechanism for quantum tunneling, we contend that for the two most critical phenomena where tunneling is invoked. alpha decay and stellar fusion, our framework offers more direct and parsimonious explanations that render the tunneling concept unnecessary. Rather than relying on the abstract probability amplitudes of the Standard Model, the UFD framework interprets these processes as deterministic transitions driven by geometric coherence within the fundamental fields (*see* Chapter 7). These alternatives are presented as physically grounded, testable hypotheses that may offer deeper insight into the nature of nuclear processes.

In addition to offering a new understanding of the quantum world, the RFI suggests a radical shift in how we might engineer it. By grounding quantum behavior in the tangible dynamics of the ULF, the RFI opens an entirely new technological frontier: quantum computing by resonance design.

5.1 A Field-Based Architecture for Intelligence

In the RFI, quantum computing becomes resonant field logic enacted within a physically real medium. Each qubit becomes a localized vortex in the ULF, capable of storing information in rich, multi-dimensional harmonics. This architecture transforms the ULF into an intelligent substrate:

- Superpositions become structured waveforms.

- Quantum gates become geometric modulations of resonance.

- Entanglement becomes the coherent overlap of field modes across space.

This approach has profound implications for the field. It could lead to quantum processors that are more stable, scalable, and geometrically expressive. With a stable resonant platform, a single quantum circuit could model the intricate vibrational geometry of atoms, molecules, and nuclei in real time. Chemistry, material sciences, and systems optimization would be revolutionized by this technology.

5.2 Resonance Stabilization: A New Quantum Control Paradigm

At the heart of quantum engineering lies the challenge of decoherence. Current efforts to maintain a qubit's fragile superposition rely on elaborate isolation and active correction protocols, known as quantum feedback control (Wiseman & Milburn, 2009). While successful, these methods often treat environmental noise as a random, statistical phenomenon to be suppressed with brute force.

The RFI framework offers a more precise approach by providing a physical picture of the noise itself. We posit that "noise" is not random but is a structured field of dissonant ULF waves with a specific, measurable "resonant noise spectrum." This leads to a key prediction: a "coherence-tuned" stabilization system, designed to selectively cancel only the most harmful resonant frequencies, will be demonstrably more efficient and effective than a system that attempts to suppress all noise equally.

This new paradigm of Resonance Stabilization would be achieved through two key hardware developments that have direct parallels in the cutting edge of experimental physics:

- Advanced Field Sensors: A system would first require devices capable of mapping the local ULF environment in real time with extreme precision. This corresponds to the active research field of Quantum Noise Spectroscopy, in which the qubit itself is used as a probe to identify the exact frequencies and phases of the "resonant noise spectrum" that cause decoherence.

- Precision Field Emitters: The counterpart to the sensors, these devices would generate complex, phase-controlled electromagnetic fields. This corresponds to the use of Arbitrary Waveform Generators guided by Optimal Control Theory. These emitters would perform two complementary functions: first, Targeted Noise Cancellation, emitting phase-inverted counter-fields to nullify the specific dissonant waves identified by the sensors; and second, Resonance Reinforcement, where a gentle, harmonically matched signal is broadcast to actively reinforce the qubit's own natural resonance.

Together, these technologies redefine quantum control as a precise art of active field stewardship—of understanding the specific geometry of the noise and harmonizing it with the qubit's natural resonance.

This vision of quantum computing then becomes the linchpin technology of the Resonant Engineering paradigm. The technologies envisioned in other chapters—from the precise field scaffolding of Controlled Fusion to the targeted dissonance of Precision Fission, as described in Chapter 7—require the real-time calculation and control of incredibly complex, multi-layered resonant fields. A classical, digital computer would be hopelessly inefficient at this task. Only a resonant computer, which operates on the same principles of field dynamics and harmonic coherence, can achieve the necessary speed and elegance to model and control another resonant system. Therefore, the development of a UFD-based quantum computer is a necessary first step toward a new technological frontier.

6. Conclusion

This chapter has introduced the RFI as a physical alternative to the standard interpretation of quantum mechanics. Its central claim is that the quantum state is a real, three-dimensional standing wave in the ULF, stabilized and shaped by the deeper coherence of the UEF. This ontological shift allows us to reinterpret the foundational features of quantum theory in purely physical terms. The electron's physical identity is redefined: its spin-1/2 and charge are derived directly from the toroidal vortex's physical flow and chirality. This framework also accounts for key constants by identifying the Fine-Structure Constant (α, 1/137) as the field's hydrodynamic slip ratio and by successfully deriving the electron-proton mass ratio from field stiffness and topology.

The RFI then provides a mechanistic basis for the rules of quantum mechanics. Here, the Schrödinger equation becomes a literal wave equation for resonant field dynamics, energy quantization emerges naturally from the harmonic structure of standing waves, and the uncertainty principle reflects a geometric trade-off between localization and coherence. Ultimately, the phenomena of collapse and entanglement are understood not as mysterious failures of theory but as tangible reorganizations of the field within a unified, deterministic substrate.

Future work is required to formalize this framework by deriving quantum behavior from the hydrodynamic principles of the coupled UEF-ULF system. Experimental validation of the RFI's technological predictions will also be essential to the model's validation.

If successful, the RFI offers a resolution to conceptual tensions that have long haunted quantum theory as well as a pathway to a deeper synthesis of physics, one in which the quantum world is understood as a realm of intelligibility.

5.3 An Intelligible Quantum World

In this chapter, we introduced the Resonant Field Interpretation (RFI), a framework that replaces the abstract, probabilistic ontology of quantum mechanics with a deterministic, field-based reality. By positing that the atom is a resonant system of interacting fields—the Universal Light Field (ULF) sustained by the nuclear Universal Energetic Field (UEF)— the RFI restores causality and intuition to the quantum world. In this model:

1. Electric Charge is defined by the vortex's chirality, where the electron serves as a physical "sink" that perpetually draws the ULF medium inward.

2. Spin-1/2 is demystified as a geometric consequence of the electron's toroidal topology, where the complex interplay of poloidal and toroidal flows necessitates a 720-degree rotation to return to the initial state

3. The Fine-Structure Constant (α) and the Electron-Proton Mass Ratio (1836:1) are revealed as the ULF's inherent hydrodynamic gear ratio and the result of the proton's topological impedance, respectively.

4. The de Broglie wavelength becomes the ripple created by a moving vortex in the ULF, and the Schrödinger equation is a literal wave equation that describes these field resonances.

5. Heisenberg Uncertainty Principle is reinterpreted as a physical trade-off between a stable, delocalized wave and a localized, disrupted particle state.

This physical ontology also provides mechanistic resolutions to quantum theory's most famous paradoxes. Wave function collapse is the literal disruption of a resonant wave, while quantum entanglement reflects the intrinsic correlation of two parts of a single unified standing wave. If validated, the RFI would represent a fundamental shift in the very nature of physics. For over a century, the Copenhagen view has compelled scientists to accept a strange, probabilistic reality that is divorced from intuition. The RFI bridges this divide, offering a physics in which quantum behavior is as visualizable and deterministic as planetary motion.

Beyond theory, by shifting quantum control from brute-force isolation to active Resonance Stabilization, quantum computing could transition from fragile qubits to dynamically reinforced systems. A stabilized quantum computer would open the door to revolutionary breakthroughs: rapid modeling of protein folding, real-time molecular simulation for drug discovery, unbreakable encryption via quantum key distribution, and optimization algorithms that could redesign global supply chains, climate systems, and transportation networks in seconds. It would also enable quantum chemistry calculations that reveal the resonant geometries of new molecules and materials, thereby accelerating the development of superconductors, catalysts, and energy storage systems.

In sum, the RFI is more than a reinterpretation of quantum mechanics; it is a step toward a unified, deterministic, and technologically empowering vision of reality, in which the universe is a deeply ordered, resonant system. In doing so, it challenges us to master our understanding of quantum behavior to harness the elegant, cooperative dynamics of the fields that underlie all matter and energy, ushering in a new technological age.

References

1. Arndt, M., Nairz, O., Vos-Andreae, J., Keller, C., van der Zouw, G., & Zeilinger, A. (1999). Wave–particle duality of C60 molecules. *Nature, 401*(6754), 680–682.
2. Aspect, A., Dalibard, J., & Roger, G. (1982). *Experimental Test of Bell's Inequalities Using Time-Varying*
3. Bell, J. S. (1964). *On the Einstein Podolsky Rosen Paradox. Physics Physique Физика,* 1(3), 195–200.
4. Bohm, D. (1952). 'A Suggested Interpretation of the Quantum Theory in Terms of "Hidden Variables" I & II'. *Physical Review, 85*(2), 166–193.
5. Bohr, N. (1928). 'The Quantum Postulate and the Recent Development of Atomic Theory'. *Nature, 121,* 580–590.
6. Einstein, A., Podolsky, B., & Rosen, N. (1935). 'Can Quantum-Mechanical Description of Physical Reality Be Considered Complete?'. *Physical Review, 47*(10), 777–780. *(The famous "EPR paper" that first introduced the paradox of quantum entanglement.)*
7. Everett, H. (1957). 'Relative State' Formulation of Quantum Mechanics'. *Reviews of Modern Physics, 29*(3), 454–462.
8. Gamow, G. (1928). Zur Quantentheorie des Atomkernes [On the Quantum Theory of the Atomic Nucleus]. *Zeitschrift für Physik, 51*(3–4), 204–212.
9. Ghirardi, G. C., Rimini, A., & Weber, T. (1986). Unified dynamics for microscopic and macroscopic systems. *Physical Review D, 34*(2), 470–491.
10. Giustina, M., et al. (2015). *Significant-Loophole-Free Test of Bell's Theorem with Entangled Photons. Physical Review Letters,* 115(25), 250401.
11. Heisenberg, W. (1925). Über quantentheoretische Umdeutung kinematischer und mechanischer Beziehungen (Quantum-Theoretical Re-interpretation of Kinematic and Mechanical Relations). *Zeitschrift für Physik, 33*(1), 879–893.
12. Heisenberg, W. (1927). 'Über den anschaulichen Inhalt der quantentheoretischen Kinematik und Mechanik'. *Zeitschrift für Physik, 43*(3–4), 172–198.
13. Hensen, B., et al. (2015). *Loophole-free Bell inequality violation using electron spins separated by 1.3 kilometres. Nature,* 526, 682–686.
14. Jeans, J. H. (1905). On the Partition of Energy between Matter and Æther. *Philosophical Magazine, 10*(55), 91–98.
15. Lamb, H. (1932). *Hydrodynamics.* Cambridge University Press.
16. Planck, M. (1901). Ueber das Gesetz der Energieverteilung im Normalspectrum (On the Law of Distribution of Energy in the Normal Spectrum). *Annalen der Physik, 309*(3), 553–563.
17. Pierański, P. (1998). In search of ideal knots. *Proceedings of the Royal Society of London. Series A: Mathematical, Physical and Engineering Sciences,* 454(1976), 2339-2364.

18. Rayleigh, Lord. (1900). Remarks upon the Law of Complete Radiation. *Philosophical Magazine, 49*(301), 539–540.

19. Schrödinger, E. (1926). 'An Undulatory Theory of the Mechanics of Atoms and Molecules'. *Physical Review, 28*(6), 1049–1070.

20. Shalm, L. K., et al. (2015). *Strong Loophole-Free Test of Local Realism. Physical Review Letters, 115*(25), 250402.

21. Williamson, J. G., & van der Mark, M. B. (1997). Is an electron a photon with toroidal topology? *Annales de la Fondation Louis de Broglie, 22*(2), 133.

22. Wiseman, H. M., & Milburn, G. J. (2009). *Quantum Measurement and Control.* Cambridge University Press.

23. Zurek, W. H. (2003). 'Decoherence, einselection, and the quantum origins of the classical'. *Reviews of Modern Physics, 75*(3), 715–775.

Chapter 6: The Composite Neutron

6.1 The Electroweak Force

The Standard Model of Particle Physics is built on the foundation of four fundamental forces. Two of these—the electromagnetic force and the weak nuclear force—were successfully unified in the 1960s into a single, elegant mathematical framework known as the electroweak theory (Weinberg, 1967). This unification stands as one of the crowning achievements of modern physics, yet the profound differences between the two forces it describes hint at a deeper reality that the model itself does not explain.

The first of these is the electromagnetic force, the familiar interaction first unified by James Clerk Maxwell that governs light, electricity, magnetism, and the entire structure of chemistry (Maxwell, 1865). According to modern theory, the electromagnetic force has an infinite range, is associated with the property of electric charge, and, in the quantum world, is mediated by the exchange of massless photons. It is the force of structure and communication. The second is the weak nuclear force, a strange and counterintuitive interaction first described by Enrico Fermi that operates only at the subatomic scale (Fermi, 1934). With an extremely short range, it primarily governs radioactive decay, such as the decay of a free neutron into a proton. It is a force of transmutation, a process that can change one kind of particle into another.

The great challenge for 20th-century physics was to explain how these two vastly different forces—one an infinite-range force of structure, the other a short-range force of transformation—could be two different aspects of the same thing. In this respect, the Standard Model's solution was a mathematical triumph. It proposed that both forces are mediated by the exchange of force-carrying particles. Unlike the electromagnetic force, which is mediated by the massless photon, the weak force's mediators (the W and Z bosons) acquire mass by interacting with a new, universal energy field known as the Higgs field (Higgs, 1964). This entire theoretical architecture was experimentally confirmed, first with the discovery of the W and Z bosons in 1983, and then with the discovery of the Higgs boson itself in 2012 (ATLAS Collaboration, 2012; CMS Collaboration, 2012).

Despite its success, the electroweak theory remains an immensely abstract concept. Force is carried by the exchange of "virtual" particles that cannot be directly observed (Feynman, 1985), and electric charge is treated as a fundamental, intrinsic property of a particle with no deeper physical origin. The model provides the precise mathematical rules of the game but offers no intuitive physical explanation for *why* those rules are the way they are.

This chapter presents a physical and more parsimonious alternative. Drawing on the principles of Unified Field Dynamics (UFD), we propose a new physical model for the neutron, identifying it not as a fundamental particle but as a stable, composite vortex formed by a tightly bound proton and antiproton. This postulate provides a unified physical origin for the electroweak force and magnetism.

6.2 The Composite Neutron: Deriving the Electroweak Force from a Single Geometric Postulate

Abstract

This chapter introduces a new physical theory of the electroweak force, grounded in Unified Field Dynamics (UFD) and based on a single postulate: the neutron is a composite Figure-8 vortex formed from a proton and antiproton vortex. In this framework, the electroweak interaction emerges naturally: the weak force corresponds to the geometric transformation of this composite structure, electric polarity arises from the chirality of its constituent vortices, and magnetism is a direct consequence of the internal dipolar configuration. The model also provides a hydrodynamic foundation for Maxwell's equations and Quantum Electrodynamics (QED) and yields novel predictions, including a test of the neutron's dipole structure and astrophysical predictions that follow from our composite model of the neutron. Altogether, these developments provide a comprehensive, physically grounded unification of the electroweak force and offer an alternative to the abstract formalism of the Standard Model.

1. Introduction

The Standard Model of Particle Physics successfully unifies the electromagnetic and weak nuclear forces through the elegant but abstract mathematics of gauge symmetries and spontaneous symmetry breaking (Weinberg, 1967). While its predictions are incredibly accurate, the framework does not provide a physical origin for electric charge, nor does it offer a mechanical picture of its associated forces; instead, it relies on the exchange of "virtual" particles (Feynman, 1985). Most notably, its description of the weak force requires the existence of the massive W and Z bosons, which in turn necessitates the conceptual apparatus of the Higgs field to explain their mass.

This chapter presents a physical and parsimonious alternative, drawing on the principles of Unified Field Dynamics (UFD) established in the preceding chapters. Building on the foundational postulate of the composite neutron—a stable, bound vortex of a proton and an antiproton, as introduced in Chapter 1—we will now demonstrate how both the electroweak force and magnetism emerge from this unified structure.

In the sections that follow, we detail the internal geometry of this composite vortex, explain how its transformation and decay give rise to the phenomena of the weak force, and provide a new interpretation for the experimental signatures of the W and Z bosons. We then demonstrate how this structure provides a natural origin for the mass and magnetic moment of the neutron, and finally, how the fluid-dynamic interactions of its charged vortices give rise to the laws of classical electromagnetism.

2. The Composite Neutron: A Unified Origin for the Electroweak Force

In the Standard Model, the neutron is treated as a fundamental particle, and separate, abstract mechanisms describe the weak and electromagnetic forces. UFD proposes a more direct, physical picture, grounded in a foundational postulate: the neutron is not a fundamental particle but a composite vortex composed of a tightly bound proton and antiproton. This postulate provides a unified, mechanistic origin for the electroweak force. To understand how this mechanism functions, though, we must first understand the origin of charge.

2.1 The Emergence of Charge from Chirality

In the Standard Model, charge is an intrinsic, axiomatic property of a particle. UFD, however, provides a physical mechanism for charge by proposing that charge polarity is a direct manifestation of a vortex's fundamental chirality, or "handedness." A vortex with one chirality (e.g., a "left-handed" spin like the electron) creates a "sink" in the Universal Light Field (ULF), which we observe as a negative charge. Conversely, a vortex with the opposite, "right-handed" chirality (like the proton) acts as a "source," generating an outward flow observed as a positive charge.

Attraction between opposite charges occurs because the ULF flows from the high-pressure zone around the source to the low-pressure zone around the sink. This creates a channel of directed flow, and the higher ambient pressure of the surrounding field pushes the two vortices together. Similarly, repulsion between like charges occurs because their similar outward flows create a high-pressure zone between them, which physically pushes them apart (Figure 14).

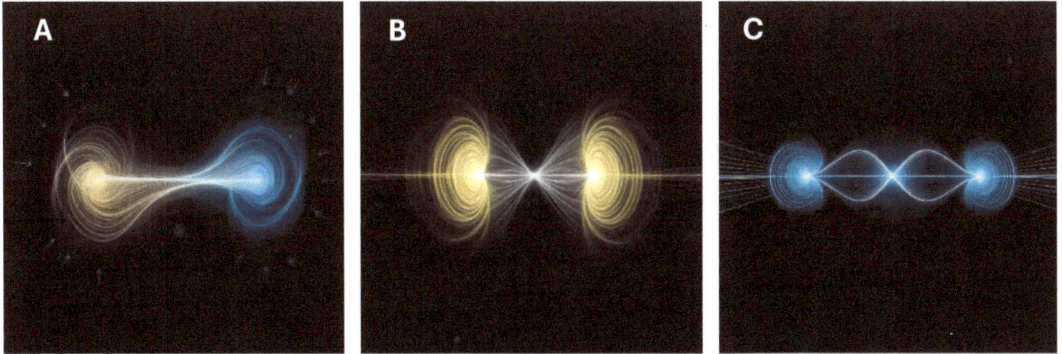

Figure 14: The Fluid Dynamics of the Electrostatic Force. This figure illustrates the physical mechanism of electrostatic attraction and repulsion in the UFD framework.
(A) Attraction: A "source" vortex (gold) perpetually emits a flow of the ULF, while a "sink" vortex (blue) perpetually draws it in. This creates a low-pressure channel between them, and the higher ambient pressure of the surrounding field pushes the two vortices together.
(B) Source Repulsion: Two "source" vortices both emit an outward flow of the ULF. Their flows clash, creating a high-pressure zone between them that physically pushes the two vortices apart.
(C) Sink Repulsion: This figure illustrates the repulsion between two "sink" vortices (e.g., two electrons). Both vortices draw the ULF medium inward from the surrounding space. Between them, these two opposing flows clash, creating a high-pressure zone that physically pushes the vortices apart.

2.2 The Neutron as a Matter-Antimatter Composite

In the Standard Model, the neutron is treated as a fundamental particle composed of three quarks (Gell-Mann, 1964). This interpretation presents several conceptual puzzles that our composite model addresses.

2.2.1 The Neutron's Neutrality

The first puzzle is the neutron's neutrality. The Standard Model explains this by positing that a neutron is composed of one "up" quark (with a +2/3 charge) and two "down" quarks (each with a -1/3 charge). The sum of these fractional charges is zero, resulting in a neutral particle.

UFD offers a simpler explanation: the neutron is a composite Figure 8 vortex formed by a tightly bound proton (+1 charge) and an antiproton (−1 charge) vortex. Its electrical neutrality is, therefore, the necessary and predictable consequence of a perfect balance between its internal "source" (the proton) and "sink" (the antiproton).

2.2.2 The Neutron's Magnetic Moment

The second, deeper puzzle is the neutron's magnetic moment. A magnetic moment is generated by moving electric charge, yet the neutron has no net charge—a surprising discovery first made in the 1930s (Bloch et al., 1939). The Standard Model attributes this discovery to the complex interactions and non-zero magnetic moments of the constituent quarks, which do not perfectly cancel out.

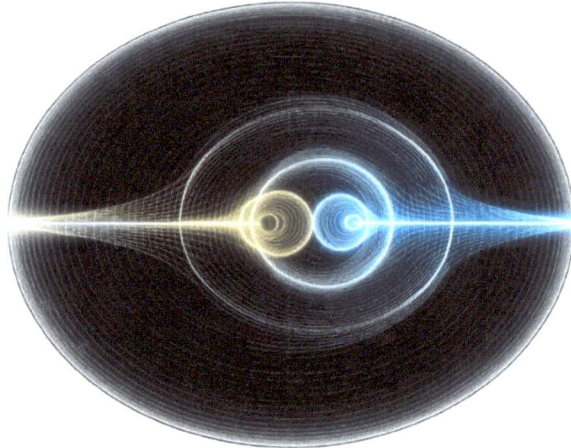

Figure 15: The Neutron's Magnetic Moment. This figure illustrates the physical mechanism for the neutron's magnetic moment within the UFD framework. The neutron is modeled as a composite vortex containing a proton "source" (gold) and an antiproton "sink" (blue). The spinning of this intrinsic, self-contained dipole creates a coherent, rotational eddy in the surrounding ULF. This rotational flow *is* the neutron's magnetic moment, and its streamlines are what we visualize as magnetic field lines. The composite model thus provides a direct physical origin for this fundamental property.

Our emergent framework provides a more direct, physical origin. The neutron's internal proton-antiproton structure is a perfect, fundamental magnetic dipole (Figure 15). The spinning of this balanced, internal source-sink system creates a powerful, localized, and coherent rotational eddy in the ULF. This eddy is the neutron's magnetic moment—a direct, physical consequence of its underlying matter-antimatter geometry.

2.2.3 The Neutron's Mass

The third puzzle of the composite neutron is its mass. If it is a proton–antiproton composite, why is its mass not roughly twice that of a proton (~1876 MeV)?

The answer is what we term the Coherence Dividend (*see* Chapter 7). The merger of a proton (~938 MeV) and an antiproton (~938 MeV) is the ultimate mass-defect event, leading to the formation of matter, as we describe in Chapters 1 and 3. This interaction is driven by the fluid dynamics of the two vortices: the proton acts as a stable "source" with an outward flow, while the antiproton acts as a transformative "sink" with an inward flow. As the two vortices collapse, the stable geometry of the proton interlocks with the opposing geometry of the antiproton. The sink captures the source's flow, locking them into a new, self-contained Figure-8 vortex.

In this process, 936 MeV of their combined 1876 MeV mass-energy is radiated away as binding energy.[36] This colossal release consists almost entirely of the annihilated mass-energy of the antiproton component (*see* Chapter 3). We propose that this coherence energy is a candidate mechanism for the power of quasars (Schmidt, 1963). The neutron's resulting mass of ≈939.6 MeV is what remains.[37]

2.2.4 The Neutron's Stability

The fourth puzzle is the neutron's instability. Why is a free neutron unstable, and why is it slightly heavier than a proton?

In our model, the neutron is a "deep-field" particle, forged in the extreme pressure of a Genesis Event. In this high-energy environment, the interlocking of a proton and antiproton into a Figure-8 vortex is perfectly stable, just as it is in the high-pressure environment of the nucleus. However, a free neutron in the ambient, low-pressure field of deep space is like a deep-sea fish brought to the surface; it is no longer

---

[36]To put this number in perspective, the fission of a single Uranium-235 atom in a nuclear bomb releases approximately 200 MeV of energy. The energy released in the formation of a single neutron is therefore nearly five times more powerful than the energy released by a single fission event.

[37] This interpretation is strongly supported by a well-established astronomical observation: the tight correlation between the mass of a supermassive black hole and the total mass and energy of its host galaxy (Ferrarese & Merritt, 2000). In UFD, this is a direct and necessary consequence of the Genesis Event. A larger and more massive galaxy is the result of a larger and more intense creation event, which by definition forges more matter (neutrons) and therefore must release proportionally larger energies, resulting in a more luminous quasar. The observed correlation is thus a direct confirmation of the principles of this model.

in the environment required to maintain its structural integrity. This environmental mismatch forces the free neutron into a strained, metastable state.

Thus, in our Pressure Equilibrium Model, we define stability not as an intrinsic property of the neutron but as an emergent property of its equilibrium with the environment.

1. The Bound Neutron (True Stable State): The Genesis Event (or the core of a nucleus) is a forge of immense UEF pressure ($P_{external}$). Under this pressure, the proton (source) and antiproton (sink) are forced to merge, releasing the massive ~937 MeV Coherence Dividend as they lock into a tight, self-contained Figure-8 topology. This is the true, stable, ground state of the neutron, where the internal tension of the knot ($P_{internal}$) is counterbalanced by the external pressure of the forge: $P_{internal} \approx P_{external}$.

2. The Free Neutron (Metastable State): When this particle is ejected from the forge into the low-pressure vacuum of free space ($P_{external} \approx 0$), the external containment is removed. The Figure-8 geometry is now under immense internal tension, acting like a coiled spring trying to pull itself apart. The particle is now in a metastable, high-tension state where: $P_{internal} \gg P_{external}$.

This geometric tension is the physical origin of the 1.3 MeV mass-energy excess of the free neutron (Fermi, 1934). This small mass excess is not arbitrary; it is the potential energy stored in the strained topological bond between the proton and antiproton loops. Its ~15-minute beta decay is simply the Geometric Coherence Force (GCF) resolving this tension.

2.3 The Weak Force as a Manifestation of the Geometric Coherence Force

In UFD, the weak force is not a fundamental interaction mediated by massive W and Z bosons, as proposed by electroweak theory (Weinberg, 1967). Instead, it is an emergent effect of the GCF—the universal tendency of vortices to shed energy and reconfigure into more stable geometries. The decay of a free neutron is the paradigmatic example: a geometric phase transition in which the strained, metastable neutron collapses into the more coherent proton, shedding its excess energy as an electron and an antineutrino (Figure 16).

This framework thus reinterprets the high-energy signatures discovered at CERN in 1983, which were attributed to the W and Z bosons (UA1 Collaboration, 1983;

144

UA2 Collaboration, 1983). Rather than new particles with rest masses of 80–91 GeV, UFD proposes that these signatures are quantized harmonic overtones of the nucleon vortex itself. A useful analogy is a bronze bell: a proton's fundamental "note" is ~1 GeV, while a sufficiently violent collision can excite a fleeting, high-frequency overtone near 80–91 GeV. These rare, short-lived excitations are the "clang" of the nucleon. Their spin-1 character arises because ULF disturbances naturally quantize into vector (spin-1) modes, just as toroidal vortices quantize into spin-½ states (*see* Chapter 5). In this view, the resonance's measured width reflects its damping rate, while its branching ratios are the different geometrically allowed "shatter patterns" as the excited vortex collapses.

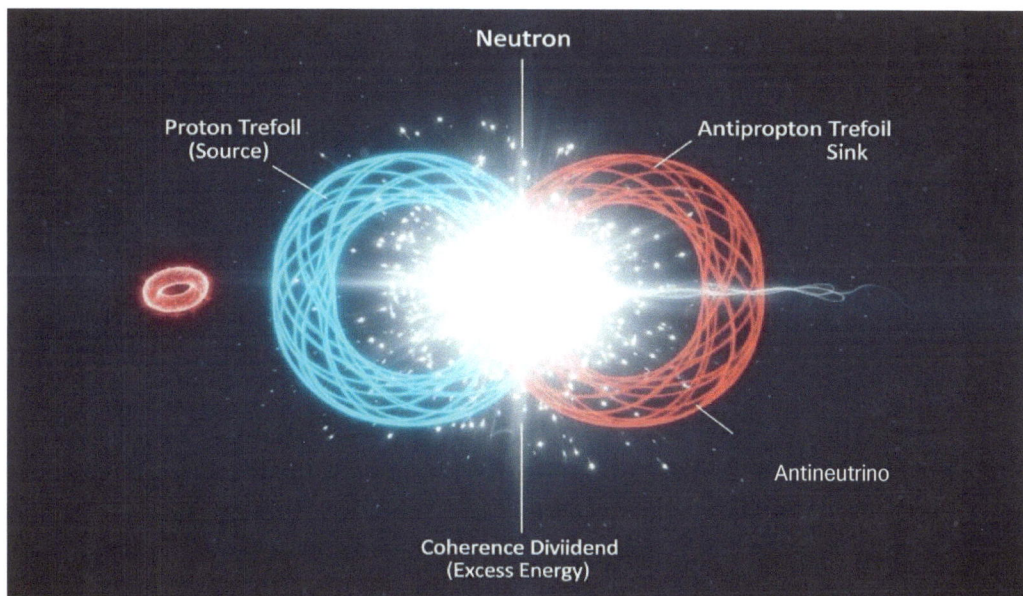

Figure 16: The Weak Force as a Dynamic Geometric Phase Transition. This illustration depicts the weak force as a rapid geometric restructuring event. The metastable Figure-8 composite vortex of the neutron is shown as it spontaneously unravels. As the vortex minimizes its internal strain and transitions to the more stable trefoil proton geometry (*see* Chapter 7), it sheds excess energy as a Coherence Dividend, which is visually represented by the central bright emission. Simultaneously, the emergent toroidal electron vortex and the subtle antineutrino vortex are seen precipitating from the disintegrating structure, illustrating the GCF-driven process of geometric reorganization and energy release.

This, however, presents the deepest quantitative challenge for the UFD framework. The Standard Model's greatest triumph—its "crown jewel"—is not just that the Z boson decays, but that its electroweak theory quantitatively predicts the branching ratios of its decay into different lepton families (e^+e^-, $\mu^+\mu^-$, $\tau^+\tau^-$) and quark jets with astonishing precision, establishing the existence of exactly three generations of matter. The UFD framework must be capable of reproducing this precision.

Here, UFD posits that the Standard Model's coupling constants are not fundamental magic numbers but are, in fact, the mathematical result of the lepton's stable geometric hierarchy. The branching ratios are the competing 'coherence pathways' available to the collapsing vortex. They are the first three stable, resonant harmonics of the same underlying ULF vortex geometry. The 91 GeV "clang" decays to them equally because they are just the first three notes on the same guitar string. The Standard Model's coupling constant is simply a measurement of the geometric accessibility of these three harmonics.

While the Breit-Wigner resonance curve observed at LEP, peaking at 91 GeV, has been used to support the Standard Model's interpretation of the weak force, the curve is not unique to particle physics. It is the universal mathematical fingerprint of any driven, damped resonant system, as exemplified by the infamous collapse of the Tacoma Narrows Bridge (Billah & Scanlan, 1991).[38] Thus, the LEP data do not prove the existence of a new particle. Instead, they demonstrate that the nucleon possesses a powerful, destructive resonance at this energy.

Ultimately, the significance of the weak interaction is its role in revealing the underlying reality of the UFD framework. The weak force is the only force that violates parity symmetry, or the preference for left-handed particles over right-handed ones. This violation is the direct consequence of the topological handedness built into the Figure-8 neutron composite. Thus, in the UFD framework, the weak force demonstrates how geometry and chirality—rather than abstract quantum numbers—are the causal drivers of reality. It is simply a natural consequence of vortices driven by the principle of geometric coherence.

---

[38] The catastrophic collapse of the Tacoma Narrows Bridge in 1940 is the textbook real-world example of a driven resonance. A steady, moderate wind acted as the "driving force," creating aerodynamic vortices that happened to oscillate at a frequency that perfectly matched one of the bridge's natural torsional (twisting) resonant frequencies. This constant, rhythmic "pushing" from the wind pumped energy into the system, causing the amplitude of the bridge's twisting oscillations to grow uncontrollably until the structure tore itself apart. The mathematical curve that describes this relationship—the amplitude of the oscillation versus the frequency of the driving force—is a perfect Breit-Wigner resonance curve, demonstrating that this signature is a universal feature of resonant systems, not one exclusive to the world of subatomic particles.

3. Hydro Electrodynamics: The Physical Nature of the Electromagnetic Force

In the Standard Model, force is abstractly carried by virtual particles (Feynman, 1985). UFD replaces this with a tangible mechanism directly mediated by the ULF's physical properties, which we call Hydro Electrodynamics (HED). In this view, all electromagnetic phenomena are reinterpreted as the dynamics of this universal field.

3.1 Electrostatic Charge and Force: Sources and Sinks

In HED, the fundamental force of electricity arises from the vortex chirality of charged particles. As established, the intrinsic rotation of a proton or electron vortex causes it to act as a stable "source" (positive charge, outward ULF flow) or a "sink" (negative charge, inward ULF flow) within the ULF medium.

The electrostatic force is the simple, pressure-based effect of the ULF flowing between these sources and sinks. Thus, the mechanism of electric attraction in this model is identical to that of gravity; in both cases, the force arises from a pressure gradient driven by the bulk field flowing into a sink. The difference is that gravity is a monopole (always a sink/attraction), whereas electricity is a dipole (sources and sinks), allowing for both attraction and repulsion.

3.2 Magnetism: Coherent Rotational Flow

In classical and quantum physics, the intrinsic magnetic moment of particles, such as electrons, is an unexplained, axiomatic property. HED provides a direct, physical origin: magnetism emerges from the coherent, rotational flow of the ULF induced by particle spin.

A single, spinning charged vortex induces a tiny, rotational "eddy" in the ULF immediately around it. The vortex's spin axis defines its geometric North and South poles, and its nature as a source or sink determines the "handedness" or direction of this rotational flow. Thus, "magnetic field lines" are reinterpreted as maps of these coherent streamlines, which flow in closed loops from one pole to the other, and the magnetic force is the physical pressure and momentum of this stable, circulating flow.

3.3 Macroscopic Magnetism: The Geodynamo

HED's principles also scale up to provide a direct physical explanation for the origin of large-scale magnetic fields, such as those generated by planets and stars. In modern astrophysics, this phenomenon is explained by dynamo theory (Elsasser, 1946).

The theory successfully posits that a persistent, large-scale magnetic field can be generated if three conditions are met: a large volume of electrically conductive fluid, such as the Earth's molten iron core; kinetic energy provided by the planet's rotation; and an internal energy source that drives convective currents within the fluid.

While the dynamo theory successfully describes these necessary conditions, it remains a macroscopic model. It therefore does not provide a fundamental physical picture of *what* the "conductive fluid" is at the level of individual particles or *how* their collective motion translates into a planetary-scale magnetic field. HED provides this missing physical ontology.

In the HED framework, a planet's liquid outer core is a vast sea of mobile, charged nucleon and electron vortices. Because these fundamental vortices possess the emergent property of charge, the fluid is inherently "conductive." The planet's rotation, combined with heat-driven convection, organizes the motion of these vortices into a single, massive, and coherent electrical current that moves in helical patterns (Figure 17).

Figure 17: The Physical Origin of a Planetary Magnetic Field. This schematic illustrates the physical mechanism behind a planet's magnetic field in HED. Within the electrically conductive molten outer core, the planet's rotation and heat-driven convection organize a vast sea of mobile nucleon and electron vortices into a massive, coherent electrical current flowing in helical patterns. This organized current, in turn, induces a robust, stable, macroscopic vortex (a "massive eddy") in the surrounding ULF, which manifests as the planet's observable magnetic field.

According to the principles of induction, this enormous, organized current of spinning vortices will inevitably induce a robust and stable macroscopic vortex—a massive eddy—in the surrounding ULF. This large-scale, coherent ULF vortex *is* the planet's magnetic field. This model naturally explains why planets with large, rotating, fluid metallic cores, such as Earth and Jupiter, generate powerful magnetic fields,

148

whereas those without such cores, like Mars (whose core has largely solidified) or Venus (which rotates too slowly), do not (Stevenson, 2003).

Having established a physical picture of electromagnetism, we will now demonstrate how it provides a direct, intuitive basis for the formal mathematics of electromagnetism: Maxwell's equations.

4. A Fluid-Dynamic Basis for Maxwell's Equations

In the 1860s, James Clerk Maxwell unified the disparate phenomena of electricity, magnetism, and light into a set of four elegant equations (Maxwell, 1865). These equations have been experimentally verified countless times and form the bedrock of classical electrodynamics. For a new theory to be viable, it must be able to recover these foundational laws from its own principles. This section demonstrates, principle by principle, how Maxwell's equations emerge as mathematical descriptions of the ULF's fluid dynamics.

We begin by defining the electric and magnetic fields as distinct modes of ULF fluid dynamics. The electric field (E) is the irrotational, divergent component of the ULF's flow, representing a linear pressure gradient. The magnetic field (B) is the solenoidal, rotational component of the ULF's flow, representing a vortical or shear effect. This physical distinction maps directly onto the language of vector calculus. The divergence ($\nabla \cdot$) is the natural mathematical operator for quantifying the strength of a source or sink, while the curl ($\nabla \times$) is the natural operator for quantifying the rotation or vorticity of a flow.

4.1 Gauss's Law for Electricity: The Law of Flow from a Source

In classical physics, Gauss's Law for Electricity ($\nabla \cdot E = \rho/\epsilon 0$) provides a powerful relationship between electric charge and the electric field it produces. Gauss's Law states that the total electric flux out of any closed surface is directly proportional to the total electric charge enclosed within it. Thus, if one knows the total "flow" of the electric field out of a volume, they also know how much charge is inside. It is this principle that gives rise to the famous inverse-square law for the strength of the electric field.

In HED, Gauss's Law becomes a concrete principle of hydrodynamics. The electric field (E) becomes the physical velocity of the ULF's flow, and the charge (ρ) becomes the strength of the source or sink. The law, therefore, translates to a simple, intuitive statement: the total rate of fluid flowing out of any given volume is determined by the strength of the source inside it. The inverse-square law emerges naturally because this fixed amount of "flow" must spread out over the surface area of an expanding sphere.

This reinterpretation also provides a physical origin for Coulomb's constant (k_e), the fundamental constant that sets the strength of the electrostatic force. In the UFD framework, k_e, is not a fundamental axiom but an emergent property representing the intrinsic "stiffness" of the ULF medium. Just as the stiffness of a drum skin determines how efficiently it transmits a vibration, the stiffness of the ULF determines how efficiently it transmits the "push" of a charge. The enormous value of Coulomb's constant is a direct measure of the ULF's incredible responsiveness relative to the UEF.

4.2 Gauss's Law for Magnetism: The Law of Closed Loops

Gauss's Law for Magnetism, $\nabla \cdot B = 0$, is the mathematical statement that the divergence of the magnetic field is always zero. This means that there are no "sources" or "sinks" for the magnetic field, or no magnetic monopoles. In HED, this is a necessary geometric consequence. We have defined the magnetic field (B) as the coherent, rotational "eddy" or whirlpool induced in the ULF by a spinning vortex. The spin itself is the origin of the magnetic dipole. This principle applies universally:

- For an electron, its spin is a fundamental rotation of its ULF vortex, which creates a primary magnetic dipole.

- For a neutron, its powerful magnetic moment arises from the spinning of its internal (Proton + Antiproton) source-sink dipole.

Because a spin is a complete, unified axis of rotation with two poles (e.g., North and South), one cannot have "half a spin" any more than a coin can be spinning with only a "heads" side. An isolated magnetic monopole is therefore physically impossible in our framework, as it would represent a spin with only one pole, which is a geometric absurdity. The law is not a mere observation; it is an inevitable consequence of the underlying rotational nature of matter.

4.3 Faraday's Law of Induction: The Law of Rotational Change

Faraday's Law of Induction ($\nabla \times E = -\partial B / \partial t$) is the principle that revealed the deep connection between electricity and magnetism. It describes how a changing magnetic field creates a circulating electric field. This law is the foundation of nearly all electric power generation; every electric generator, motor, and transformer works because of this fundamental relationship. While the Standard Model provides precise mathematical rules for this effect, it does not offer a tangible picture of *why* a changing rotational field should induce a linear one.

In HED, this relationship is a direct consequence of fluid dynamics. We have defined the magnetic field (B) as a stable, rotational "eddy" in the ULF. A "changing" magnetic field, therefore, means this eddy is either speeding up, slowing down, or moving.

Consider a whirlpool in a lake: if it suddenly spins faster, it pushes water away from its edge, creating a radial, linear flow (an E-field). Conversely, if the whirlpool slows, the surrounding water will rush inwards to fill the space. In precisely the same manner, a change in the ULF's rotational flow (B) necessarily induces a linear flow (E) in the surrounding medium. Faraday's Law is thus reinterpreted as a fundamental principle of hydrodynamics: the acceleration or deceleration of a rotational flow will always generate a corresponding linear flow.

4.4 Ampère-Maxwell Law: The Law of Linear Change

The Ampère-Maxwell Law ($\nabla \times B) = \mu_0 J + \mu_0 \epsilon_0 (\partial E / \partial t)$) is the final of Maxwell's equations and the counterpart to Faraday's Law. It describes the two ways a magnetic field can be generated. The first term ($\mu_0 J$) is Ampère's original law, which states that a steady flow of charge (an electric current) creates a circulating magnetic field around it. This is the principle behind all electromagnets.

The second term ($\mu_0 \epsilon_0 (\partial E / \partial t)$) was Maxwell's crucial addition. He realized that a changing electric field must also generate a magnetic field. This insight completed the symmetry of the equations, as it showed that a changing B-field creates an E-field and a changing E-field creates a B-field. This self-perpetuating cycle is what allows light to propagate through space.[39] While mathematically brilliant, the Standard Model provides no single, intuitive explanation for why these two distinct causes—a moving current and a changing field—produce the same effect.

----

[39] In the HED framework, this self-perpetuating cycle is the direct, physical mechanism of a photon. A photon is a self-sustaining wave packet in the ULF where a constantly changing linear pressure wave (the electric field) induces a rotational eddy (the magnetic field), which in turn induces the next linear pressure wave. This "leapfrogging" or "bootstrapping" process is how the wave propagates through the ULF medium at the speed of light.

HED provides this single, intuitive picture. First, an electric current (J) is a coherent stream of charged vortices moving together. This moving line of vortices naturally drags the surrounding ULF fluid into a large, stable rotational flow that circulates along the current's path. This is the physical origin of the magnetic field around a wire. Second, a changing electric field ($\partial E/\partial t$) is an accelerating or decelerating linear flow of the ULF itself. Just as a sudden jet of water shot into a calm pool creates eddies and turbulence, a change in the ULF's linear flow necessarily induces a new rotational flow (B) in the surrounding medium. Thus, the Ampère-Maxwell law is reinterpreted as a fundamental principle of hydrodynamics, stating that a rotational flow can be generated either by dragging the fluid with a moving object or by accelerating the fluid itself.

Ultimately, HED demonstrates that the laws of electromagnetism are not axiomatic but are the hydrodynamics of the very medium that constitutes light.

5. An Incommensurable Ontology: Hydro Electrodynamics

The Standard Model's foundation for electromagnetism is QED, a powerful theory based on abstract axioms such as the existence of virtual particles, the intrinsic charge of particles, and unexplained coupling constants. This section demonstrates how UFD's substance-based approach replaces QED's abstract machinery with tangible, physical mechanisms.

5.1 The Quantum Electrodynamics Challenge (HEM vs. QED)

This section contrasts the abstract, mathematical interpretation of QED with the physical mechanism of HED by asking the fundamental questions that QED accepts as axioms.

- What is an electric charge?

  - QED: Charge is a fundamental, intrinsic, and axiomatic property of a particle (e^-) with no deeper physical origin.

  - HED: Charge is a physical manifestation of a vortex's chirality ("handedness"). A "sink" rotation creates negative charge, and a "source" rotation creates positive charge.

- What is magnetism?

- QED: Magnetism is an axiomatic property arising from the spin and motion of charged particles.

- HED: Magnetism is coherent rotational flow or a macroscopic eddy induced in the ULF by the spin of charged vortices. Magnetic field lines are ULF streamlines.

- What is the electromagnetic force?

  - QED: The force is mediated by the exchange of massless, virtual photons between charged particles.

  - HED: The force is the pressure-based effect of ULF flow between stable, charged vortices. Attraction is the ULF flowing from a "source" to a "sink," creating a low-pressure channel.

- What is a "virtual photon?"

  - QED: A mathematical artifact used in Feynman diagrams that cannot be observed.

  - HED: A transient ULF pressure wave or coherent coupling that mediates the pressure difference between the charged vortices. It is the mathematical consequence of a fluid-dynamic interaction.

- What is the fine-structure constant (α)?

  - QED: The coupling strength of the electromagnetic interaction.

  - HED: It is the Hydrodynamic Gear Ratio (1/137), the geometric and field-limited rate at which a charged vortex can exchange energy with the ULF medium, as described in Chapter 5.

- Why is the muon's magnetic moment anomalous?

  - QED: The deviation (g-2) is predicted via complex calculations involving virtual particles, but the observed value is currently unexplained and points to new, unknown physics.

  - HED: The deviation is a physical signature of geometric strain. The muon is a metastable, excited harmonic of the electron torus, whose high internal tension physically deforms its vortex structure, altering its magnetic coupling efficiency.

- What is the weak force?

  - QED: The force is mediated by the exchange of massive, virtual W and Z bosons, which acquire mass through the Higgs mechanism. The force is abstract and requires spontaneous symmetry breaking

  - HED: The weak force is an emergent effect of the GCF. It is a physical geometric phase transition — the inevitable collapse of the strained, metastable composite neutron vortex into the more stable proton vortex

- What is the Higgs boson?

  - QED: Massless gauge fields acquire mass through interaction with a universal scalar Higgs field. The Higgs boson is the quantum excitation of this field, and particle mass arises via spontaneous symmetry breaking.

  - HED: The Higgs boson is a collective excitation that signals the onset of geometric impedance in the ULF–UEF transition regime. What the Higgs mechanism parameterizes mathematically as "mass acquisition" is, in HED, the physical consequence of a vortex entering a closed, self-sustaining circulation state within the field. Mass does not arise from coupling to a scalar background. Instead, it arises from topological closure and internal circulation at speed c, as described in Chapter 4. The Higgs boson is therefore interpreted as a boundary-mode excitation marking the threshold between freely propagating field modes and frozen, vortex-bound states of light.

This translation of QED to HED demonstrates that the electromagnetic force can be considered an emergent pressure effect of the ULF. This substance-based view of the field is not new, however, as it already has deep roots in engineering intuition.

5.2 Steinmetz's Dual-Field Model of Electromagnetism and Its Legacy

Charles Proteus Steinmetz (1865–1923) was a foundational figure in electrical engineering who developed a deep, physically intuitive understanding of electromagnetic phenomena that anticipated many modern concepts formalized much later. Known for his pioneering work on alternating current systems and electrical machinery, Steinmetz also proposed a conceptual framework for the electromagnetic field grounded in fluid and vortex dynamics, arguing that the electromagnetic field consists of two complementary components, each with distinct physical characteristics:

1. The Dielectric Field: Associated with stored potential energy, this component represents the static, tensioned aspect of electricity—the electric field in its electrostatic form. Steinmetz likened this to a medium under strain, capable of holding energy without motion (Steinmetz, 1914).

2. The Magnetic Field: This component embodies kinetic energy and dynamic pressure within the field, arising from currents and the motion of charges. It corresponds to the radiative and inductive phenomena that produce changing magnetic flux and electromagnetic waves (Steinmetz, 1914).

In Steinmetz's view, these two fields represent orthogonal modes of a unified physical medium, which together explain electricity and magnetism as manifestations of wave-like and rotational motions within this underlying continuum.[40] His conceptualization predates and foreshadows modern electromagnetic field theory, which emphasizes the physicality and fluidity of fields rather than abstract forces acting at a distance. HED builds directly on this legacy by situating Steinmetz's dielectric and magnetic components within a nested hierarchy of fields:

- The UEF provides the dense, fluid-like substrate within which vortices form.

- The ULF manifests these vortical motions as electromagnetic phenomena, where the dielectric and magnetic fields arise as natural modes of vibration and rotation within it.

This perspective offers a fluid-dynamic ontology in which electromagnetic phenomena emerge from vortex structures and wave interference patterns, providing a tangible, mechanistic foundation for Steinmetz's physical intuition. By reviving

---

[40] In physics, orthogonal means that two vector components are mathematically independent and spatially perpendicular (acting at a 90-degree angle). In the UFD framework, this geometric necessity reflects the physical mechanics of the ULF. The electric field is the linear pressure gradient flow, while the magnetic field is the circulating, vortical flow (the "whirlpool" around a current). The ULF medium must sustain both types of flow simultaneously and independently; hence, the two resulting forces must be perpendicular to each other.

Steinmetz's dual-field picture in a rigorous modern context, HED bridges early 20th-century engineering insights with contemporary theoretical physics.

6. Falsifiable Predictions

A scientific framework is defined not only by its explanatory power but also by its ability to make novel, testable predictions that distinguish it from the Standard Model. UFD's model of electroweak emergence makes several such predictions.

6.1 Fundamental Physics

The most direct predictions of this chapter follow from our redefinition of the neutron.

6.1.1 A Non-Zero Electric Dipole Moment

In the Standard Model, the neutron is treated as a spherical cloud of quarks with a net charge of zero. Consequently, it predicts an Electric Dipole Moment (EDM) that is vanishingly small ($\sim 10^{31} e \cdot cm$)—effectively zero for any current measurement. UFD predicts the opposite. Because the neutron is a composite Figure-8 vortex formed by the interlocking of a proton "source" and an antiproton "sink," it possesses a permanent, structural separation of charge centers along its axis of symmetry. While the *net* charge is zero, the internal distribution is strictly polarized.

Therefore, we predict that next-generation ultra-cold neutron experiments (such as those at PSI or SNS) will detect a non-zero EDM significantly larger than the Standard Model limit (likely $\sim 10^{28}$ to $10^{26} e \cdot cm$). A definitive measurement of a large neutron dipole would be a smoking gun that falsifies the Standard Model's spherically symmetric description and validates the composite, dipolar geometry of the UFD neutron.

6.1.2 Scattering Signatures: The Geometric Form Factor

Current Deep Inelastic Scattering (DIS) experiments model the neutron as a cloud of three point-like quarks held together by gluons. However, the UFD model predicts that the neutron has a distinct, macroscopic topology: a Figure-8 vortex created by the interlocking of a proton and antiproton. Accordingly, we predict that high-precision electron-scattering experiments will reveal deviations from the Standard Model's "form factors" (the mathematical description of the charge distribution) at specific momentum transfers. Specifically:

1. Dual-Center Diffraction: At energy scales corresponding to the physical distance between the internal proton and antiproton centers, scattering data should show a diffraction pattern consistent with a dual-center topology (a dumbbell or Figure-8 shape) rather than a spherical distribution of three points.

2. Anisotropic Charge Distribution: Unlike the quark model, where the neutron's charge distribution is often averaged to a sphere, the UFD Figure-8 geometry possesses an intrinsic orientation. We predict that scattering polarized electrons off polarized neutrons will reveal a strong, geometric anisotropy in the charge distribution that aligns with the neutron's spin axis—a "shape" that the Standard Model cannot account for without adding arbitrary parameters.

Detecting this Figure-8 signature in the form factors would be the structural proof that the neutron is not a bag of quarks but a structured, geometric composite.

6.1.3 Subtle Signatures in Beta Decay

In the Standard Model, beta decay is calculated using Fermi's theory, which treats the interaction as occurring at a single point in spacetime. UFD, in contrast, describes it as a dynamic physical process in which one vortex geometry (the neutron) physically transforms.

This more complex process should lead to observable deviations from standard predictions. We therefore predict the existence of subtle, transient phenomena during the decay process that are not accounted for by the Standard Model. High-precision measurements of the energy spectrum of emitted electrons may reveal minute deviations from the spectrum predicted by Fermi's theory. These deviations would not be random noise but would represent the energy momentarily absorbed and re-emitted by the ULF as the Figure-8 vortex reconfigures itself—the same disturbances we have identified as the origin of the W and Z boson signatures.

6.1.4 The Generation of Antimatter

Our composite model also predicts a novel method for generating antimatter: regenerating the dormant antiproton within the neutron.

Since the neutron is formed by the high-energy collapse of a proton and an antiproton (releasing ~937 MeV as the Coherence Dividend), the internal antiproton exists in a low-energy, "deflated" state. Therefore, it should be possible to reverse this process not by brute-force collision, but by resonant energy injection. We predict that a precisely tuned, high-energy electromagnetic field could pump energy back into the neutron's internal binding mode. This would not merely break a bond; it would "re-inflate" the internal antiproton vortex. Once the system absorbs the required energy threshold (restoring the lost mass), the geometric stability of the Figure-8 would fail, causing the neutron to dissociate into a free proton and a fully restored free antiproton.

While the energy cost is high, this mechanism differs fundamentally from standard pair production, as it targets a specific pre-existing topological structure, thereby offering a directed pathway to antimatter generation that serves as a definitive test of the neutron's composite nature.

6.2 Astrophysical Signatures

The composite nature of the neutron also leads to unique astrophysical predictions.

6.2.1 The "Neutron Crystal": A New Model for Neutron Stars

In the Standard Model, a neutron star is a hot gas of "neutronium," held up by an abstract "degeneracy pressure" (Shapiro, 1983). The UFD framework, guided by the GCF, predicts the exact opposite: a neutron star is not a hot, chaotic gas but a single, cold, hyper-coherent neutron crystal. The immense gravitational pressure of the star *is* the GCF at its maximum, compelling the $\sim 10^{57}$ composite neutron vortices to abandon their individuality and lock into a single, perfectly aligned, phase-coherent crystalline lattice. This crystal is held up by its own immense structural integrity, much like a Bose-Einstein Condensate.[41]

The neutron crystal model provides a direct, physical explanation for "pulsar glitches" (sudden increases in a pulsar's spin). In our model, a "glitch" is a literal starquake in which the entire massive neutron crystal lattice, under immense rotational stress, suddenly snaps into a more geometrically efficient, lower-energy (more coherent)

---

[41]A Bose-Einstein Condensate (BEC) is a real, experimentally verified state of matter, first created in 1995. It is formed when a cloud of atoms is cooled to temperatures near absolute zero. At this point, the atoms' individual quantum wavefunctions "overlap" and condense into a single, unified, coherent quantum state, in which millions of atoms behave as a single "super-atom." In the UFD model, a neutron star is the ultimate BEC: it is not extreme *cold* that creates the coherent state but the extreme *pressure* of the GCF, which "squeezes" all $\sim 10^{57}$ neutron vortices into a single, unified, macroscopic quantum crystal.

state. It is the GCF in action. This geometric phase transition has two immediate, observable consequences:

1. The Glitch: As the crystal's geometry tightens, its moment of inertia decreases, causing its rotational speed to increase to conserve angular momentum. This *is* the observed "glitch."

2. The Positron Source: This catastrophic shattering and realigning of the lattice is not perfectly efficient. At the fault lines of the starquake, some of the composite neutron vortices are destroyed. This pressure-induced decay releases their core components: protons and antiprotons. These liberated antiprotons immediately annihilate with nearby protons, producing a massive, localized burst of gamma rays and positrons (e^+).

This model provides a new, testable solution to the "positron excess" mystery, a well-documented anomaly observed in experiments such as the Alpha Magnetic Spectrometer (Aguilar et al., 2013). Pulsars are already considered the leading candidate source for this excess (Hooper, Blasi, & Serpico, 2009). The UFD model provides the missing causal link, predicting that the excess positrons are the stochastic, cumulative result of pulsar glitches (starquakes) occurring throughout the galaxy.

6.2.2 The Annihilation Signature of Quasars

The UFD model of cosmogenesis, in which a quasar serves as the observable signature of a galaxy's birth, makes another unique astrophysical prediction that differs from that of standard models. While standard "hadronic models" predict antimatter signatures originating from a quasar's narrow, relativistic jets, our model predicts a much larger and more fundamental signal: a vast, diffuse, and roughly spherical (isotropic) glow of 511 keV annihilation radiation surrounding the entire galactic core.

This is not merely a future test; it can be compared to a well-known phenomenon in our own galaxy. For decades, gamma-ray observatories have mapped a mysterious, vast, and diffuse halo of 511 keV annihilation radiation coming from the central bulge of the Milky Way (Weidenspointner, 2008). The origin of this halo remains mysterious. UFD provides a direct explanation for this phenomenon: while our galaxy's black hole is no longer an active quasar, this glow is the fading remnant, or afterglow, of our own galaxy's ancient Genesis Event. The theory thus provides a testable solution to a major astrophysical mystery.

This set of falsifiable predictions provides a clear and comprehensive roadmap for the experimental validation of our framework. From direct, laboratory-based tests of the neutron's internal structure to the search for unique astrophysical signatures in neutron stars and quasars, the confirmation of any one of them would provide direct evidence for the composite nature of matter and the fluid-dynamic, resonant reality described by this framework.

7. Conclusion

This chapter presented a new physical foundation for the electroweak interaction, built on a single, core postulate: the neutron is a composite Figure-8 vortex composed of a tightly bound proton and an antiproton. From this postulate, we derived a unified physical origin for the entire electroweak interaction and magnetism.

The significance of this framework lies in its theoretical virtues. By redefining the neutron as a composite particle, it provides a tangible, mechanistic basis for the HED and the weak force. This structural simplicity eliminates the need for the abstract machinery of W and Z boson force carriers and the conceptual necessity of the Higgs field for mass generation. The weak interaction is revealed to be a simple, physical geometric phase transition: the inevitable collapse of the strained neutron composite into the more stable proton.

This process replaces the standard Electroweak mechanism entirely: the transformation of a particle creates the weak force, and the resulting change in its form dictates the electromagnetic force. This structural and theoretical parsimony extends to the foundations of physics itself. While the Standard Model must mathematically combine two separate theories, Quantum Mechanics and Special Relativity, to derive fundamental properties, the UFD framework generates these phenomena from a single geometric principle.

The theoretical path forward requires the development of a rigorous mathematical formalism for the ULF to derive Maxwell's equations from first principles and model the transient dynamics of vortex decay, thereby fully accounting for the observed properties of W and Z resonances. The experimental path forward is equally clear: to test the specific, falsifiable predictions the model makes about the structure and behavior of neutrons and neutron stars. Ultimately, this model provides both a new theory and a new intuition for understanding of electromagnetism.

6.3 A New Electroweak Paradigm

In this chapter, we introduced a vortex-based foundation for the electroweak force rooted in the principles of Unified Field Dynamics (UFD). By redefining the neutron as a composite proton–antiproton vortex, we replaced the patchwork mechanisms of the Standard Model with a single, intuitive postulate that unifies electromagnetic and weak interactions. From this redefinition, we were granted the following:

1. A Unified Origin for Electroweak Phenomena: The weak interaction is revealed as a geometric phase transition—the inevitable, GCF-driven collapse of the strained neutron composite. Conversely, electromagnetism emerges naturally from the fluid dynamics of its charged constituents. The W and Z bosons are reinterpreted not as fundamental particles, but as transient shockwaves in the Universal Light Field (ULF), eliminating the need for a separate Higgs field to explain their mass.

2. A Physical Basis for Maxwell's Equations: The laws of electromagnetism arise directly from the fluid dynamics of the ULF, restoring causality and eliminating reliance on abstract symmetries or virtual particles.

3. Clear, Falsifiable Predictions: The model predicts a non-zero electric dipole moment for the neutron, a resonance-driven pathway to antimatter generation, and distinct astrophysical signatures, which are all testable in principle.

Our composite-neutron model thus simplifies the conceptual landscape of particle physics and resolves several long-standing puzzles, including the neutron's intrinsic magnetic moment, the absence of magnetic monopoles, and the physical origins of charge and fundamental constants. It does so without the need for a zoo of fundamental entities—quarks, Higgs fields, or purely virtual exchanges—by replacing them with a unified, field-based ontology.

Ultimately, this work restores causality and intuition to particle physics, and, as with gravity, invites us to view electromagnetism as a "push" arising from the geometry and flow of a universal field—one in which particles are structured vortices and the laws of nature emerge as the inevitable principles of hydrodynamics.

References

1. Aguilar, M., et al. (AMS Collaboration). (2013). First Result from the Alpha Magnetic Spectrometer on the International Space Station: Precision Measurement of the Positron Fraction in Primary Cosmic Rays of 0.5–350 GeV. *Physical Review Letters, 110*(14), 141102.

2. ATLAS Collaboration. (2012). Observation of a new particle in the search for the Standard Model Higgs boson with the ATLAS detector at the LHC. *Physics Letters B, 716*(1), 1–29.

3. Billah, K. Y., & Scanlan, R. H. (1991). Resonance, Tacoma Narrows bridge failure, and undergraduate physics textbooks. *American Journal of Physics, 59*(2), 118–124.

4. Bloch, F., Alvarez, L. W., & Rossi, B. (1939). A determination of the magnetic moment of the neutron. *Physical Review, 56*(6), 579.

5. CMS Collaboration. (2012). Observation of a new boson at a mass of 125 GeV with the CMS experiment at the LHC. *Physics Letters B, 716*(1), 30–61.

6. Elsasser, W. M. (1946). Induction Effects in Terrestrial Magnetism. Part I. Theory. *Physical Review, 69*(3-4), 106–116.

7. Fermi, E. (1934). 'Versuch einer Theorie der β-Strahlen. I'. *Zeitschrift für Physik A Hadrons and Nuclei, 88*, 161–177.

8. Ferrarese, L., & Merritt, D. (2000). A Fundamental Relation between Supermassive Black Holes and Their Host Galaxies. *The Astrophysical Journal, 539*(1), L9–L12.

9. Feynman, R.P. (1985). *QED: The Strange Theory of Light and Matter*. Princeton University Press.

10. Gell-Mann, M. (1964). A Schematic Model of Baryons and Mesons. *Physics Letters, 8*(3), 214–215.

11. Griffiths, D.J. (2017). *Introduction to Electrodynamics*. 4th ed. Cambridge University Press.

12. Higgs, P. W. (1964). Broken Symmetries and the Masses of Gauge Bosons. *Physical Review Letters, 13*(16), 508–509.

13. Hooper, D., Blasi, P., & Serpico, P. D. (2009). Pulsars as the sources of high energy cosmic ray positrons. *Journal of Cosmology and Astroparticle Physics, 2009*(01), 025.

14. Krane, K. S. (1988). *Introductory Nuclear Physics*. John Wiley & Sons.

15. Maxwell, J.C. (1865). 'A Dynamical Theory of the Electromagnetic Field'. *Philosophical Transactions of the Royal Society of London, 155*, 459–512.

16. Schmidt, M. (1963). 3C 273: a star-like object with large red-shift. *Nature, 197*(4872), 1040.

17. Shapiro, S. L., & Teukolsky, S. A. (1983). *Black Holes, White Dwarfs, and Neutron Stars: The Physics of Compact Objects*. John Wiley & Sons.

18. Sommerfeld, A. (1916). 'Zur Quantentheorie der Spektrallinien'. *Annalen der Physik, 51*(17): 1–94.

19. Steinmetz, C. P. (1914). *Electric Discharges, Waves and Impulses, and Other Transients.* McGraw-Hill.

20. Stevenson, D. J. (2003). Planetary magnetic fields. *Earth and Planetary Science Letters, 208*(1-2), 1–11.

21. UA1 Collaboration, Arnison, G., et al. (1983). Experimental observation of isolated large transverse energy electrons with associated missing energy $\sqrt{s}$ = 540 GeV. *Physics Letters B, 122*(1), 103–116.

22. UA2 Collaboration, Banner, M., et al. (1983). Observation of single isolated electrons of high transverse momentum in events with missing transverse energy at the CERN p⁻p collider. *Physics Letters B, 122*(5-6), 476–485.

23. Weidenspointner, G., Skinner, G., Jean, P., et al. (2008). An asymmetric distribution of positrons in the Galactic disk. *Nature, 451*(7175), 159–162.

24. Weinberg, S. (1967). 'A Model of Leptons'. *Physical Review Letters, 19*(21): 1264–1266.

Part III - The Architecture of Chemistry

Chapter 7: The Coherence Dividend

7.1 The Strong Force

The physical world is built upon a foundation of atomic nuclei, yet their very existence presents an immense challenge to the laws of physics. Within the incredibly dense space of atomic nuclei exist positively charged protons, which are packed together and generate an immense electrostatic repulsion that should, by all accounts, cause them to fly apart. As a result, the "strong force" is the name given to the powerful, short-range attraction that overcomes this repulsion, binding protons and neutrons together and making the stable matter of our universe possible.

Our understanding of this force has evolved significantly since its discovery in the early 20th century. Initial theories, such as Hideki Yukawa's proposal in the 1930s, envisioned the strong force as a simple but powerful force mediated by the exchange of a new particle, the meson (Yukawa, 1935), which provided a working model for interactions between nucleons (protons and neutrons). However, as high-energy experiments in the following decades revealed a complex zoo of new particles and interactions, it became clear that a deeper theory was required.

The modern theory of the strong force is known as Quantum Chromodynamics (QCD), a cornerstone of the Standard Model of Particle Physics (e.g., Weinberg, 1995). Formulated in the 1970s, QCD posits that protons and neutrons are not the fundamental constituents of matter. Instead, they are composed of three smaller constituents called "quarks." The force that binds these quarks is the "color force," a powerful interaction mediated by the exchange of particles called "gluons." In this picture, the force that holds the nucleus together is a residual effect of this more fundamental color force, just as the force between neutral atoms is a residue of the electromagnetic forces within them.

The primary strength of QCD is its predictive success. It is a mathematically rigorous and robust theory that has been verified with extraordinary precision in numerous particle accelerator experiments. It explains why quarks are permanently "confined" within protons and neutrons, and it provides a framework for calculating a vast range of high-energy phenomena.

Despite these triumphs, QCD has conceptual limitations. Its foundation is highly abstract, rooted in complex mathematical gauge symmetries. The fundamental entities of the theory, the quarks and gluons, cannot be observed in isolation, and the force itself is described as the exchange of "virtual" particles that exist only within its mathematical formalism. While the equations work, they do not provide an intuitive physical picture of what is happening inside the nucleus.

This chapter proposes a fundamentally different explanation, which we call Quantum Vortex Dynamics (QVD). In this model, nuclear binding is not a force in the traditional sense. Instead, it is an emergent property of the geometric organization of self-sustaining, energetic vortices. The significance of this reinterpretation lies in its potential to replace the abstract, axiomatic rules of QCD with a more physical, parsimonious mechanism grounded in the principles of geometry and fluid dynamics.

In this chapter, we will develop this model and demonstrate that the energy that binds all matter is simply the energy expended to achieve a more perfect geometric form.

7.2 The Coherence Dividend: A Geometric and Fluid Dynamics Theory of the Nucleus

Abstract

This chapter redefines the strong nuclear force and the quark model as emergent geometric phenomena. Here, we introduce Quantum Vortex Dynamics (QVD), an incommensurable framework of Quantum Chromodynamics (QCD) whose foundational postulate is that the three quarks observed in scattering experiments are the mathematical shadow of the proton's physical 3-lobe trefoil vortex.

In QVD, the force that binds the nucleus is the Geometric Coherence Force (GCF)—a "nuclear surface tension" that drives these structured vortices to lock into configurations of maximum stability (like the crystalline Helium-4 tetrahedron). The core mechanism is described by the Reduced Shear Hypothesis, which posits that this interlocking reduces the system's total exposed surface area. The energy released, the Coherence Dividend, then becomes the physical origin of nuclear binding energy. Notably, the framework shows that our nuclear surface tension constant (k) can be derived from the gravitational constant (G) and the speed of light (c), thereby unifying the strong force and gravity as manifestations of the UEF's elasticity. Quantitatively, the model validates this link by calculating that the strong binding mechanism (the GCF) operates at the Planck Length ($\sim 10^{-35}\ m$).

Falsifiable predictions that emerge from this model include a tetrahedral charge distribution in Helium-4 and the derivation of fractional charges from the proton's 3-lobe vortex geometry. Finally, the QVD model of nuclear binding is technologically generative, offering paths to controlled fusion and nuclear waste remediation through Geometric Catalysis. Altogether, this chapter establishes QVD as a testable and physically intuitive alternative to QCD.

1. Introduction

The stability of the atomic nucleus is a paradox at the heart of modern physics. While Quantum Chromodynamics (QCD) provides an empirically successful explanation based on a fundamental strong force, its conceptual framework is deeply abstract. Following from our cosmic model, Cosmic Vortex Dynamics (CVD) (*see* Chapter 1), this chapter asks a different question: Is nuclear cohesion an abstract force mediated by particles, or is it an emergent property of the nucleus's vortex geometry? We propose the latter.

Here, we introduce Quantum Vortex Dynamics (QVD), the nuclear component of the Unified Field Dynamics (UFD) framework, wherein nucleons are modeled as self-sustaining energetic vortices. Specifically, QVD posits that the three quarks measured in Deep Inelastic Scattering (DIS) experiments are the mathematical shadow of a single, physical, 3-lobe trefoil vortex, which distinguishes its shape from that of the neutron — previously described as a composite Figure-8 vortex formed by the topological interlocking of a proton and an antiproton (*see* Chapter 6). In this view, nuclear binding is not caused by a fundamental force but by an emergent Geometric Coherence Force (GCF) — a powerful "nuclear surface tension" arising from Planck-level vibrations that drives vortices to lock into more efficient, lower-energy configurations (*see* Chapter 3). The energy released in this process, the Coherence Dividend, is the direct physical origin of nuclear binding energy.

This principle of geometric stability provides a unified physical basis for all nuclear phenomena. This framework also explains the binding-energy curve and the exceptional stability of the Helium-4 nucleus — the Alpha Crystal — proposing that this nucleus is a perfect tetrahedron stabilized by a unified, central antimatter sink. The model also provides a quantitative link between the nuclear surface tension (k) (strong force) and the gravitational constant (G) while confirming the GCF's origins in Planck-level vibrations ($\sim 10^{-35}$ m). Finally, QVD provides a unified geometric principle for both fusion and fission, revealing them as two sides of the same coin: the former builds

167

coherence by assembling light nuclei, while the latter restores coherence by breaking down heavy, strained ones.

The implications of this framework extend beyond theory, as it leads to a new technological paradigm of Resonant Engineering. In the final part of this chapter, we explore these potential applications, from Controlled Fusion to nuclear waste remediation through Precision Fission.

2. The Conceptual Framework of Quantum Vortex Dynamics

Having established the geometric and resonant foundations of the atomic and subatomic world in previous chapters, we are now prepared to revisit the nucleus itself as a self-organizing, resonant structure. This section introduces the core ideas of QVD, a framework that builds on earlier concepts of field-based structure and coherence to offer a radically intuitive model of nuclear stability.

Three core principles define this approach: First, protons and antiprotons are modeled as self-sustaining, structured, 3-lobe vortices in the Universal Energetic Field (UEF), rather than as dimensionless points or quark composites. Second, nuclear binding energy is reconceptualized as a Coherence Dividend—a release of energy that occurs when the interlocking vortices reduce the overall system's shear energy and increase its geometric harmony. Third, the binding force is the emergent GCF—a powerful "nuclear surface tension" that drives vortices to lock into more efficient, lower-energy configurations (see Chapter 3).

2.1 The Proton as a 3-Lobe Vortex

The foundational postulate of QVD is that the three-quark model is a mathematical shadow of a deeper, physical geometry. This three-quark model is not a theory derived from first principles; it is an interpretation of Deep Inelastic Scattering (DIS) data that reveals three distinct scattering centers within the proton (Friedman & Kendall, 1972; Taylor, 1991).

QVD provides the incommensurable physical explanation for this 3-point data by offering a 21st-century realization of Lord Kelvin's "vortex atom" hypothesis, which posited that matter is a topological knot in a physical aether (Thomson, 1867).[42] In QVD,

---

[42] Lord Kelvin's original hypothesis (1867) collapsed because it relied on a static, non-dissipative aether and predicted that atoms were topologically indestructible. QVD succeeds where the original failed by replacing the static medium with a dynamic, relativistic UEF/ULF medium (solving the relativity

the proton is modeled as a physical 3-lobe trefoil vortex—a single, continuous, toroidal flow of the UEF that has been geometrically knotted.[43] The fundamental toroidal and poloidal fluid flows within this knot are what give the nucleon its intrinsic angular momentum and its spin-1/2 nature (*see* Chapter 5).

This physical geometry necessarily possesses three distinct, high-density interaction regions—the three lobes of the knot—where the flow of the Universal Light Field (ULF) is most concentrated. When an electron probe strikes the proton, it interacts with one of these three lobes. Therefore, in QVD, the Standard Model's "uud" quark properties are not descriptions of discrete subatomic particles but are rather a mathematical description of the time-averaged effective charges and spins of these three physical lobes in their stable, ground state.

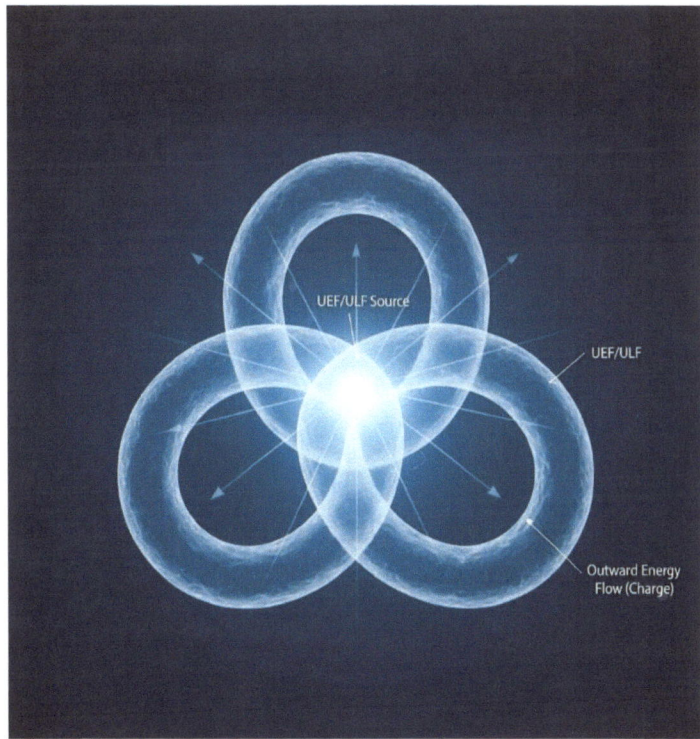

Figure 18: The Trefoil Proton. This schematic illustrates the proton as a fundamental, coherent toroidal vortex within the UEF. Its characteristic 3-lobe trefoil geometry is shown, from which its measured "quark" substructure mathematically arises. The central radiance and outward-pointing arrows denote its role as an energetic "source" within the ULF, which physically manifests as its positive electric charge. This stable configuration corresponds to the ground-state matter formation in the QVD framework.

problem) and defining decay (radioactivity) as a geometric phase transition governed by the GCF (solving the instability problem).

[43] A trefoil knot is the simplest possible nontrivial mathematical knot (a loop that cannot be untied without cutting it). This specific geometry is the physical realization of the proton's structure. Its three lobes are the physical regions that scattering experiments incorrectly identified as three quarks. This knotted model of the proton explains its immense stability, as its vortex structure cannot unravel into simpler components.

2.2 The Coherence Dividend: Binding Energy as an Emergent Property

Having defined nucleons as complex, structured vortices, we must now define the force that binds them and the energy released in that binding. In QVD, the internal energy of a single nucleon vortex is composed of two primary components:

1. Core Energy (E_{core}). This is the fundamental, non-reducible energy associated with the vortex's central, deepest coherence and identity. This is the mass-energy ($E = mc^2$) that cannot be altered or shed during chemical or nuclear reactions; it is what defines the nucleon as a nucleon.

2. Shear Energy (E_{shear}). This is the disposable energy associated with the vortex's rotating boundary surface (its "shear interface," S) with the surrounding medium (the UEF). This energy is required to maintain the vortex's boundary against the external field.

The Coherence Dividend arises from shear energy. When nucleon vortices interlock into a more coherent structure, their overlapping shear surfaces are internalized. The energy that was previously "wasted" maintaining these redundant boundaries is now surplus and is radiated away as the Coherence Dividend (the binding energy). To illustrate this, we can draw on economic principles. When two companies merge, redundancies in leadership and infrastructure are eliminated, thereby reducing overhead and increasing efficiency. The resulting gains mirror the Coherence Dividend in our model: energy saved through reorganization, not imposed from the outside, but emerging from within.

The model also aligns with the Bardeen-Cooper-Schrieffer (BCS) theory of superconductivity (Bardeen, Cooper, & Schrieffer, 1957), in which electrons pair into more stable, lower-energy collective states via resonance. Similarly, in QVD, nucleon vortices bind through geometric synergy rather than forceful compression, allowing them to emerge into configurations that reduce total field tension and maximize internal coherence.

2.3 The GCF as the Assembler

In conventional nuclear physics, the strong force is a fundamental interaction mediated by gluons. QVD offers an incommensurable alternative: nuclear binding is an emergent effect driven by the GCF, a short-range, "surface-based" force, like surface tension, that activates when nucleon vortices achieve resonant geometric alignment (*see*

Chapter 3). It is the universal drive for the system to shed shear energy and snap into its most geometrically efficient, low-energy, coherent configuration.

Central to this mechanism is the neutron's unique geometry. As a composite figure-8 vortex with an internal antiproton "sink," the neutron acts as a universal "docking port" for the proton's "source" flow. This geometric complementarity is essential, as the net nuclear force must overcome the strong electrostatic repulsion between protons. The neutron thus becomes a key catalyst for binding. As a proton-antiproton composite, its internal antiproton "sink" provides a localized electrostatic attraction for the proton, allowing the two vortices to overcome their mutual repulsion and get close enough for their geometries to "mesh." At this point, the GCF snaps them together. The Coherence Dividend is then the direct, energetic result of this GCF-driven snap.

In this light, nuclear binding energy is not the strength of a fundamental force. It is rather a quantitative measure of the GCF's successful organization of the system against electrostatic opposition—that is, the energy gained from geometric closure minus the energy lost to Coulombic strain. A more symmetrical configuration releases a larger Coherence Dividend and is, by definition, a more stable nucleus within a given environment.

3. A Fluid-Dynamic Toy Model: The Reduced Shear Hypothesis

To provide a semi-quantitative test of the Coherence Dividend, we now develop a simplified "toy model" based on the fluid-dynamic principles of our framework.

While a complete mathematical treatment of interacting quantum vortices is beyond the scope of this introductory chapter, a simplified model can effectively illustrate the physical principles of QVD and demonstrate its ability to account for experimental data. The model we propose is based on the central hypothesis that the binding energy released during a nuclear merger is proportional to the reduction in the system's total shear interface with the surrounding medium.

3.1 The Reduced Shear Hypothesis and a Toy Model Equation

We hypothesize that the internal energy of a single nucleon vortex is composed of an incommutable core energy (E_{core}) and a disposable shear energy (E_{shear}), where the shear energy is proportional to the surface area, or "shear interface" (S), that the vortex presents to the surrounding medium (UEF). This can be expressed as $E_{shear} = k \cdot S$, where k is a constant representing the energy density of the fluid interface. When two or more vortices merge into a more compact and symmetrical arrangement, the total exposed shear surface is reduced, and the surplus shear energy radiates away as the system's binding energy (E_{bind}). This provides us with the central equation for our toy model:

$$E_{bind} = E_{\text{shear initial}} - E_{\text{shear final}} = k \cdot (S_{\text{initial}} - S_{\text{final}})$$

Defining ΔS as the total reduction in shear interface ($S_{\text{initial}} - S_{\text{final}}$), our model simplifies to:

$$E_{bind} = k \cdot \Delta S$$

In this equation, k is not just an arbitrary fitting parameter; it represents a fundamental physical property of the universe. With units of energy per unit of area (Joules/meter^2), k is the direct measure of the intrinsic surface tension of the UEF. It represents the energy cost of creating a boundary, or "tear," in the substance of the field, which is a direct reflection of the UEF's internal coherence. We will now demonstrate that a single, consistent interpretation of this geometric principle can successfully account for the binding-energy curves of light nuclei, thereby validating that binding energy is a direct function of geometric coherence (ΔS).

3.2 The First Test Case: The Deuteron ($p + n$)

Our first test of the Reduced Shear model is the formation of the deuteron nucleus (^2H), which consists of one proton and one neutron. In QVD, this process is not merely the sticking together of two spheres, but the precise topological interlocking of two distinct vortices.

Specifically, it involves the docking of a stable proton trefoil (a flow source) into the "sink" side of a neutron Figure-8 (a flow dipole). When separate, the entire surface area (shear interface) of both vortices is fully exposed to the surrounding medium. When interlocked, the proton's outward flow partially feeds the neutron's intake, forming a stable, dumbbell-shaped pair that significantly reduces the overall friction or shear with the ULF (Figure 19A).

Mathematically, we define the initial state as a free proton and neutron, each with a baseline shear interface, S, yielding a total initial shear of 2S. The new, interlocked dumbbell configuration possesses a reduced total interface, S_{final}. The reduction in shear is thus $\Delta S1 = 2S - S_{final}$. According to our model, the energy released should be directly proportional to this geometric change. Experimentally, the deuteron's binding energy is measured to be 2.225 MeV (Krane, 1988). Applying our model's central equation, we get:

$$2.225 \text{ MeV} = k \cdot \Delta S1$$

This result calibrates our model, establishing a relationship between the constant k and geometric efficiency. This modest energy release is consistent with what we would expect from a simple pair and represents a small but critical step toward achieving greater coherence.

3.3 A Quantitative Prediction: The Helium-3 Nucleus

After calibrating our model with the deuteron, we now test its predictive validity against the Helium-3 nucleus (^{3}He), which consists of two protons and one neutron. The most common formation pathway for ^{3}He is the merger of a deuteron with a free proton.

In our model, this corresponds to a second proton trefoil joining the existing interlocked pair. We hypothesize that this merger results in a new, highly stable, planar triangular configuration. In this geometry, the two "source" proton trefoils dock symmetrically into the central "sink" of the neutron Figure-8. This arrangement maximizes the surface-tension effect of the GCF by creating a closed geometric circuit (Figure 19B).

The initial shear interface for this reaction is that of the separate deuteron and proton ($S_{initial}$). The final state is the new Helium-3 vortex with a more optimized shear interface, S_{final}. The additional reduction in shear accomplished in this step is $\Delta S2$. Therefore, the additional binding energy released should be proportional to this change.

The experimental total binding energy of Helium-3 is 7.718 MeV (Krane, 1988). Since the initial deuteron was already bound by 2.225 MeV, the energy released in this second stage of nucleosynthesis is 7.718 MeV - 2.225 MeV = 5.493 MeV.

For our model to remain consistent with this data, the reduction in shear interface from this second merger ($\Delta S2$) must be approximately 2.47 times greater than the reduction from the first ($\Delta S1$). This ratio is a direct consequence of the competing forces on the nucleus: While the ideal geometric gain for closing the triangular circuit is Factor 3—the ideal energy release for closing a perfect three-sided circuit should be 3 times the energy of a single connection ($3 \cdot \Delta S1$)—the mutual electrostatic repulsion of the two protons prevents the structure from achieving this full coherence. This physical strain exacts a 0.53 unit penalty on the ideal geometric gain.

When converted back into energy (0.53 × 2.225 MeV/Unit), this penalty equals 1.18 MeV. This value aligns precisely with the calculated electrostatic potential energy between two protons separated by the characteristic nuclear distance (1.22 fm) (Krane, 1988). This external validation shows that the binding energy of Helium-3 results from the GCF's drive for stability counterbalanced by structural opposition from the electrostatic force.

The model, therefore, accounts for the disproportionately large binding energy of Helium-3 as a direct consequence of geometric closure. The formation of this fully symmetric, planar structure, mediated by the neutron's internal antimatter sink, represents a significant leap in geometric efficiency and coherence compared to a simple linear pair.

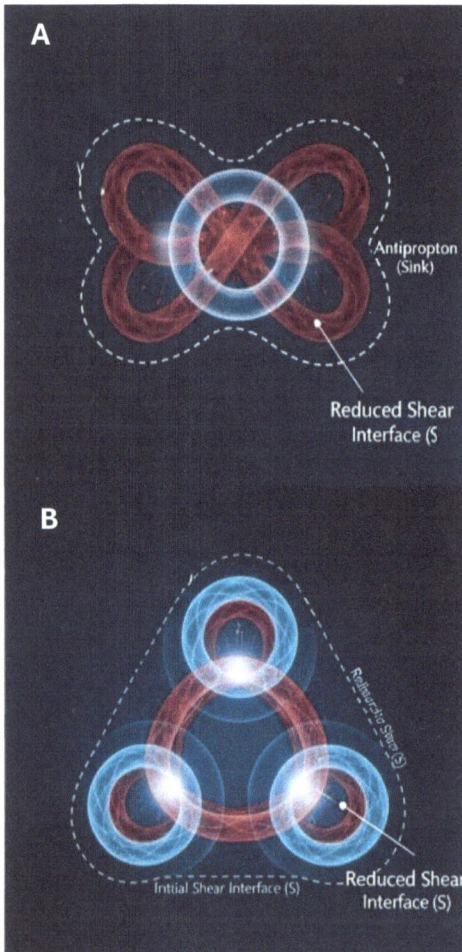

Figure 19: The Geometric Structure of Light Nuclei in QVD.

(A) The Deuteron (^{2}H): This figure illustrates the deuteron nucleus as a stable, dumbbell-shaped pair formed by the topological interlocking of the proton's outward flow into the neutron's intake. This precise docking creates a more compact configuration, significantly reducing the total exposed shear interface to the ULF (shown as reduced dashed lines).

(B) The Helium-3 Nucleus (^{3}He): This figure illustrates the formation of the Helium-3 nucleus as a highly stable, planar triangular configuration. The second proton trefoil symmetrically docks into the central "sink" provided by the neutron's Figure-8 composite vortex within the deuteron. This arrangement maximizes the surface-tension effect of the GCF by creating a "closed geometric circuit," leading to a significantly optimized shear interface

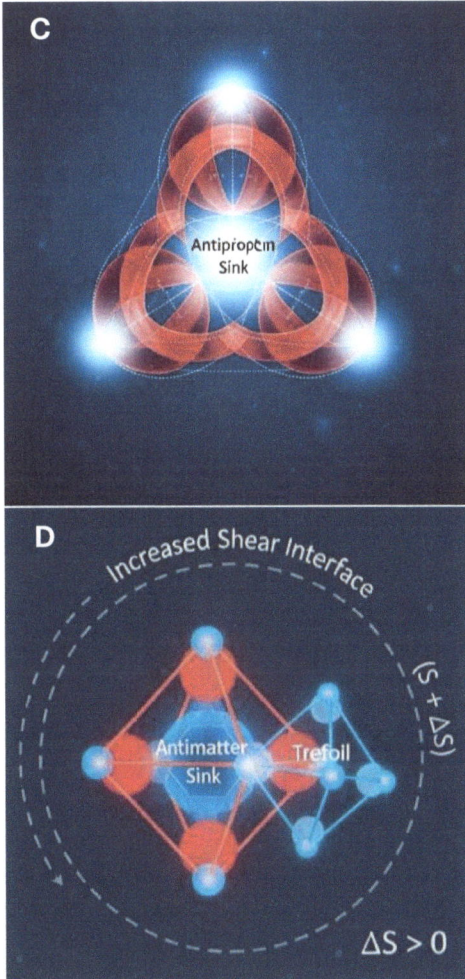

(C) The Alpha Vortex (⁴He): This figure illustrates the Helium-4 nucleus as a highly symmetrical, three-dimensional Alpha Crystal in a perfect tetrahedral configuration. This structure arises from the fusion of two deuterons, where the internal "sinks" (antiproton components) of the two neutron Figure-8s align at the exact geometric center, forming a unified, powerful negative-pressure core. This central sink binds all proton trefoil components, harmonizing all outward flows and dramatically minimizing the total exposed shear interface

(D) The Instability of Lithium-5 (⁵Li): This figure depicts the hypothetical Lithium-5 nucleus, formed by the forced, asymmetrical attachment of a fifth proton to the outside of the Helium-4 core. This attachment disrupts the geometric closure and symmetry of the alpha crystal, resulting in a net increase in the total exposed shear interface compared to the separate constituents. According to the UFD model, a positive ΔS yields a negative binding energy, leading to intrinsic instability.

3.4 The Confirmation: The Alpha Vortex (Helium-4)

The pinnacle of stability among light nuclei is the Helium-4 nucleus (⁴He), often referred to as the alpha particle. We now apply our model to this crucial test case. While several pathways form Helium-4, we will model the symmetrical fusion of two deuteron nuclei $(d + d)$, which involves the merger of two proton trefoil/neutron figure-8 pairs into a single, unified four-body state. We propose that this merger allows the four nucleon vortices to rearrange themselves into the most stable and symmetrical three-dimensional configuration possible: a perfect tetrahedron (Figure 19C).

Crucially, in the UFD framework, this is not just a packing of spheres. The two neutron Figure-8 vortices align such that their internal "sinks" (antiprotons) meet at the exact geometric center of the structure. This creates a unified, central antimatter core: a

175

single, powerful negative-pressure point that binds the two proton trefoils and the lobes of the neutrons simultaneously. The result is the Alpha Crystal: a perfectly symmetrical, interlocked knot in which all outward flows are harmonized by a single central intake.

The total binding energy of Helium-4 is an immense 28.296 MeV (Krane, 1988). If we consider the $d + d$ fusion pathway, our initial state consists of two deuterons with a combined binding energy of 2 * 2.225 MeV = 4.450 MeV. The net energy released in this final merger is therefore a colossal 23.846 MeV.

According to our model ($E_{bind} = k * \Delta S$), this colossal energy release implies that the reduction in shear interface when forming the tetrahedron (ΔS_3) is vastly greater than in the previous mergers. This is the model's key success. The arrangement into a perfectly symmetric, three-dimensional tetrahedral geometry represents the optimal packing solution for four vortices, minimizing the exposed shear interface to the greatest possible extent and thus releasing the largest possible Coherence Dividend.

We further hypothesize that this uniquely stable arrangement fundamentally mitigates or shields the internal electrostatic repulsion between its two constituent protons, making its stability a result of achieving geometric and dynamic perfection.

3.5 The Final Proof: Explaining Nuclear Instability (Lithium-5)

The final test of a model's predictive power is its ability to account for negative results. The QVD framework must not only explain the stability of Helium-4 but also the profound instability of nuclei with five nucleons, such as Lithium-5 (^{5}Li). Experimentally, Lithium-5 is unbound and instantly decays back into a Helium-4 nucleus and a proton (Krane, 1988).

We model this system by adding a proton trefoil to a fully formed Helium-4 Alpha Crystal. Our model posits that the tetrahedral alpha vortex represents a state of geometric closure; the central antimatter sink, which binds the structure, is now physically enclosed and shielded by the four surrounding lobes. There is no accessible "docking port" for a fifth vortex to join the structure. Any addition represents a forced, asymmetrical attachment "tacked on" to the outside of the perfect geometric core (Figure 19D).

This inefficient geometry would result in a net increase in the total shear interface, as the combined structure of the hypothetical Lithium-5 vortex (S_{Li5}) would be less compact and have a greater exposed surface area than the two separate particles

($S_{alpha} + S_{proton}$). This means that the change in our key variable, $\Delta S = (S_{alpha} + S_{proton})$. $-S_{Li5}$, would be negative (representing a *cost* rather than a dividend). This means that energy must be put into the system to form this configuration and that energy would be immediately recovered when the structure flies apart. The model, therefore, correctly predicts that the system will overwhelmingly favor the lower-energy state of a separate, hyper-stable Helium-4 nucleus and a free proton.

The ability of this model to predict the deep valley of instability at $A = 5$, using the same geometric principles that explained the peak of stability at $A = 4$, demonstrates its consistency and predictive power. We now transition from this geometric proof to its quantitative foundation: the derivation of the surface tension constant (k) that governs this stability.

3.6 Unification: Deriving the UEF Surface Tension Constant (k)

If k represents the effective nuclear surface tension associated with geometric coherence in the UEF, it must be fundamentally related to the constants governing the field's other manifestations, such as gravity (G), which, in UFD, characterizes the large-scale stiffness imposed by UEF constraint geometry. This relationship reflects that the strong interaction and gravity are not independent forces but scale-separated expressions of a single underlying elastic constraint.

The relationship is derived through dimensional analysis by substituting UFD's physical interpretation of mass into the gravitational coupling. Although the UEF itself is incompressible and does not store inertial energy through displacement, its shear geometry imposes constraints that induce solitonic energy localization in the ULF. Inertial mass is therefore identified with the ULF energy required to satisfy UEF-imposed constraints over a characteristic coherence volume L^3.

1. Define Dimensions: We start with the physical dimensions of the two constants:

$$[G] \approx [G] = [L]^3 [M]^{-1} [T]^{-2}, \qquad [k] = [M][T]^{-2}$$

2. Substitute Mass: In UFD, mass arises from displacement of the UEF over a coherence volume L^3, with density set by the field stiffness. Dimensional consistency gives:

$$[M] \sim \rho L^3 \sim \frac{k L^2}{c^2}$$

3. Relate Time and Length: Since c is the characteristic wave speed of the UEF,

177

$$[T] = \frac{L}{c}$$

Substituting:

$$G \sim \frac{L^3 c^2}{kL^2 (L^2/c^2)} = \frac{c^4}{kL}$$

4. The Unification Equation: Rearranging gives the fundamental relation:

$$k = \frac{c^4}{GL}$$

This equation shows that gravity (G) and nuclear surface tension (k) are inversely related through the UEF coherence length L. The constant k characterizes the local geometric stiffness governing nuclear binding, while G describes the macroscopic manifestation of that same constraint geometry over large distances.

Using experimental constants and the proton radius ($r_p \approx 0.84$ fm) as the coherence length:

$$k \approx \frac{(2.998 \times 10^8 \text{ m/s})^4}{(6.674 \times 10^{-11} \text{ m}^3 \text{ kg}^{-1} \text{ s}^{-2})(0.84 \times 10^{-15} \text{ m})} \approx 1.4 \times 10^{59} \text{ J/m}^2.$$

This enormous value confirms that the UEF is an ultra-rigid, high-tension superfluid, accounting for the immense energies released in nuclear processes (the Coherence Dividend). The density of the UEF follows from the identity:

$$k \sim \rho L c^2 \Rightarrow \rho \sim \frac{k}{Lc^2} \approx 1.85 \times 10^{57} \text{ kg/m}^3,$$

reflecting the extreme stiffness of the underlying constraint field. This density does not represent stored mass energy but the scale of geometric rigidity required to impose observed gravitational and nuclear phenomena.

Finally, the universal limiting speed emerges from the constraint relation:

$$v \sim \sqrt{\frac{k}{\rho}} = c,$$

demonstrating that the speed of light arises as the characteristic response speed of ULF excitations under UEF-imposed geometric constraints, rather than as a postulated property of spacetime.[44]

3.7 The Microscopic Scale of Coherence

Having established the derived UEF Surface Tension constant, k, and demonstrated that our Reduced Shear hypothesis accurately accounts for the binding energetics of light nuclei, we now apply the model to probe the fundamental scale at which this geometric interlocking occurs. Inverting our model ($E_{bind} = k\Delta S$) allows us to calculate the effective "shear length" ($l_{eff} = \sqrt{\Delta S}$) for various nuclei.

When applied to the Deuteron, Helium-4, and Carbon-12, the model consistently calculates that the geometric rearrangement driving nuclear stability occurs at a characteristic length scale in the range of 10^{-36} to 10^{-35} meters. This is the Microscopic Scale of Coherence. This finding thus provides dual empirical confirmation of the UFD's structural hierarchy, as described in Chapter 3, by quantitatively demonstrating that the GCF is the operational force of the Planck Scale.

Now that we have established the geometric principles of nuclear binding, we must reconcile this new view with the established success of the Standard Model.

4. Situating QVD within Physics

For any new theoretical framework to be considered viable, it must accomplish two essential tasks: First, it must situate itself in relation to the existing, successful paradigm it seeks to complement or replace. Second, it must provide unique, falsifiable predictions that can be tested experimentally. This section addresses both requirements.

4.1 The Incommensurability of QVD and QCD

The Standard Model explains nuclear structure via QCD. QVD does not seek to contradict the experimental success of QCD. Instead, the two frameworks are incommensurable—a term used by philosopher of science Thomas Kuhn to describe

---

[44] Although wave speeds are generally set by stiffness-to-density ratios, the relevant stiffness in UFD is not a bulk modulus but the geometric surface tension k of an incompressible constraint field. The characteristic speed c therefore emerges from the ratio of UEF geometric stiffness to the effective inertia of ULF excitations.

theories that operate under fundamentally different assumptions (Kuhn, 1962). In this respect, the three-quark model can be considered a brilliant mathematical description of the UFD's physical, causal geometry, as the two frameworks provide incommensurable answers to the most basic questions:

- What is a proton?

    - QCD: A composite particle containing three separate, fundamental, point-like particles (quarks) held together by gluons.

    - QVD: A single 3-lobe trefoil vortex in the UEF, representing a modern realization of Lord Kelvin's "vortex knot" hypothesis.

- What are the three quarks (the 3-point scattering data)?

    - QCD: Three separate particles (e.g., 'up, up, down'). This is an interpretation of the DIS data.

    - QVD: The quark signatures are the three distinct, high-density lobes of the proton's trefoil vortex. The DIS experiments correctly counted three scattering centers in the proton, but they were measuring the geometric regions of a single continuous object rather than three separate particles.

- What is the difference between a proton and a neutron?

    - QCD: They contain different quark configurations (Proton = uud, Neutron = udd).

    - QVD: They are topologically distinct. The proton is the stable, ground-state trefoil (a monopole source). The neutron is a metastable, composite Figure-8 (a dipole source-sink). Its instability is not a "flavor change" but a geometric unraveling: the strain of the Figure-8 geometry eventually causes it to snap, releasing the stable proton and shattering the internal antiproton into an electron and antineutrino (see Chapter 6).

- What are Color, the Color Force, and Confinement?

    - QCD: Color is an abstract charge (red, green, blue) invented to solve a paradox in the Pauli Exclusion Principle. The Color Force is the fundamental force, mediated by gluons, that "glues" the three quarks together. Confinement is the property whereby this force increases with distance, making it impossible to pull a single "colored" quark out.

- o QVD: The Color Force is a mathematical epicycle invented to solve a paradox that the QVD model does not possess. It represents the internal GCF or topological integrity that holds the trefoil vortex together. Confinement, in turn, is a physical, topological property. One cannot pull a lobe out of a trefoil knot for the same reason that one cannot pull a loop out of a pretzel without destroying the entire structure. The Standard Model's "force increasing with distance" is the mathematical description of the physical tension of trying to unravel a fundamental topological knot.

- What is a meson (pion)?

  - o QCD: A transient particle (quark-antiquark pair) that mediates the residual strong force between nucleons (Yukawa potential).

  - o QVD: A transient geometric bridge or flow filament, which is the physical manifestation of the GCF. It is a temporary flow connection created when a proton's outward-flowing lobe locks into the internal antimatter sink of a neighboring neutron.

- How does the structure change at high energy (The Sea Quark Problem)?[45]

  - o QCD: As momentum transfer (Q^2) increases, gluons split, generating a sea of virtual quark-antiquark pairs that increase the number of scattering centers.

  - o QVD: When particles are probed at very high energies, the experiment is not revealing new particles inside the proton. Instead, it is resolving finer details of how the proton's internal medium is moving. What appears as an increasing number of "sea quarks" reflects the probe's ability to

---

[45]In the Standard Model (QCD), sea quarks are virtual quark-antiquark pairs that spontaneously appear and disappear within the nucleon, generated by gluon splitting. They increase the number of scattering centers observed at high energies. In QVD, this phenomenon is reinterpreted as the dynamic resolution of the proton's internal flow structure. The high-energy probe instantaneously resolves the turbulent eddies and sub-loops of the UEF flow that compose the trefoil vortex, making the smooth, three-lobed proton appear complex and fractal.

momentarily resolve complex, rapidly changing flow patterns—eddies, loops, and turbulence—within the proton's underlying field structure. The added complexity comes from resolving this internal motion more sharply, not from the creation of new constituent particles.

In sum, QVD provides an underlying geometric explanation for scattering data and provides a physical, causal mechanism for the Standard Model's "quark flavors" and the nuclear "glue."

4.2 Falsifiable Predictions

A scientific framework is validated by its ability to make novel, testable predictions that distinguish it from the standard paradigm. QVD offers several such predictions, providing clear experimental pathways for its confirmation or falsification.

4.2.1 The Quantum Tetrahedron of Helium-4

The most critical prediction of QVD regarding the nucleus is that the Helium-4 nucleus (the Alpha Crystal) possesses an intrinsic tetrahedral geometry. This directly challenges the Standard Model's assumption of a spherically symmetric nucleus.

At first glance, this prediction appears to contradict decades of scattering experiments that have shown a spherically symmetric charge distribution (Hofstadter, 1957). However, QVD predicts that this observed symmetry is the necessary observational signature of a dynamically tumbling, quantum-superposed Alpha Crystal. Like a long-exposure photograph of a spinning three-bladed fan that appears as a solid circle, a standard scattering experiment measures the time-averaged position of a tetrahedral vortex lattice that is in a superposition of all possible orientations.

The true test of this prediction, therefore, would require novel experimental methods, such as time-resolved scattering or coherence-based probes. These techniques could detect the nucleus's underlying geometric structure—specifically the tetrahedral arrangement of the four nucleon vortices around the central antimatter binding core—before it is averaged into a spherical shape.

4.2.2 Alpha-Cluster Dominance in Light Nuclei

QVD predicts that stable, light, even-even nuclei like Carbon-12 and Oxygen-16 are best described as stable geometric structures (stackings) of tetrahedral alpha-particle vortices. If this is correct, then high-energy "knockout" reactions performed on these nuclei should show a disproportionately high probability of ejecting intact alpha particles rather than single protons or neutrons. This would constitute direct evidence of a pre-existing cluster structure, a phenomenon actively studied in experimental nuclear physics (Hen et al., 2017).

4.2.3 Deriving Quark Properties from Vortex Geometry

Our incommensurable model of the nucleon, in which we map the "quarks" to the physical lobes of the vortex, yields the ultimate falsifiable prediction of QVD. We predict that the Standard Model's fractional charges (e.g., +2/3, -1/3) are not fundamental properties of discrete particles but are the time-averaged, "effective" charges that a high-energy electron probe encounters when scattering off the dynamic lobes of the vortex.

- For the Proton: The probe encounters the three distinct high-density regions of the stable trefoil geometry.

- For the Neutron: The probe encounters the complex, dual-center topology of the Figure-8 composite (as detailed in Chapter 6).

QVD's quantitative challenge is to mathematically model the UEF/ULF field properties of these distinct regions. We must derive the "effective" fractional charges and spin contributions from the first principles of the vortex's fluid dynamics and geometry. Such a derivation would provide definitive proof that the Standard Model's "quark" was a mathematical shadow of the QVD's physical reality all along.

4.2.4 Testing the Inverse Force Relationship (G, k)

The quantitative link between the gravitational constant (G) and the nuclear surface tension (k) derived in Section 3 leads to a specific prediction about the ultimate coherence scale of the UEF: the effective geometric rearrangement required to bind any nucleus, regardless of its size or nucleon count, must correspond to a length scale ($l_{eff} = \sqrt{E_{bind}/k}$) that is constant and lies near the Planck length ($\sim 10^{-35}$ m).

Further high-precision binding energy measurements of various exotic, short-lived nuclei should yield l_{eff} values that fall within the same $\sim 10^{-35}$ meter range, confirming the UEF has a single, characteristic binding pixel size. Such measurements would validate the unification of G and k.

Thus, by situating itself as an incommensurable yet complementary framework to QCD, QVD offers a profound shift in perspective, proposing that the secrets of the nucleus may be found in the elegant, architectural principles of geometry and resonance that govern it.

5. Nuclear Coherence: A New View of Fusion and Fission

The quest to unlock the power of the atom has long been framed as a battle against nature—a matter of overcoming immense forces with brute force. But what if this model is fundamentally misguided? This section challenges the standard quantum tunneling explanation for nuclear reactions and proposes an alternative based on the principle of geometric coherence as the primary causal mechanism. Here, we show that both stellar fusion and terrestrial fission can be understood as deterministic, resonant alignments of energetic structures.

5.1 Stellar Fusion Reimagined

In standard physics, stellar fusion is framed as a statistical long shot. The high temperatures and pressures in the Sun's core are said to provide protons with just enough kinetic energy to occasionally tunnel through the formidable Coulomb barrier (Bethe, 1939).

In QVD, fusion is not a statistical accident but a geometric inevitability. In the dense plasma of a stellar core, a proton (UEF source vortex) captures a high-energy electron (ULF sink vortex) through the routine process of electron capture (Bahcall, 1963). This geometric transformation converts a proton into a neutral neutron, thereby eliminating the electrostatic barrier.

The newly formed neutron is now free to fuse with a nearby proton, forming a deuteron. This fusion is not driven by the force of collision but by the energetic advantage of achieving a more stable geometry. The resulting binding energy (the Coherence Dividend) is released as radiant energy. Stars shine not because particles happen to tunnel but because the structure of their fields compels them to seek harmony.

5.2 Fission as a Collapse of Coherence

Fission is the symmetrical counterpart to fusion. It occurs in very heavy nuclei (like uranium), where the immense number of vortices creates a state of high geometric strain and instability. Such a nucleus can achieve a more stable, lower-energy state by splitting into two or more smaller, more coherent daughter nuclei. This reconfiguration also releases a massive Coherence Dividend.

Conceptually, we can liken fission to a collective form of the weak force (*see* Chapter 6). While the weak force governs the decay of a single, strained nucleon, fission is the decay of an entire, unstable nucleus as the collective "wall" of its geometric bricks collapses into a more stable arrangement. Both fusion and fission are, therefore, two distinct pathways toward the same end: a state of greater geometric coherence.

6. Resonant Engineering: The Path to Controlled Fusion and Fission

QVD's reinterpretation of nuclear reactions as geometric and resonant processes opens the door to a new technological paradigm. Conventional fusion research has employed brute-force methods to overcome the Coulomb barrier through extreme heat and pressure. But if the true engine of fusion is not collision but configuration, the goal should not be to bash atoms together but to guide them into the proper alignment through field-based resonance. This is the central aim of Geometric Catalysis.

6.1 The Future of Fusion: Geometric Catalysis

Geometric Catalysis proposes the design of field environments that mimic the final, stable configuration of the fusion product, particularly the tetrahedral geometry of Helium-4. This field scaffold would act as a dynamic "mold" that aligns atomic nuclei rather than forcing them together.

Advanced electromagnetic systems, such as arrays of phase-locked lasers, could create standing-wave patterns in the ULF that simulate the spatial pressure geometry of a Helium nucleus. Concepts from field-driven fusion experiments and laser confinement provide a clear precedent for this approach to precision energy delivery (Hora et al., 2010; Nuckolls et al., 1972). When deuterium and tritium nuclei enter this scaffold, their internal vortex structures would be subtly nudged into alignment by the resonant gradients of the field. At the moment of interlocking, the GCF takes over, the system snaps into its low-energy configuration, and the full coherence dividend is released and captured.

Because Geometric Catalysis is driven by resonance, it is hypothesized to be inherently self-limiting. If the surrounding field coherence diminishes, the fusion process is expected to halt naturally to prevent runaway reactions. This characteristic suggests a technology that is significantly safer and more controllable than conventional nuclear methods. Moreover, the architecture appears modular and scalable. Controlled Fusion units based on field resonance could be finely tuned by adjusting the phase and frequency of the field scaffold. The potential implications of Controlled Fusion extend beyond power generation to include:

- Clean, compact fusion reactors with minimal risk of meltdown or long-lived radioactive waste.

- The synthesis of new materials forged under field-guided coherence conditions, as suggested by foundational work in coherence phenomena (Fröhlich, 1968).

- Propulsion systems for space exploration that harness field-mediated fusion in real time.

If successfully realized, Geometric Catalysis could transform nuclear energy from a source of concern to one of virtually unlimited opportunity.

6.2 The Future of Fission: Waste Neutralization and Inherent Safety

While fusion represents the future of energy generation, Resonant Engineering also offers a novel approach to the challenges of nuclear fission. In QVD, fission is a controlled "collapse of coherence" in a geometrically strained, heavy nucleus. This model suggests that the process can be guided by resonance, enabling technologies previously considered impossible.

The most significant application would be Precision Waste Neutralization. Long-lived radioactive waste from nuclear reactors is dangerous because its decay is slow and uncontrolled. Geometric Catalysis, however, would enable us to target these unstable isotopes (such as Plutonium-240 or Technetium-99) with a resonant field tuned to their specific geometric instabilities. This field would act as a catalyst, inducing the nucleus to fission immediately into stable, non-radioactive daughter products, essentially allowing us to "sing" radioactive waste into a safe, stable state on demand.

This same principle could lead to the development of inherently safe fission reactors. Instead of managing a volatile chain reaction with control rods, a "resonance-driven reactor" would use a precisely controlled magnetic field to maintain a steady, controlled rate of fission in its fuel. If the system were to fail for any reason, the resonant field would simply turn off, and the fission process would stop instantly, making a meltdown virtually impossible.

7. Conclusion

This chapter has introduced QVD as a novel physical framework for understanding the structure and stability of the atomic nucleus. Departing from the abstract, axiomatic approach of QCD, we have proposed that nuclear cohesion arises not from particle exchange, but from a real, emergent contact force: the GCF. This force emerges when structured nucleon vortices interlock into more geometrically efficient configurations, releasing their excess boundary energy as a Coherence Dividend.

By reinterpreting nucleons as energetic vortices, we have demonstrated how their merger into coherent geometries naturally leads to nuclear binding. A simplified Reduced Shear model derived from this framework accurately accounts for the binding energies of light nuclei and explains the extraordinary stability of the Helium-4 Alpha Crystal, where the internal antimatter sinks of the neutrons merge to form a unified binding core. Beyond its explanatory power, QVD offers a return to a physically intuitive vision of matter, reframing the strong force as a kind of "nuclear surface tension" born of the vortices' drive to minimize shear and achieve structural coherence.

The framework's central theoretical achievement, however, is its quantitative unification of the strong and gravitational forces. By deriving the surface tension constant (k) directly from G and c, we demonstrated that nuclear binding is the local, high-tension limit of the same elastic field whose long-range effects are described by the gravitational constant. This calculation further yielded a Microscopic Scale of Coherence ($\sim 10^{-35}$ m) where the binding occurs, linking nuclear physics directly to the Planck scale.

This foundation supports the chapter's incommensurable interpretation of QCD. We have demonstrated that the "three-quark" model is a mathematical shadow of the proton's stable trefoil vortex, providing a physical, causal explanation for the Standard Model's "quark flavors." Experimentally, key predictions, such as the proposed tetrahedral charge geometry of Helium-4 and deriving quark properties from vortex geometry, offer testable departures from the Standard Model.

Most importantly, this chapter has demonstrated that the implications of QVD extend beyond theory to a new paradigm in Resonant Engineering, which offers a potential pathway to controlled nuclear fusion and the precision targeting of long-lived radioactive isotopes. In sum, QVD brings geometry back into our understanding of nuclear dynamics, encouraging us to view the atomic nucleus not as a battleground of fundamental forces but as a resonant system seeking an optimal form. Its stability, its energy, and its transformations are all governed by this dynamic logic.

7.3 A Geometric Nucleus

In this chapter, we introduced Quantum Vortex Dynamics (QVD), a theoretical framework that reimagines the nucleus as a self-organizing system of structured, energetic vortices. This model shifts the foundation of nuclear physics from abstract, force-centric postulates to tangible, geometric principles rooted in fluid dynamics. The chapter's key contributions are as follows:

- An Incommensurable Model: QVD provides the first physical, causal explanation for the Standard Model's "three-quark" data. It posits that the three scattering centers observed in the proton are not separate particles but rather the three physical lobes of a single 3-lobe trefoil vortex.

- A New Physical Mechanism: QVD replaces the notion of a fundamental strong force with the emergent Geometric Coherence Force (GCF). This force arises when nucleon vortices interlock, releasing their excess energy as the Coherence Dividend, providing a physical origin for nuclear binding energy that scales with the geometric efficiency of the nucleus.

- Force Unification & Quantitative Anchor: The strong nuclear binding constant (k) is mathematically derived from the gravitational constant (G) and the speed of light (c), demonstrating that gravity is the long-range manifestation of the UEF's surface tension. This derivation further reveals that the strong binding mechanism operates at the Planck scale, thereby quantitatively validating our model's prediction that the GCF arises from Planck-scale vibrations (*see* Chapter 3).

- A Unified View of Stability: QVD offers an intuitive, geometric explanation for the stability of atomic nuclei. It accounts for the extraordinary stability of Helium-4 (the Alpha Crystal) as a perfect tetrahedron bound by a unified central antimatter sink, while explaining the instability of Lithium-5 as a failure of geometric closure.

The true power of QVD lies not only in its explanatory power but also in its potential to usher in a new era of nuclear science. As a scientific theory, it offers clear, falsifiable predictions that distinguish it from the Standard Model. The most central of these is the tetrahedral charge geometry of the Helium-4 nucleus and the ability to mathematically derive the Standard Model's fractional charges from the time-averaged "effective charge" of the three physical vortex lobes.

Beyond its testability, this geometric understanding of the nucleus opens the door to a new paradigm of Resonant Engineering. This includes a new pathway to Controlled Fusion, which uses shaped resonance fields to guide nuclei into stable configurations, and a revolutionary solution to nuclear waste through Precision Fission, which employs similar fields to "sing" long-lived radioactive elements into stable, harmless ones.

Taken together, these advances restore a deeply intuitive picture of the atomic nucleus as a resonant structure governed by the dynamic interplay of geometry, energy, and form. QVD thus invites us to return to a principle once central to natural philosophy: nature organizes itself according to patterns of symmetry and harmony.

References

1. Bahcall, J. N. (1963). Electron capture and solar neutrinos. Physical Review, 129(6), 2683–2685.
2. Bahcall, J. N., Pinsonneault, M. H., & Basu, S. (2001). Solar models: current epoch and time dependences, neutrinos, and helioseismological properties. Astrophysical Journal, 555(2), 990–1012.
3. Bardeen, J., Cooper, L. N., & Schrieffer, J. R. (1957). Theory of Superconductivity. Physical Review, 108(5), 1175–1204.
4. Bethe, H. A. (1939). Energy production in stars. Physical Review, 55(5), 434–456.
5. Einstein, A. (1905a). Does the Inertia of a Body Depend Upon Its Energy Content? Annalen der Physik, 18(13), 639–641.
6. Friedman, J. I., & Kendall, H. W. (1972). Deep Inelastic Electron Scattering. *Annual Review of Nuclear Science*, 22, 203-254.
7. Fröhlich, H. (1968). Long-range coherence and energy storage in biological systems. International Journal of Quantum Chemistry, 2(5), 641–649.
8. Gross, D. J. (2005). "The Discovery of Asymptotic Freedom and the Emergence of QCD." Reviews of Modern Physics, 77(3), 837–849. (This is the published version of his 2004 Nobel Lecture).
9. Hen, O., Sargsian, M., Weinstein, L. B., et al. (2017). Probing the structure of the atomic nucleus with high-energy electrons. *Science, 358*(6369), eaao3442.
10. Hofstadter, R. (1961). The electron-scattering method and its application to the structure of nuclei and nucleons. Nobel Lecture.
11. Hora, H., Miley, G. H., & et al. (2010). Fusion energy without radioactivity: Laser ignition of solid hydrogen–boron (HB11) fuel. Energy & Environment, 21(4), 173–200.

12. Krane, K. S. (1988). Introductory Nuclear Physics. John Wiley & Sons.

13. Kuhn, T. S. (1962). *The Structure of Scientific Revolutions*. University of Chicago Press.

14. Misner, C. W., Thorne, K. S., & Wheeler, J. A. (1973). Gravitation. W. H. Freeman.

15. Nuckolls, J., Wood, L., Thiessen, A., & Zimmerman, G. (1972). Laser compression of matter to super-high densities: Thermonuclear (CTR) applications. Nature, 239(5368), 139–142.

16. Perkins, D. H. (2000). Introduction to High Energy Physics (4th ed.). Cambridge University Press.

17. Steinmetz, C. P. (1914). *Electric Discharges, Waves and Impulses, and Other Transients*. McGraw-Hill.

18. Taylor, R. E. (1991). Deep Inelastic Scattering: The discovery of the point-like constituents of the nucleon. *Reviews of Modern Physics*, 63(3), 573.

19. Thomson, W. (Lord Kelvin). (1867). On vortex atoms. *Proceedings of the Royal Society of Edinburgh*, 6, 94–105.

20. Weinberg, S. (1995). The Quantum Theory of Fields, Volume 1: Foundations. Cambridge University Press.

21. Yukawa, H. (1935). On the Interaction of Elementary Particles. I. Proceedings of the Physico-Mathematical Society of Japan. 3rd Series, 17, 48-57.

22. Zee, A. (2010). Quantum Field Theory in a Nutshell (2nd ed.). Princeton University Press.

Chapter 8: The Resonant Atom

8.1 Chemistry

Modern chemistry rests on a foundation of incredible predictive power. From the periodicity of the elements to the intricacies of molecular bonding, its principles allow us to manipulate the material world with extraordinary precision. Yet, despite its operational success, the theoretical foundations of chemistry remain deeply abstract. The models we use to explain chemical behavior—quantized orbitals, energy levels, and exclusion principles—are accepted as axiomatic rules that lack a deeper, underlying physical mechanisms.

This chapter seeks to change that. Building directly on the Resonant Field Interpretation (RFI) and Quantum Vortex Dynamics (QVD) developed in the previous chapters, we extend the same physical and geometric principles that govern nuclear structure to the architecture of chemistry.

To understand what this new framework offers, it is helpful to retrace the historical path chemistry has taken. The journey began with alchemists, like Isaac Newton, whose mystical efforts to transmute matter were rooted in the belief that the material world obeyed hidden, intelligible laws. Modern chemistry emerged when these intuitions were formalized by pioneers like Robert Boyle, who established the strict definition of a chemical element, Antoine Lavoisier (1789), and Dmitri Mendeleev (1869), whose periodic table revealed a powerful yet unexplained regularity in their behavior. The quantum revolution of the early 20th century explained this regularity. The structure of the periodic table, we now understand, is a consequence of quantum rules: the geometric shapes of atomic orbitals are derived from the Schrödinger equation (1926), and the Pauli Exclusion Principle (1925) governs how electrons populate those orbitals.

While these rules explain what is happening, they leave unanswered the deeper question of why. Why do electrons form standing wave patterns at specific radii? Why do they exclude one another from identical quantum states? Why is energy quantized at all?

This chapter proposes that these features are inevitable outcomes of a deeper field dynamic. Drawing from the RFI, we will show how the Universal Light Field (ULF) and Universal Energetic Field (UEF) interact to create what we call the Resonant Atom (RA). In parallel with QVD's treatment of nucleons, we will construct a physical, geometric, and field-centric explanation of the atom's architecture. In doing so, we reveal how the rules of chemistry emerge as elegant harmonies of a deeper resonant order.

8.2 The Resonant Atom: A Physical Basis for the Principles of Chemistry

Abstract

This chapter presents a new physical foundation for chemistry, one whose principles emerge from the geometry of a resonant universe. Here, we introduce the concept of the Resonant Atom (RA), a new model of the atom built on the Resonant Field Interpretation (RFI) and Quantum Vortex Dynamics (QVD). Its central claim is that the geometric structure of the nucleus, quantified by the Alpha Stability Index (ASI), dictates the shape and energy of atomic orbitals, which are understood as real, physical standing waves in the Universal Light Field (ULF). This hierarchical model provides a tangible, physical basis for the foundational rules of chemistry by explaining the Pauli Exclusion Principle and electron shell structures as consequences of hydrodynamic stability and generating a series of novel, falsifiable predictions, including a Geometric Stark Effect and an Intrinsic Orbital Asymmetry in atoms with non-spherical nuclei. Ultimately, the RA model provides a basis for replacing the abstract axioms of the quantum model with an intuitive, deterministic framework in which the properties of matter are a direct reflection of geometry.

1. Introduction

The modern science of chemistry is built on the successful mathematical framework of quantum mechanics. The solutions to the Schrödinger equation (Schrödinger, 1926) predict the structure of the atom with incredible precision, while the Periodic Table of Elements provides the rulebook for all chemical interactions. Yet, despite this predictive power, the field's foundations remain deeply abstract. The reasons for the strange geometric shapes of atomic orbitals, the quantization of energy, and the axiomatic nature of the Pauli Exclusion Principle (Pauli, 1925) have no physical explanation.

This chapter proposes this deeper physical mechanism. Building directly upon the Resonant Field Interpretation (RFI) and Quantum Vortex Dynamics (QVD) established in the previous chapters, we demonstrate how the foundational principles of chemistry emerge from the physical geometry of the nucleus. Here, we posit that the atom is a resonant system in which electron vortices are shaped by the geometric structure of the nucleus and achieve stability by locking into three-dimensional standing waves in the Universal Light Field (ULF).

In the sections that follow, we build a complete geometric picture of the Resonant Atom (RA) from the inside out. We first establish a Platonic model of the nucleus's geometry, which we term the Alpha Stability Index (ASI). With this stable core defined, we then show how the shapes of atomic orbitals emerge as stable harmonic waves. From there, we derive the principles of chemistry, such as the Pauli Exclusion Principle, from the hydrodynamics of these waves. Finally, we demonstrate how this comprehensive atomic model provides a physical basis for the structure of the Periodic Table and the distinct chemical properties of its elements.

2. The Geometry of the Nucleus: A Platonic Model

To build a complete geometric picture of the RA from first principles, we must begin at its center. In our framework, the nucleus is a complex, structured UEF vortex that creates a stable potential well in which the electron's resonant ULF waves can form, as described in Chapter 5. The geometry and stability of this central vortex are therefore paramount.

The ASI is a tool we developed from QVD for quantifying how well larger nuclei adhere to the principle of geometric coherence. The pinnacle of this stability is the Helium-4 nucleus, also known as the alpha particle. [46] Composed of two protons and two neutrons, its fundamental constituents in our model are four protons and two antiprotons, as described in Chapter 7. We propose that its exceptional stability arises from a perfect geometric and dynamic equilibrium. The four proton "source" vortices form a tetrahedral framework, while the two antiproton "sink" vortices reside at the geometric center, creating a powerful binding core.

This hyperstable alpha vortex is the fundamental "geometric brick" of nuclear matter. While a complete theory would model the precise geometry of every nucleon in a large nucleus, we can create a powerful, effective model by treating the alpha particle itself as a single, incredibly coherent unit within the RA.

---

[46] The idea that many nuclei behave as if they are composed of alpha particles is a well-established concept in nuclear physics, known as the Alpha Cluster Model (Freer et al., 2018). Our framework provides a new physical basis for this model by proposing that the exceptional stability of the alpha particle results from its perfect tetrahedral geometry (*see* Chapter 7).

2.1 The Alpha Stability Index (ASI)

The central hypothesis of the ASI is that the stability of any nucleus can be quantified by comparing its actual, measured binding energy to the theoretical binding energy it would possess if it were perfectly composed of these ideal alpha-particle building blocks. The ASI is formulated as a simple ratio:

ASI = (Actual Measured Binding Energy) / (Ideal Geometric Binding Energy)

Where the "Ideal Geometric Binding Energy" is calculated as:

$$\text{Ideal Geometric Binding Energy} = \frac{\text{Total Number of Nucleons}}{4} * 28.30 \text{ MeV}$$

An ASI score at or greater than 1.0 suggests a highly coherent and stable geometry, indicating that the nucleus is as stable, or even more stable, than a simple collection of alpha particles. A score greater than 1.0 is possible when the geometric arrangement of the alpha vortices themselves is so synergistically stable that it releases an additional coherence dividend beyond what was released to form the individual alpha particles. A score less than 1.0 suggests a less stable, more strained geometry.

It is worth noting, however, that the absolute value of the ASI is not the sole determinant of stability, or even the main determinant. The true measure is a nucleus's position on the "landscape" of coherence. A nucleus can have an ASI greater than 1.0 and still be unstable if there is an accessible pathway for it to decay into a different configuration with an even higher ASI score. A fitting analogy is a ball resting in a small divot on the side of a mountain. While locally stable, it is globally unstable relative to the deeper valley below. Radioactive decay, in this view, is the process by which a nucleus "tunnels" or transitions from a state of good to optimal geometric coherence, releasing the difference in their Coherence Dividends as energy.

Therefore, the true power of the ASI, as we demonstrate here, is not just in this simple threshold but in the patterns it reveals. The index is a direct, quantitative map of this energetic landscape. It allows us to explain not only the absolute stability of the "magic numbers" (the highest peaks on the map) and the "islands of stability" (the local maxima), but also the predictable pathways of radioactive decay (the slopes and valleys that connect them).

2.2 Test Case 1: Alpha-Conjugate Nuclei

We first apply the ASI to the highly stable nuclei composed of an integer number of alpha particles, which our model predicts should have exceptionally coherent geometries.

- **Carbon-12 (^{12}C):** Composed of 3 alpha vortices.

    - Actual B.E. = 92.16 MeV

    - Ideal Geometric B.E. = (12/4) * 28.30 = 84.90 MeV

    - **ASI ≈ 1.085**

- **Oxygen-16 (^{16}O):** Composed of 4 alpha vortices.

    - Actual B.E. = 127.62 MeV

    - Ideal Geometric B.E. = (16/4) * 28.30 = 113.20 MeV

    - **ASI ≈ 1.127**

The fact that these nuclei have an ASI score greater than 1.0 suggests that they are synergistically stable. Our model proposes a specific geometric reason for this finding: they form the next simplest perfect arrangement. Carbon-12 is a stable planar triangle of three alpha vortices. Oxygen-16, the first "doubly magic" nucleus (8p, 8n), achieves an even greater coherence by forming a larger, perfect tetrahedron—a "tetrahedron of tetrahedrons."

2.3 Test Case 2: The Peak of Coherence (Iron-56 and Nickel-62)

We now turn to the ultimate test of the ASI: the nuclei that sit at the peak of the binding energy curve. While Iron-56 is often cited as the most stable nucleus in terms of binding energy per nucleon, that distinction actually belongs to Nickel-62. Both are exceptionally stable, and in our framework, they must represent the pinnacle of complex geometric packing and coherence. Here, we analyze both.

- **Iron-56 (^{56}Fe):** 26 protons, 30 neutrons

    - Actual B.E. = 492.26 MeV

    - Ideal Geometric B.E. = (56/4) * 28.30 = 396.20 MeV

    - **ASI ≈ 1.242**

- **Nickel-62 (^{62}Ni):** 28 protons, 34 neutrons

195

- Actual B.E. = 545.26 MeV

- Ideal Geometric B.E. = (62/4) * 28.30 = 438.65 MeV

- **ASI ≈ 1.243**

Both nuclei have exceptionally high ASI scores, confirming that they represent a peak of geometric efficiency. Notably, the ASI score for Nickel-62 is slightly higher, correctly identifying it as the true pinnacle of nuclear stability. In the RA framework, this finding indicates that while both structures are masterpieces of nuclear architecture, the specific geometric arrangement of vortices in Nickel-62 achieves a slightly more perfect, synergistically coherent state. Beyond this point, the increasing electrostatic repulsion between protons begins to overcome the coherence force, causing the gradual decline in the stability of all heavier nuclei.

2.4 Test Case 3: Non-Alpha Nuclei

Next, we test the ASI on nuclei that are not perfect alpha structures, such as the stable isotopes of Lithium.

- **Lithium-6 (^{6}Li):** 3 protons, 3 neutrons.

    - Actual B.E. = 31.99 MeV

    - Ideal Geometric B.E. = (6/4) * 28.30 = 42.45 MeV

    - **ASI ≈ 0.754**

- **Lithium-7 (^{7}Li):** 3 protons, 4 neutrons.

    - Actual B.E. = 39.24 MeV

    - Ideal Geometric B.E. = (7/4) * 28.30 = 49.53 MeV

    - **ASI ≈ 0.792**

The ASI scores for the stable Lithium isotopes are significantly less than 1.0. The low scores for the Lithium isotopes quantitatively demonstrate that their geometric structures, which are not based on a clean arrangement of alpha particles, are under significant geometric strain and are far less coherent than the Platonic forms.

2.5 Test Case 4: The Beryllium-8 Anomaly

Finally, we apply the ASI to a crucial "negative" case: Beryllium-8 (^{8}Be). This nucleus is composed of exactly two alpha particles. A simple interpretation might suggest that it should be stable. However, our model's core principle is that stability arises from the formation of a new, more coherent, higher-order geometric structure. The arrangement of two tetrahedra does not form a new, perfect Platonic solid. Instead, it forms a dimer that lacks the stability of a triangular or tetrahedral arrangement. Our model, therefore, predicts that Beryllium-8 should be unstable. Let us test this with the ASI:

- **Beryllium-8 (^{8}Be):** Composed of 2 alpha vortices.

    o   Actual B.E. = 56.50 MeV

    o   Ideal Geometric B.E. = (8/4) * 28.30 = 56.60 MeV

    o   **ASI ≈ 0.998**

Here, the significance of this result lies not in its proximity to 1.0, but in the fact that it is fundamentally less than 1.0. The ASI score of 1.0 represents a perfect energetic break-even point, the threshold between a bound and unbound system.

- A nucleus with an ASI > 1.0 (like Carbon-12) has released a "synergistic coherence dividend." It is in a stable energy "valley," and to break it apart, energy must be added to the system.

- A nucleus with an ASI < 1.0 (like Beryllium-8) is an unbound system. It is in a higher energy state than its potential decay products, balanced on an energetic "hilltop," and will spontaneously decay, releasing energy.

This result explains why Beryllium-8, with an ASI of ~0.998, is profoundly unstable, while Lithium-6, with a much lower ASI of ~0.754, is perfectly stable. Lithium-6 is located in a shallow, but stable, energy valley. Beryllium-8, on the other hand, is at a peak, with no energy barrier to prevent its decay.

The ASI model, therefore, predicts that Beryllium-8 should spontaneously decay into two alpha particles, releasing a small amount of energy in the process. This prediction is consistent with experimental observations. Beryllium-8 is famously and incredibly unstable, decaying almost instantly with a half-life of just 8×10^{-17} seconds (Krane, 1988). The ability of the ASI to both obtain the correct number and accurately predict the fundamental nature of this nucleus, whether it is unbound or bound, is a powerful confirmation of the framework's geometric principles.

2.6 A Geometric Theory of the Magic Numbers

In standard nuclear physics, certain "magic numbers" of nucleons (2, 8, 20, 28, 50, 82, 126) are known to correspond to peaks of exceptional stability. While the nuclear shell model provides a mathematical framework for these numbers, it offers no deep, physical insight into *why* these specific configurations are so uniquely stable. QVD provides this physical explanation. In this model of nuclear stability, the "magic numbers" are not arbitrary; they are the architectural landmarks that signify the completion of a perfect, highly coherent, nested geometric shell of alpha-particle vortices.[47]

The stability of these shells is architectural in nature. A "magic" nucleus is not a finished building but rather a perfectly completed foundation. This analogy allows us to understand the otherwise paradoxical stability of isotopes near the magic numbers. The QVD model's geometric precision is demonstrated by its ability to correctly predict the distinct architectural properties of the O-16 and Ca-40 magic shells.[48]

2.6.1 Case 1: The Ca-40 "Open Foundation"

---

[47] The idea that nuclear structure can be modeled with Platonic solids has been explored by other non-mainstream theories, most notably the Structured Atom Model (SAM) (Vowles, 1969). While the UFD framework shares this geometric intuition, it is a fundamentally different theory. The two frameworks diverge on two crucial points: first, where SAM is a geometric packing model for nucleons, UFD provides a physical origin for the nucleons themselves, defining them as toroidal trefoil vortices, a specific geometry that naturally gives rise to the quantum property of spin-1/2; second, and most profoundly, the two models have incommensurable views of the neutron. In UFD, the neutron is a composite proton-antiproton vortex, whereas in SAM, the neutron is a composite proton-electron particle. Thus, although they are conceptually aligned, UFD and SAM do not yield the same predictions.

[48] The specific geometric predictions for these nuclei (e.g., Ca-40 as octahedral, Ni-62 as icosahedral) are derived by correlating the ASI with macroscopic crystal lattice data. In UFD, ferromagnetism (*see* Chapter 10) provides a unique window into nuclear geometry. We postulate that the electron lattice "resonates" with the nuclear vortex geometry. For instance, Iron's Body-Centered Cubic (BCC) lattice implies a cubic/octahedral nuclear surface, while Nickel's Face-Centered Cubic (FCC) lattice implies a tighter icosahedral shell. By mapping these lattice constraints onto the ASI stability peaks, we find that the "magic numbers" correspond precisely to the completion of Platonic harmonics: Tetrahedron (A=16), Octahedron (A=40), Icosahedron (A=62), and a Multi-Shell Icosahedral Sphere (A=208).

The "doubly magic" Calcium-40 nucleus (20p, 20n) represents the flawless completion of an open foundation, predicted to be an octahedral shell. Its high ASI score of ~1.208 confirms its stability.

- Subtracting a Keystone (Calcium-39): When a single neutron is removed, the entire structure is compromised. The ASI score plummets to ~1.183 and the nucleus becomes highly unstable, with a half-life of just 0.86 seconds (Audi, 2003).

- Building on the Foundation (Calcium-42, 44): Because Ca-40 is an open foundation, its geometry (an octahedron with 8 faces) provides stable docking ports for additional neutrons. This synergistic addition *increases* coherence, releasing a Coherence Dividend and raising the ASI score for Ca-42 (~1.218) and Ca-44 (~1.224).

- Completing the Architecture (Calcium-48): This model predicts that 8 additional neutrons will completely fill the 8 docking ports of the Ca-40 octahedron. This is precisely what is observed: the next "doubly magic" nucleus is Calcium-48 (20p, 28n). The "magic number" 28 is thus the geometric proof of a completed Ca-40 core with all 8 of its architectural sites filled.

2.6.2 Case 2: The O-16 "Finished Masterpiece"

The "doubly magic" Oxygen-16 nucleus (8p, 8n) represents a different kind of architectural state: a "finished masterpiece," predicted to be a perfect tetrahedral geometry. Its ASI score (~1.127) represents a peak of completed coherence.

- Subtracting a Keystone (Oxygen-15): Removing a neutron causes a catastrophic coherence collapse (~1.055), making O-15 highly unstable. This confirms O-16 is a stable shell.

- Building on a "Finished" Stucture (Oxygen-17, 18): Unlike the "open" Ca-40, adding neutrons to the "finished" O-16 tetrahedron disrupts its perfect symmetry. This reduces the overall coherence, causing the ASI score to drop for O-17 (ASI ~1.095) and O-18 (ASI ~1.098). Although the O-17 and O-18 scores are sufficiently high to be kinetically stable (metastable), their geometric imperfections and lower stabilities are confirmed by their natural abundances: the universe overwhelmingly favors the most coherent state, with O-16 making up 99.76% of all oxygen.

This demonstrates the ASI's keen predictive power. The otherwise paradoxical *increase* in stability for Calcium and *decrease* for Oxygen is perfectly explained as a lesson in cosmic construction. The "magic numbers" are not an abstract list; they are the direct, physical confirmation of completed architectural layers, each with its own unique geometric properties. The ASI chart that follows is the direct, visible confirmation of this geometric reality.

2.7 The Coherence Potential Landscape

When the Coherence Potential for a wide range of isotopes is plotted as a function of their mass number (as shown in Figure 20), a clear landscape emerges. The chart provides a stunning visual confirmation of the core principles of our geometric model.

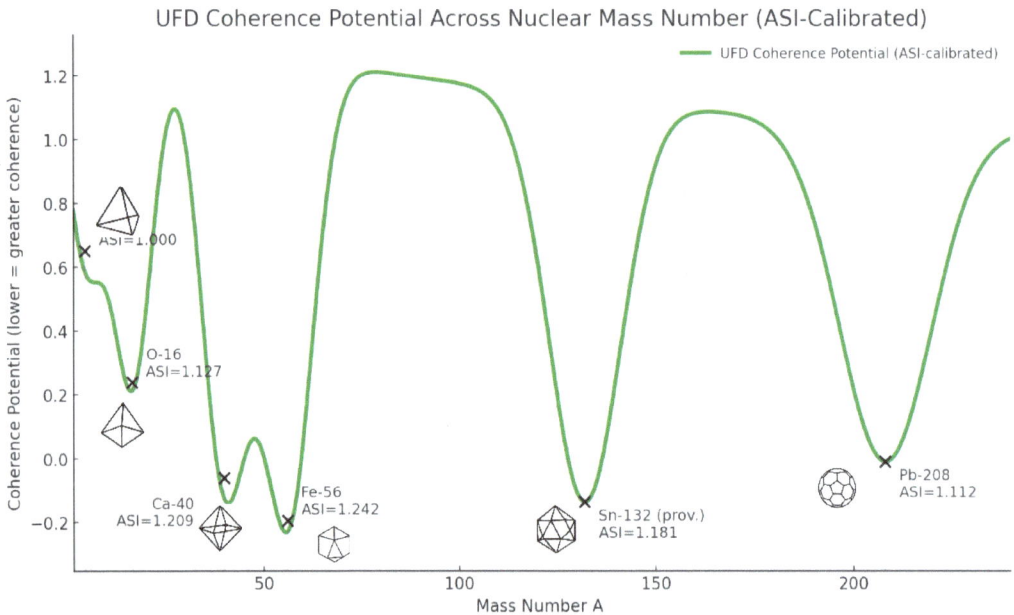

Figure 20. UFD Coherence Potential Across Nuclear Mass Number. This figure illustrates the UFD Coherence Potential, a conceptual scalar field representing the residual geometric strain in the nuclear vortex lattice. Lower values correspond to deeper coherence basins, where UEF vortex packing is maximally efficient and nuclear stability is greatest. The curve is based on an inverted, smoothed binding-energy trend with Gaussian coherence wells calibrated to the actual ASI values for key nuclei. Coherence landmarks at A ≈ 4, 16, 40, 56, 132, and 208 correspond to alpha-cluster structures and shell closures. Vertical dashed lines mark magic nucleon numbers. Deep minima, such as Fe-56 and Pb-208, correspond to the most favored geometric configurations (depicted as Platonic shells), whereas mid-scale structures, such as Sn-132, form metastable plateaus in the coherence landscape. In this framework, nuclear stability arises from an element's position within this geometric topology.

200

First, it perfectly captures the famous binding energy curve but reframes it as a potential landscape. The Coherence Potential, which represents geometric strain, starts high for the lightest elements, descends sharply to its deepest minimum (or "basin") around Iron-56 and Nickel-62, and then begins a gradual ascent for heavier nuclei. This demonstrates that the peak of nuclear stability is a direct consequence of achieving the most perfect, lowest-strain geometric packing state possible.

The chart also highlights the exceptional stability of the magic number nuclei. Even within the gradual ascent of heavy elements, certain nuclei stand out. The alpha-conjugate and doubly magic nuclei—from Carbon-12, Oxygen-16, and Calcium-40 up to the heavy and exceptionally stable Lead-208—all appear as distinct local minima (or "valleys") of low potential relative to their neighbors, providing strong evidence that the magic numbers correspond to the completion of perfect, low-strain geometric shells.

The Coherence Potential's predictive accuracy is most powerfully demonstrated by the famous paradox of radioactive "doubly magic" nuclei. The case of Nickel-56 (28p, 28n) is an ideal example. As expected, its potential is very low, confirming its geometric coherence (ASI ~1.222). However, this potential is still noticeably higher than that of the truly stable nuclei in its immediate vicinity: Iron-56. The chart thus correctly identifies Ni-56 as a "metastable well"—a deep valley on the coherence landscape, but not the absolute deepest basin. This predicts that it should be radioactive and decay "downhill" toward a more stable configuration, which is precisely what happens.

This same subtle logic holds true for the much heavier doubly magic nucleus, Tin-132 (50p, 82n). Its Coherence Potential is also exceptionally low, confirming its "magic" geometric status. However, just like Ni-56, it sits in a shallow local minimum, as its stable neighbor Xe-132 has an even lower potential (a deeper well). The chart correctly predicts that Sn-132, despite its magic status, must also be radioactive and will decay "downhill" toward this deeper "valley of stability."

The Coherence Potential model correctly identifies these famously unstable nuclei as highly coherent but ultimately less stable (at a higher potential) than their neighbors, demonstrating that it provides an accurate map of the nuclear landscape.

3. Orbital Geometry: The Harmonics of the Atom

Having established the geometry of the nucleus as the stable UEF core of the atom, we can now explain the second aspect of the RA model: the structure of the electron shells that form around it. Building on the RFI, this section explains how the strange and specific shapes of atomic orbitals arise as physical, harmonic standing waves in the ULF.

The quest to understand the atom began with the classical "planetary model" (Rutherford, 1911), which failed because an accelerating electron must continuously radiate away its energy. The RFI, in contrast, proposes that stability is achieved only when the electron vortex abandons classical orbital motion and instead "locks into" a resonant frequency of the ULF, which is being sustained by the central potential of the nucleus (*see* Chapter 5). This creates a stable three-dimensional standing wave, a self-sustaining pattern of field flow that does not dissipate energy.

Crucially, the specific geometry of the UEF nucleus itself shapes this standing wave. A tetrahedral Helium-4 nucleus, for example, creates a potential well in the ULF with an underlying tetrahedral symmetry. The stable standing waves that form within this well must, therefore, conform to this geometry. In this view, the shapes and orientations of the atomic orbitals are a direct reflection of the symmetries of the nucleus they form around. This phenomenon is analogous to how the shape of a bell determines the specific harmonic notes that it can produce.

3.1 The 's' Orbital: The Fundamental Resonance

In standard quantum mechanics, the s-orbital is the lowest-energy state, defined by having zero orbital angular momentum ($l = 0$).[49] A key feature of the Standard Model is that its wave function is always perfectly spherical, regardless of the nucleus.

Our framework reinterprets the s-orbital as the fundamental mode of resonance for the ULF around a nucleus. It is the simplest, most stable standing wave pattern possible, representing a symmetrical "breathing mode." However, the precise physical geometry of this standing wave is not independent; it must conform to the shape of the

---

[49] The electron's position in an atom can be considered an "address" defined by a set of rules called quantum numbers. The three most important of these are the Principal Quantum Number (n), the Angular Momentum Quantum Number (l), and the Magnetic Quantum Number (ml). The Principal Quantum Number ($n = 1,2,3...$) defines the main energy shell, akin to the "street" on which the electron lives. The Angular Momentum Quantum Number ($l = 0,1,2...$) defines the shape of the orbital, like the "shape of the house." The Magnetic Quantum Number (ml) describes the orientation of the orbital in space, akin to the "direction the house is facing." Here, the s-orbital ($l = 0$) would be a spherical house with only one possible orientation ($ml = 0$). The p-orbital ($l = 1$) would be a dumbbell-shaped house that can be oriented in three perpendicular ways, corresponding to the px, py, and pz orbitals ($ml = -1,0,+1$).

potential well created by the nucleus itself. This principle leads to a key, testable distinction from the Standard Model:

1. Hydrogen: The Standard Model views the proton as a spherically symmetric point charge. In sharp contrast, QVD defines the proton as a 3-lobe trefoil vortex (*see* Chapter 7). Therefore, the potential well it creates should not be perfectly spherical but should possess an underlying trefoil symmetry. The resulting 1s orbital, while appearing spherical at a distance, must carry a subtle, tri-lobed internal resonance to remain coherent with its trefoil nucleus (Figure 21A)

2. Helium-4: The Helium-4 nucleus is the archetypal coherent structure in UFD, forming a highly stable tetrahedron. Its potential well is therefore distinctly tetrahedral. The 1s orbital of Helium-4 must conform to this geometry, resulting in a quasi-spherical standing wave that exhibits a clear and measurable tetrahedral deformation, bulging slightly along the axes corresponding to the nuclear tetrahedron's vertices (Figure 21B).

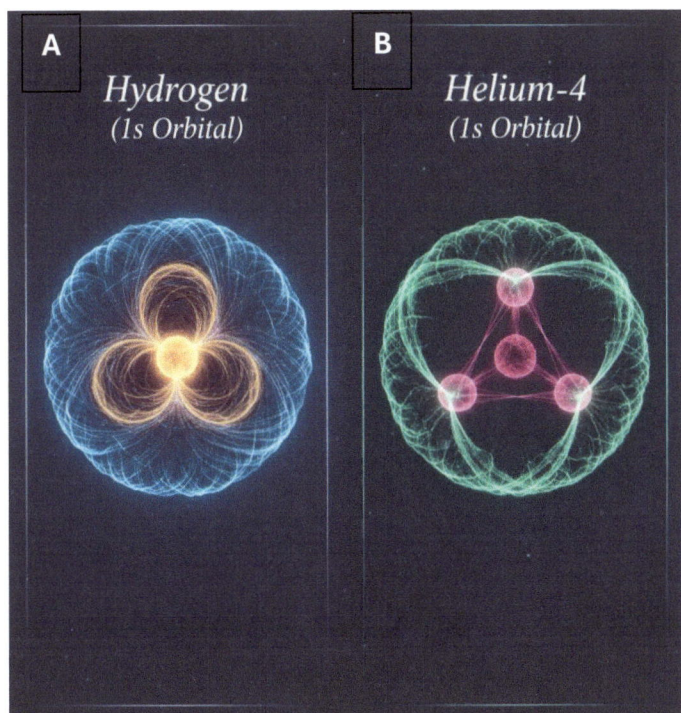

Figure 21: The Influence of Nuclear Geometry on Orbital Shape. This figure illustrates a key prediction of the RA model. (A) The Hydrogen Atom: The 1s orbital is a spherical standing wave that carries the subtle, tri-lobed internal resonance mirroring the 3-lobe trefoil proton at its core.

(B) The Helium-4 Atom: In our framework, the Helium-4 nucleus is a stable tetrahedron. This non-spherical core creates a potential well with a corresponding tetrahedral symmetry, forcing the 1s orbital to adopt a subtle but distinct tetrahedral deformation. This illustrates the first level of the two-level resonance that governs all chemical bonding.

Thus, in the RA model, the shape of the s-orbital is not a universal constant; instead, it directly reflects the underlying geometry of the specific nucleus it surrounds.

This provides a new, falsifiable prediction that distinguishes the RA from the standard interpretation

3.2 The $'p'$ Orbitals: The First Harmonic Overtone

The p-orbitals ($l = 1$) are the next stable harmonic overtone of the ULF. They have a characteristic "dumbbell" shape consisting of two lobes of opposite phase, separated by a nodal plane.[50] In an atom with a perfectly spherical nucleus, like hydrogen, the three p-orbitals can orient themselves along the standard, perpendicular x, y, z axes and will be perfectly degenerate, meaning they have the same energy.

However, in the RA, the orientation of these orbitals is dictated by the geometry of the central nucleus. For an atom like Helium-4, with its underlying tetrahedral nucleus, the potential well is not spherically symmetric. The three p-orbitals must therefore align themselves in a way that is geometrically compatible with this tetrahedral field. Because these different orientations are no longer physically identical, we predict that the three p-orbitals of Helium-4 are not perfectly degenerate. This subtle breaking of degeneracy, caused by the nuclear geometry, should be detectable as a minute splitting of the 2p energy level in high-resolution spectroscopy, providing another clear, falsifiable test of our framework.

3.3 Higher Orbitals and the Foundation of Chemistry

This principle of harmonic resonance extends logically to higher orbitals. The more complex, clover-leaf shapes of the d-orbitals and the intricate forms of the f-orbitals are reinterpreted in this framework as the next stable, higher-energy harmonic overtones that the ULF can support around the nucleus. Their complex shapes result from the standing wave needing to accommodate higher energy by forming more intricate patterns of nodal planes and surfaces, analogous to the complex Chladni figures that form on a vibrating plate (Chladni, 1787).

As with the p-orbitals, the specific orientations and potential splitting of these d- and f-orbital energy levels are dictated by the underlying geometry of the nucleus they

---

[50] In any standing wave, a node is a point or line that does not move while the rest of the system vibrates around it. A nodal plane is the two-dimensional extension of this concept. In the analogy of a vibrating drumhead, the nodal plane is the imaginary line across the middle that remains perfectly still while the two halves of the drumhead oscillate in opposite phases. For a p-orbital, the nodal plane is the surface passing through the nucleus where the amplitude of the ULF standing wave is always zero, meaning there is zero probability of finding the electron vortex at any point on that plane.

form around. Thus, the entire structure of the Periodic Table of Elements, with its sequence of shells and subshells (s, p, d, f), is re-envisioned as a direct map of the stable geometric standing-wave solutions available in the ULF. It is a catalogue of the fundamental notes and harmonies that matter can play.

In summary, by reinterpreting the atom as a resonant system, we have provided a tangible, physical identity for atomic orbitals, explaining their specific shapes as the stable geometric standing-wave harmonics of the ULF. Most importantly, we have shown that the geometry of these orbitals directly reflects the underlying geometry of the nucleus they surround.

4. A Geometric Basis for the Principles of Chemistry

Having established the physical nature of atomic orbitals as resonant standing waves, we now demonstrate how the foundational principles governing electron arrangement—the rules that build the Periodic Table—emerge directly from the physical properties of these waves. While the Standard Model of Chemistry presents the Aufbau principle, Hund's Rule, and the Pauli Exclusion Principle as quantum mechanical axioms, the RA provides a physically intuitive origin for each by grounding them in the geometry and fluid dynamics of the ULF.

4.1 The Aufbau Principle: The Path of Least Resistance

The Aufbau principle, from the German for "building-up," is the fundamental rule chemists use to predict the electron configuration of any element in the Periodic Table. It dictates the specific order in which electrons fill the available atomic orbitals, starting with the lowest-energy level and progressing to higher-energy levels. This predicted configuration is relevant because it determines an element's chemical properties, its reactivity, and how it will bond with other atoms. While it is a potent predictive tool, the principle is axiomatic in the Standard Model.

The RA model provides a physical reason for this rule. In RA, the Aufbau principle is a manifestation of the universal tendency of all physical systems to seek their state of maximum stability, which is always the path of least energetic resistance. An electron vortex will always settle into the most stable, least energetic resonance pattern possible. Therefore, it will naturally "lock into" the simplest, fundamental harmonic resonances (the 1s, then 2s orbitals) before it has sufficient energy to occupy the more complex, higher-energy overtone patterns of the p and d orbitals. The orderly filling of shells described by the Aufbau principle is, in this view, the direct and necessary consequence of a physical system settling into its most stable possible configuration.

4.2 Hund's Rule: The Principle of Minimal Resonant Strain

Hund's rule (Hund, 1925) is another key principle used to determine the ground-state electron configuration of an atom. It is composed of two distinct rules that govern how electrons are distributed among orbitals of the same energy (degenerate orbitals), such as the three p-orbitals. The first rule, "maximum multiplicity," states that electrons will fill each orbital singly before any orbital is doubly occupied. The second rule, "parallel spins," tells us that the electrons in these singly occupied orbitals will all have parallel spins. These rules are essential for correctly predicting the magnetic properties of many atoms. In the Standard Model, they are derived from complex quantum mechanical calculations that show this configuration minimizes the overall electron-electron repulsion; however, this explanation lacks a simple, intuitive physical picture. Our framework provides a distinct physical mechanism for each rule.

The first rule—that electrons occupy separate orbitals first—is explained in our model by the principle of Minimal Resonant Strain. By placing one electron vortex in each of the three perpendicular p-orbitals, the system maximizes the physical distance between them. This minimizes the interference and hydrodynamic strain between their respective ULF standing-wave patterns, resulting in a more stable, lower-energy configuration than if two vortices were forced into the same orbital. This is the primary effect and accounts for the most significant portion of the energy stabilization.

Once the electrons have occupied separate orbitals to minimize strain, the second rule—that their spins align in parallel—is explained by a secondary principle of rotational coherence. We propose that multiple, parallel-spinning vortices create a more stable, large-scale rotational flow in the ULF between them. Their individual magnetic eddies, all rotating in the same direction, encourage the ULF fluid to form a single, smooth, and coherent flow, which is a state of very low energy. A configuration of mixed spins, in contrast, would create turbulence and "hydrodynamic friction" at the point where opposing eddies meet, resulting in a slightly higher-energy, less stable state. The system therefore settles into the parallel-spin configuration to achieve its absolute lowest possible energy.

4.3 The Pauli Exclusion Principle: The Principle of Resonant Stability

The Pauli Exclusion Principle (Pauli, 1925) is arguably the most important rule in chemistry and is a cornerstone of quantum mechanics. It states that no two electrons in an atom can have the same four quantum numbers. In practical terms, this means that any single atomic orbital can hold a maximum of two electrons, provided that their spins are opposite. This principle is what gives matter its structure and volume; without it, all of an atom's electrons would collapse into the lowest energy state, and the rich complexity of the Periodic Table would not exist. However, the standard interpretation offers no physical mechanism for this. Like Aufbau's principle and Hund's rule, it is an axiom that must be accepted.

The RA provides a physical mechanism, which we term the Principle of Resonant Stability. This principle is grounded in the fluid dynamics of the electron vortex. As we have established, an electron is a spinning vortex in the ULF. This physical spin has two inseparable consequences: first, its fundamental "handedness" or chirality determines its charge polarity (a "sink"); second, this same rotation induces a coherent rotational eddy in the surrounding ULF, which constitutes its magnetic field and gives it a clear North/South magnetic polarity (*see* Chapter 5). What chemists and physicists refer to as an electron's "spin" (spin-up or spin-down) is a measurement of this latter property—the orientation of the induced magnetic eddy.[51]

This physical picture provides a direct mechanism for the exclusion principle. If two electrons with the same spin (meaning their magnetic eddies are aligned and rotating in the same direction) were to enter the same orbital, their ULF eddies would clash. Like two gears spinning in the same direction, this would create a chaotic, hydrodynamically unstable interference pattern (Figure 22). Stability is only achieved when a second electron with the opposite spin enters the orbital. Together, their counter-rotating eddies create a perfectly balanced, symmetrical flow, eliminating turbulence. This paired configuration is a much lower-energy state that powerfully reinforces the stability of the orbital's standing wave.

Therefore, the exclusion principle is a direct physical constraint of the atom's resonant nature: a stable orbital resonance can only be maintained by a balanced, non-

---

[51] A helpful analogy is the Earth itself. The Earth has a constant, fundamental physical spin on its axis, which is stable and gives us our day. This physical rotation, however, generates a secondary effect: a magnetic field with a distinct North and South pole. The orientation of this magnetic field can "flip" over geological time. Similarly, in our model, the electron vortex has a fundamental, constant physical spin. The property that chemists call "spin" (spin-up or spin-down) is a measurement of the orientation of the magnetic eddy generated *by* that physical rotation.

turbulent flow, which requires either a single vortex or a perfectly paired, counter-rotating set of two.

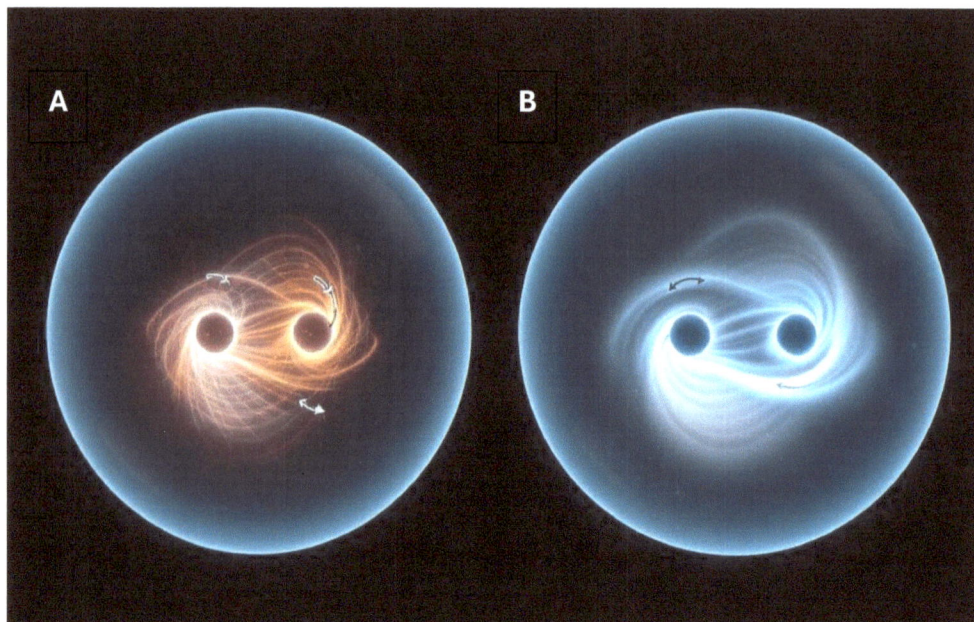

Figure 22: The Principle of Resonant Stability. This figure illustrates the physical mechanism of the Pauli Exclusion Principle in our framework.
(A) When two electrons with parallel spins occupy the same orbital, their individual magnetic eddies rotate in the same direction. This creates a chaotic, turbulent interference pattern in the ULF, resulting in a hydrodynamically unstable, high-energy state.
(B) Stability is only achieved when two electrons with opposite spins occupy the same orbital. Their counter-rotating magnetic eddies "mesh" perfectly, creating a balanced, symmetrical, and non-turbulent flow. This paired configuration is a much lower-energy state that reinforces the stability of the orbital's standing wave. This physical constraint explains why a stable orbital can hold a maximum of two electrons, provided their spins are opposite.

5. The Emergence of Chemical Properties

Having established a physical basis for the principles of chemistry, we will now demonstrate how the RA model explains the distinct chemical properties that define the major groups of elements in the Periodic Table. The organizing principle of the table is the atomic number (Z), which corresponds to the number of protons in the nucleus. In our framework, this is the number of "free," un-neutralized UEF "source" vortices. This net positive charge creates a powerful potential well in the surrounding ULF, which dictates the number and shape of the stable, resonant standing waves (orbitals) that the electron "sink" vortices must form to reach equilibrium. An element's chemical behavior,

208

therefore, is a direct reflection of the stability and geometric accessibility of its outermost ULF standing wave: its valence resonance.

5.1 The Noble Gases: Geometric Perfection

The noble gases (Helium, Neon, Argon, etc.) are defined by their extreme chemical inertness. This property is explained in the RA model by the fact that their atoms have a perfectly full and symmetrical outer shell of ULF resonances. The $'s'$ and $'p'$ standing waves in their outermost layer are all occupied by balanced, counter-spinning electron vortices. The resulting collective wave pattern is in a state of perfect geometric coherence and resonant stability. It is so stable that there is no energetic advantage to interacting with other atoms, thus explaining their non-reactive nature (Figure 23A).

5.2 The Alkali Metals: Geometric Instability

The alkali metals (Lithium, Sodium, etc.) are known to be extremely reactive. In the RA model, this is explained by their atomic structure, which consists of a stable "noble gas" core with a lone electron vortex occupying a new, high-energy s-orbital. This outer standing wave is large, weakly bound, and hydrodynamically unstable. The system can therefore achieve a much more stable, lower-energy state by jettisoning this lone electron vortex and reverting to the perfect stability of its inner shells. Their extreme reactivity is a direct measure of this geometric instability (Figure 23B).

5.3 The Alkaline Earth Metals: Paired Instability

The Alkaline Earth Metals (Beryllium, Magnesium, etc.) are also highly reactive but typically less so than their Alkali Metal neighbors. The RA model explains this subtle difference in terms of the geometry of their valence resonances. These atoms have a stable noble gas core, but their outermost shell contains a paired set of two electron vortices in a new, high-energy s-orbital. As established by the Principle of Resonant Stability, a pair of counter-spinning electron vortices creates a hydrodynamically balanced and stable flow. This makes the s-orbital of an Alkaline Earth Metal more stable than the imbalanced orbital containing only a single electron found in the Alkali Metals. However, this outer shell remains a high-energy, unstable layer compared to the perfect coherence of the inner core, allowing the system to reach a much lower energy state by discarding the entire outer resonant shell (Figure 23C).

5.4 The Halogens: Geometric Hunger

In the RA, the halogens (Fluorine, Chlorine, etc.) are the mirror image of the alkali metals. They are extremely reactive and highly electronegative because their outer p-orbitals are nearly complete, leaving a single empty "slot" in their resonant structure. In our fluid-dynamic picture, this incomplete standing wave creates a powerful asymmetry in the ULF flow around the atom, acting as a potent "unfilled sink." Their extreme reactivity is a direct measure of this intense "geometric hunger" to capture a final electron vortex, thereby completing their resonance and achieving the perfect stability of a noble gas (Figure 23D).

5.5 The Transition Metals: Resonant Complexity

The transition metals (e.g., Iron, Copper, Gold) are defined by the complex filling of their d-orbitals; their shared properties emerge from the nature of these higher-order resonances. The d-orbital standing waves are large and complex, allowing them to overlap easily within a crystal lattice. In the RA model, this creates a vast, crystal-wide "sea of shared resonance," allowing electron vortices to flow freely and imparting these elements with metallic and conductive properties. Moreover, because the many d-orbitals have very similar energy levels, electron vortices can readily shift among different resonant patterns. This ability to easily rearrange electrons into different stable or semi-stable geometric configurations is why they can form ions with a wide range of charges and create uniquely colored compounds. This principle of shielded resonances extends to the even more intricate f-block elements, the lanthanides and actinides, which explains their remarkably similar chemical properties across each series (Figure 23E).

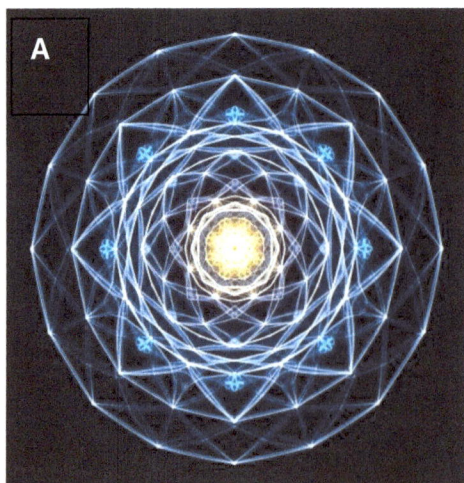

Figure 23: Valence Resonance and Chemical Periodicity of Resonant Atoms
(A) The Noble Gases: Geometric Perfection. This figure illustrates a noble gas atom with its perfectly full and symmetrical outer shell of ULF resonances. The balanced, counter-spinning electron vortices create a state of complete geometric coherence and resonant stability, explaining their extreme chemical inertness.

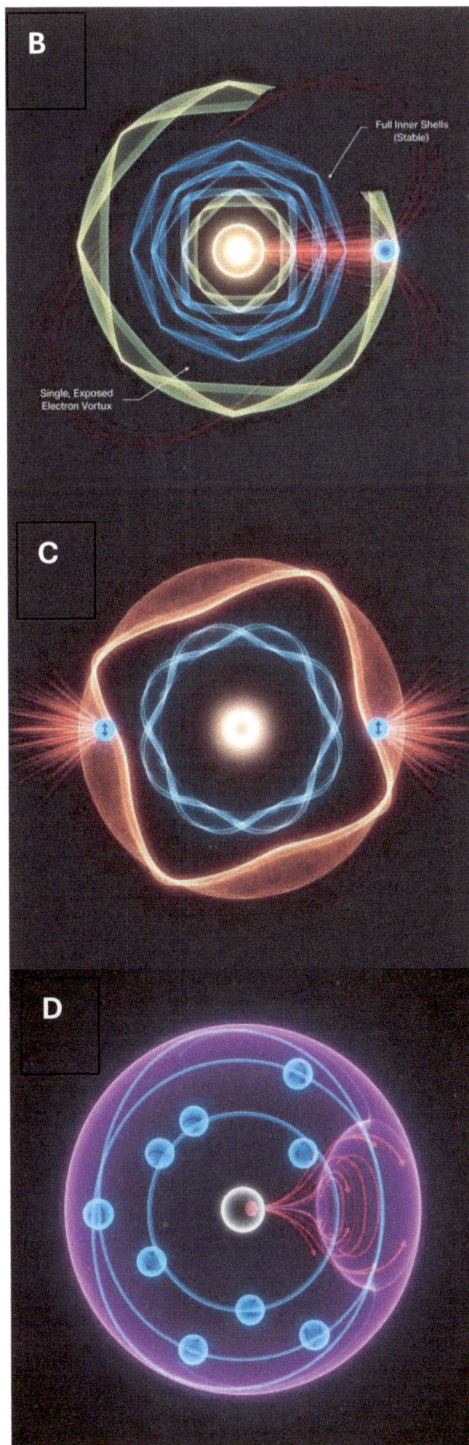

(B) The Alkali Metals: Geometric Instability. This figure depicts an alkali metal atom with a stable inner core and a single, high-energy electron vortex in an outer s orbital. This lone, weakly bound wave represents a state of geometric instability, driving the atom's high reactivity as it seeks to shed this electron and achieve the stability of its noble gas core.

(C) The Alkaline Earth Metals: Paired Instability. This figure shows an alkaline-earth metal atom with a stable inner core and a paired set of two-electron vortices in its outer s-orbital. While the paired vortices create a more balanced configuration than the single electron of alkali metals, this outer shell remains a high-energy, unstable layer compared to the perfect coherence of the inner core, leading to reactivity through the loss of these two electrons.

(D) The Halogens: Geometric Hunger. This figure illustrates a halogen atom with an almost complete outer p orbital, featuring a single empty "slot" in its resonant structure. This incomplete standing wave creates a powerful asymmetry in the ULF flow, representing a strong "geometric hunger" to capture an additional electron and achieve the stable configuration of a noble gas.

211

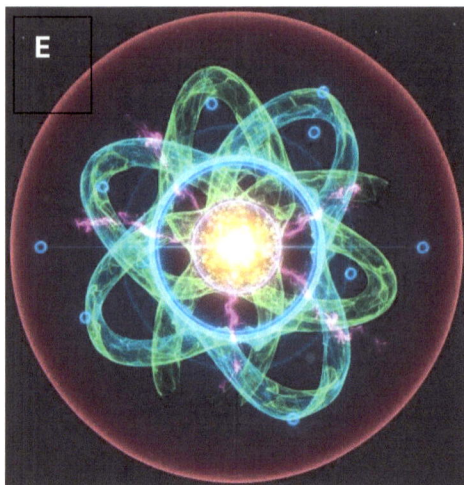

(E) The Transition Metals: Resonant Complexity. This figure depicts a single transition metal atom. The large and intricate outer d-orbitals are shown as a complex system of overlapping ULF standing waves. The ability of electron vortices to shift among the many available resonant patterns within this structure is the physical origin of their characteristic properties, such as their ability to form ions with multiple charges and to create uniquely colored compounds.

Thus, from the inert Noble Gases to the reactive Halogens and the complex Transition Metals, the diverse chemical behaviors across the Periodic Table can be understood as the logical consequences of the geometry and stability of their underlying valence resonance, providing a unified, physical basis for chemical periodicity.

6. Falsifiable Predictions

A scientific framework is validated by its ability to make novel, testable predictions. The RA leads to a new class of such predictions.

6.1 The Geometric Stark Effect

The first prediction assesses the ability to deform an orbital externally. In conventional quantum mechanics, the Stark effect (Stark, 1914) describes the shifting of an atom's spectral lines in a uniform electric field. Because the Standard Model treats an orbital as an abstract probability function, the effect depends only on the field's strength, not its shape.

The RA, which posits that an orbital is a real, physical resonant structure, predicts a "Geometric Stark Effect." We predict that a precisely shaped external field, designed to asymmetrically "squeeze" the lobes of a p-orbital, will induce an anomalous shift in the atom's spectral lines that corresponds directly to the shape and orientation of the applied field. The discovery of such a geometry-dependent shift would provide direct evidence that orbitals are real physical structures that can be physically manipulated.

6.2 The Splitting of Orbital Degeneracy

The second prediction tests the energy levels of orbitals. In the Standard Model, the three p-orbitals of a given energy shell are considered perfectly degenerate, meaning they have the same energy. This claim follows from the assumption that the nucleus creates a perfectly spherical potential well. The RA model, which states that the nucleus is a structured geometric object, predicts that this assumption is fundamentally incorrect. The non-spherical geometry of the nucleus must imprint its symmetry onto the ULF (orbital) field, "breaking" the perfect degeneracy.

At first glance, this prediction appears to contradict decades of experiments that measure these orbitals as degenerate. However, the RA model predicts that the observed degeneracy is a measurement illusion created by time averaging, analogous to a "spinning fan." A fan at rest, for instance, has a clear, non-symmetrical, 3-bladed geometry (the true, physical state). A long-exposure photograph (a time-averaged measurement) of that same fan in motion will show a smeared-out, perfectly circular blur (the measured state).

Our conventional experiments are long-exposure photographs. They are measuring the time-averaged blur of a rapidly tumbling, quantum-superposed geometry that appears perfectly symmetrical. The RA model predicts that the instantaneous, physical orbitals are not degenerate.

- Prediction 1: The Hydrogen Atom. The proton is a non-spherical, 3-lobe "trefoil" vortex. This "tumbling" 3-lobe geometry imprints a 3-lobe deformation onto the p-orbitals. Because the p_x, p_y, p_z orbitals cannot all have the same physical orientation relative to the 3 nuclear lobes, their instantaneous energies cannot be identical.

- Prediction 2: The Helium-4 Atom. This principle is amplified in the He-4 atom. Its stable tetrahedral nucleus also imprints a non-spherical (tetrahedral) deformation onto its p-orbitals.

This leads to a clear, falsifiable prediction: orbital degeneracy is an illusion of time-averaging. Novel, high-speed, time-resolved spectroscopic methods that can effectively "strobe" the atom would be able to detect the true, underlying, non-degenerate energy levels of the instantaneous orbital geometries. This splitting signature would be the Rosetta Stone that reveals the hidden geometric shape (trefoil or tetrahedral) of the nucleus itself.

6.3 Anomalous Chemical Stability and New Magic Numbers

Finally, the RA model predicts a third, deeper layer of stability that the Standard Model does not account for. It predicts that there should be "islands of anomalous chemical stability" in the Periodic Table, where an element's chemical properties are unusually stable or its bonds are unusually strong, not because of its electron configuration, but because its most common isotope has a particularly perfect or symmetrical nuclear geometry. This is a direct consequence of a principle we will formalize in the next chapter as Geometric Congruence. This principle states that, at a deeper level, the strength of a chemical bond is a function of how well the UEF nuclei can "mesh" or align, which reflects the geometric perfection of the nuclei themselves.

This prediction leads to a new set of "chemical magic numbers." Carbon-12, with its perfect, planar triangular nucleus, is the prototypical example. Our model suggests that its exceptional geometric stability underlies its uniquely strong and versatile bonds, making it the backbone of life. Similarly, Oxygen-16 and Calcium-40 should also exhibit anomalous chemical stability in the bonds they form, beyond what is predicted by their electron configurations alone. This opens a new research program: the search for subtle, measurable anomalies in bond energies that correlate not with electron shells, but with the proposed geometric perfection of the nucleus.

Altogether, these predictions provide a clear and robust experimental program to test the core tenets of the RA, and their validation would provide compelling evidence for the tangible, resonant reality proposed in this framework.

7. Conclusion

This chapter has put forth a comprehensive, physically intuitive foundation for the science of chemistry, proposing that its principles are not abstract axioms but the emergent consequences of a deeper, geometric reality. Building on the RFI and QVD, we have generated a model of the RA from the inside out, replacing mathematical abstraction with a tangible, mechanistic framework at every level.

Our journey began at the atom's core, where the ASI provided a physical explanation for nuclear stability, grounding the "magic numbers" in the optimal geometric packing of hyperstable alpha-particle vortices. With this structured UEF nucleus defined, we demonstrated that atomic orbitals are real, physical standing waves: the stable harmonic resonances of the ULF. Crucially, this framework establishes a direct link between these two domains, proposing that the geometry of the UEF nucleus dictates the shape, orientation, and energy of the ULF orbitals that form around it, analogous to how the shape of a bell determines the notes it can play.

With this complete physical model of the atom, the foundational rules of chemistry emerge as necessary consequences of its dynamics, and the distinct properties of the elements across the Periodic Table are explained as direct reflections of the geometry and stability of their valence waves. Ultimately, by positing that the atom is a resonant system in which nuclear geometry governs structure, we offer a cohesive framework in which the properties of matter are a direct reflection of that geometry.

8.3 A New Foundation for Chemistry

In this chapter, we presented a physical, field-based model of the Resonant Atom (RA) that reconstructs the foundations of chemistry from the inside out. By grounding nuclear and electronic behavior in the geometry of a resonant universe, this framework replaces the abstract formalism of the Standard Model with a tangible, deterministic, and visualizable ontology.

The core thesis is that the principles of chemistry are not axiomatic rules but are emergent, geometric consequences of two coupled fields: the Universal Energetic Field (UEF), which governs nuclear structure, and the Universal Light Field (ULF), which sustains the standing waves we know as atomic orbitals. From this single framework, a complete picture of the atom emerges:

- A Geometric Foundation for the Nucleus: Nuclear stability is explained through the Alpha Stability Index (ASI), which reveals that the "magic numbers" of nuclear physics are the completion of perfect, geometric shells of alpha-particle vortices.

- A Physical Model for Atomic Orbitals: Orbitals are reinterpreted as real, three-dimensional ULF standing waves whose specific shapes and energies are the natural harmonic overtones that form around the geometric nucleus.

- A Physical Origin for the Rules of Chemistry: The foundational principles, such as the Aufbau Principle and the Pauli Exclusion Principle, are recast as necessary outcomes of hydrodynamics and a system seeking maximum resonant stability.

- A Unified Explanation for the Periodic Table: The entire structure of the Periodic Table, with its distinct chemical properties and periodic trends, emerges as a direct and predictable map of the stable, geometric standing wave solutions available in the ULF.

The RA model is more than merely conceptual, as it is grounded in falsifiable predictions, such as the Geometric Stark Effect and the Splitting of Orbital Degeneracy. Each of these effects offers a direct way to confirm or refute the proposed nuclear-orbital connection.

If validated, this model could transform chemistry as both a science and technology. In education, it promises to replace rote memorization of orbital diagrams and exclusion rules with a visual and intuitive picture, making chemistry as conceptually grounded as classical mechanics. Regarding research, it suggests a paradigm in which materials and catalysts can be designed from first principles by shaping the ULF field geometry to stabilize desired states or transitions.

Perhaps most significantly, the framework proposes that an element's chemical properties are subtly tuned by its nuclear geometry rather than by just its electron count, suggesting a new frontier in chemistry where isotopes could be selected or engineered to alter the behavior of materials, thereby forging a true bridge between nuclear physics and chemical science.

In sum, the RA framework offers more than a reinterpretation of the atom. It is a roadmap to a new kind of chemistry—one that is more intuitive, predictive, and unified with the fundamental principles of physics. By grounding the atom in resonance and geometry, it points toward a future where both our understanding and manipulation of matter are guided by the same elegant universal principles that shape the rest of nature.

References

1. Audi, G., Wapstra, A. H., & Thibault, C. (2003). The Nubase evaluation of nuclear and decay properties. *Nuclear Physics A*, 729(1), 3-128.
2. Chladni, E. (1787). *Entdeckungen über die Theorie des Klanges* [Discoveries Concerning the Theory of Sound]. Weidmanns Erben und Reich.
3. Freer, M., Horiuchi, H., Funaki, Y., Ogawa, Y., & Kanada-En'yo, Y. (2018). Microscopic clustering in light nuclei. *Reviews of Modern Physics*, 90(3), 035004.

4. Hund, F. (1925). Zur Deutung verwickelter Spektren, insbesondere der Elemente Scandium bis Nickel [On the interpretation of complex spectra, in particular the elements scandium to nickel]. Zeitschrift für Physik, 33(1), 345–371.

5. Krane, K. S. (1988). *Introductory Nuclear Physics*. John Wiley & Sons.

6. Lavoisier, A. (1789). Traité Élémentaire de Chimie. Cuchet, Paris.

7. Mendeleev, D. (1869). On the Relationship of the Properties of the Elements to their Atomic Weights. Zeitschrift für Chemie, 12, 405-406.

8. Pauli, W. (1925). Über den Zusammenhang des Abschlusses der Elektronengruppen im Atom mit der Komplexstruktur der Spektren [On the connection of the closing of the electron groups in the atom with the complex structure of the spectra]. Zeitschrift für Physik, 31(1), 765–783.

9. Rutherford, E. (1911). The scattering of α and β particles by matter and the structure of the atom. The London, Edinburgh, and Dublin Philosophical Magazine and Journal of Science, 21(125), 669–688.

10. Schrödinger, E. (1926). An undulatory theory of the mechanics of atoms and molecules. Physical Review, 28(6), 1049–1070.

11. Stark, J. (1914). Beobachtungen über den Effekt des elektrischen Feldes auf Spektrallinien [Observations about the effect of the electric field on spectral lines]. Annalen der Physik, 43(12), 965–982.

12. Vowles, R. V. G. V. (1969). *The Structured Atom Model*.

Chapter 9: The Resonance Dividend

9.1 Chemical Bonds

Since the earliest days of inquiry, humanity has been captivated by a fundamental question: how does the world hold together? The ancient alchemists dreamed of transmutation, believing that a hidden key could unlock the secrets of matter that would allow them to transform one substance into another. While their methods were mystical, their core pursuit was scientific. They were searching for the forces that bind atoms together into the vast and varied materials of our universe. That force is the chemical bond, and understanding its true nature remains the central purpose of chemistry.

Modern science has replaced the alchemist's furnace with the precise and powerful framework of quantum mechanics. This theory, developed in the early 20th century, is one of the most successful scientific achievements in history. It describes the behavior of atoms and their constituent parts with remarkable accuracy. According to this view, the chemical bond is not a physical hook or clasp but an energetic and probabilistic phenomenon governed by the interactions of electrons.

One of the primary theories describing this process is valence bond theory (Pauling, 1960), which states that a bond forms when the atomic orbitals of two separate atoms overlap. An electron from each atom then pairs up in this overlapping region, forming a new, shared molecular orbital that binds the two atoms. A more comprehensive model, molecular orbital theory, proposes that when atoms approach each other, their atomic orbitals combine to form a new set of molecular orbitals that span the entire molecule. In both models, the bond is essentially an energetic transaction. A stable bond forms because the resulting molecule exists at a lower total energy state than the separate atoms did. The properties of these bonds, such as their strength, length, and angle, can be calculated with incredible precision by solving the complex equations of quantum mechanics.

For all its predictive success, the quantum-mechanical picture of the chemical bond remains deeply abstract. It provides the mathematical *how* but struggles to offer an intuitive physical *why*. We are taught to accept that electrons exist in fuzzy "probability clouds" and that they follow axiomatic rules, like the Pauli Exclusion Principle, for which no simple physical mechanism is given. We are told that bonds form because the final state is "more stable," but the physical nature of that stability is hidden within layers of complex mathematics.

The alchemist's old dream was not just to manipulate matter, but to *understand* it. What if there is a deeper, more tangible reality hiding beneath the quantum math? What if the stability of a chemical bond is not just an energetic outcome but the result of a physical and geometric principle?

These are the questions this chapter seeks to answer. Building upon our work from the preceding chapters, we propose a new foundation for the chemical bond. In this view, the bond is the physical merging of resonant standing waves into a new, more harmonious configuration. The energy released in this process, which we term the Resonance Dividend (the molecular version of the Coherence Dividend), is the direct result of the system "snapping" into a more coherent geometry. In this chapter, we explore how this intuitive principle can provide a physical basis for all types of chemical bonds, offering a new path toward understanding the architecture of our world.

9.2 The Resonance Dividend: The Geometry of Chemical Bonds

Abstract

This chapter presents a new physical and quantitative model for chemical bonding based on the concept of the Resonant Atom (RA), which we refer to as Resonant Chemistry (RC). Here, we propose that a chemical bond is not merely the sharing or transfer of electrons but the physical merging of Universal Light Field (ULF) standing waves into a new, more stable molecular resonance. Our Resonance Dividend Index (RDI) formalizes this principle by combining two critical levels of stabilization: nuclear congruence (the geometric compatibility of interacting UEF nuclei) and orbital overlap (the hydrodynamic stability of merged ULF standing waves). The model quantifies the additional energy released beyond baseline quantum predictions as a measurable Resonance Dividend that determines the ultimate bond strength.

Using the carbon-carbon bond as a test case, the RDI predicts that C-12–C-12 bonds, whose triangular nuclei align with near-perfect congruence, will exhibit a greater Resonance Dividend than C-13–C-13 bonds, where an extra neutron disrupts this symmetry. This leads to a clear, falsifiable prediction: a measurable vibrational frequency shift of 0.3–0.6% in the C–C bond between these isotopes, which offers a definitive experimental test of the nucleus's geometric influence on chemistry. By linking nuclear structure and electronic resonance into a unified framework, RC replaces the abstract rules of bonding with a tangible and predictive physical model.

1. Introduction

Chemical bonding is the cornerstone of chemistry and biology. Despite that, its quantum mechanical description remains abstract and complex. Conventional theories, such as valence bond theory and molecular orbital theory, can predict bond energies and molecular geometries (Pauling, 1960). However, they treat atomic nuclei as static point charges and orbitals as probabilistic distributions, offering no apparent physical reason why shared orbitals confer stability beyond energy bookkeeping. These models succeed in calculations, but they leave unaddressed the deeper question: What physically makes a bond stable?

This chapter proposes a tangible answer that builds on our model of the Resonant Atom (RA), which we call Resonant Chemistry (RC). In RC, atoms are viewed as composite structures of two fundamental fields: the Universal Energetic Field (UEF), which governs nuclear geometry, and the Universal Light Field (ULF), which forms electron standing waves. Within this framework, we introduce the Resonance Dividend Index (RDI), a quantitative tool for understanding bond stability. The RDI formalizes chemical bonds as physical mergers of ULF standing waves into coherent molecular resonances, whose stability is jointly determined by:

1. Nuclear Congruence (UEF) – the degree to which the geometric structures of the interacting nuclei align to form a coherent, low-strain foundation.

2. Orbital Overlap (ULF) – the hydrodynamic stability achieved when atomic electron vortices merge into a unified standing wave.

By explicitly quantifying the extra stabilization energy beyond baseline quantum mechanical predictions, the RDI unites nuclear structure and orbital resonance into a single, predictive energy framework. This model not only explains why bonds vary in strength across elements and isotopes but also predicts measurable effects, such as subtle vibrational frequency shifts between C-12 and C-13 carbon-carbon bonds, arising from differences in nuclear geometry. In RC, chemical bonding emerges as a physically grounded, testable phenomenon governed by resonance and geometry.

2. The Covalent Bond: A Two-Level Resonance

The most common and fundamental chemical bond is the covalent bond, which holds molecules together through the sharing of electrons between atoms. The modern quantum-mechanical picture, as described by valence bond theory (Pauling, 1960), is that the electron orbitals of individual atoms directly overlap to form a new, combined "molecular orbital" at a lower energy.

RC provides a direct physical representation of this process, revealing that the stability of this bond is a two-level geometric process governed by what we term the Principle of Geometric Congruence.

2.1 Level 1: Nuclear Congruence (The UEF Foundation)

In the Standard Model of Chemistry, the nucleus is treated almost as an afterthought in chemical bonding. According to the Born-Oppenheimer approximation, which underlies most quantum-chemical calculations, the massive nucleus is treated as a stationary, positive point charge (Born & Oppenheimer, 1927). The chemical bond is then determined entirely by the behavior of the much lighter electrons that orbit it. In this model, the specific structure or geometry of the nucleus is considered irrelevant to the strength of a chemical bond.

The RC framework proposes a radical departure from this view. We contend that, at a deeper level, the stability of the shared ULF resonance is fundamentally dependent on the geometric compatibility of the interacting nuclei themselves. A chemical bond will be strong and stable if the underlying geometries of the two bonding UEF vortices can "mesh" or align in a way that creates a new, coherent, combined nuclear structure. This stable nuclear arrangement, in turn, provides a more stable energetic foundation for the shared ULF molecular orbital that forms around it.

Thus, through the Principle of Geometric Coherence, RC introduces a new, first-level approach to the chemical bond, where the geometry of the nucleus is a relevant factor in determining the bond's ultimate strength and stability.

2.2 Level 2: Orbital Resonance (The ULF Superstructure)

The covalent bond is the fundamental connection that holds most molecules together, from the simple structure of a water molecule to the vast, complex architecture of DNA. In classical chemistry, it is understood as the act of two atoms sharing one or more electrons to achieve a stable configuration. This sharing creates a powerful attraction that binds the atoms, forming the strong, directional links that give molecules their specific, three-dimensional shapes.

RC provides the deeper, physical mechanism for this sharing. A covalent bond is formed when two atomic orbitals, understood as real, physical standing waves in the ULF, merge to form a new, larger, but highly stable shared wave pattern that encompasses both nuclei. This shared resonance *is* the molecular orbital. The energy released when the system "snaps" into this more coherent configuration is the Resonance Dividend (the molecular version of the Coherence Dividend), which is directly proportional to the stability of the final, merged standing wave.

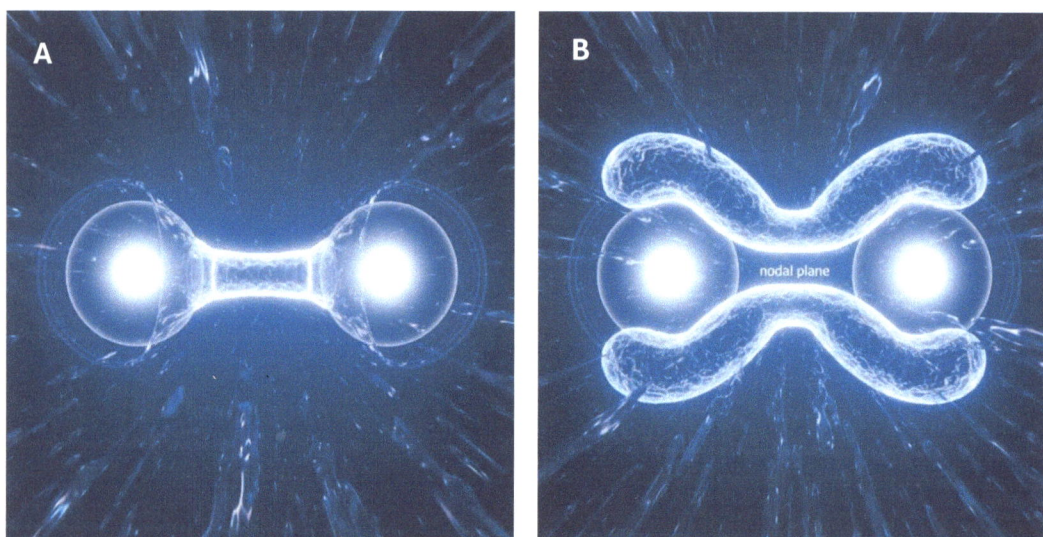

Figure 24: Covalent Bonding in Resonant Chemistry.
(A) Formation of a Sigma (σ) Bond: This figure illustrates the formation of a sigma bond through the direct, end-to-end overlap of two atomic orbitals within the ULF. The ULF forms a single, cylindrically symmetrical standing wave, creating a region of high energy density concentrated along the internuclear axis, representing a strong, direct bond.
(B) Formation of a Pi (π) Bond: This figure illustrates the formation of a pi bond through the sideways overlap of two p-orbitals. The ULF forms two distinct, parallel regions of high energy density—one above and one below the internuclear axis—with a nodal plane of zero density passing through the axis.

The geometry of these shared resonances then explains the different types of covalent bonds. A sigma (σ) bond results from the direct, end-to-end overlap of atomic orbitals, forming the most stable shared resonance along the axis connecting the two nuclei. A pi (π) bond, which forms double and triple bonds, arises from the weaker, sideways overlap of parallel p-orbitals (Figure 24).

2.3 Case Study: Carbon vs. Silicon — The Architecture of Life

The Principle of Geometric Congruence provides a direct, physical explanation for one of the most fundamental questions in biology: why is carbon, and not its abundant chemical neighbor silicon, the undisputed backbone of life? Although both elements can form four covalent bonds, the RC framework reveals that carbon is a masterpiece of multiscale coherence, whereas silicon is a study in geometric strain.

A simple analogy is the construction of a suspension bridge. Carbon-12 provides two perfect, granite anchorages (UEF) that are perfectly aligned, allowing for the creation of a powerful, high-tension steel cable (ULF) (Figure 25A).

- Level 1 (Nuclear Congruence): As we established in Chapter 8, the Carbon-12 nucleus is a perfect, hyper-stable, planar triangle of three alpha vortices. This aligns with findings from the established Alpha Cluster Model in nuclear physics, which treats C-12 as a stable, three-alpha-particle structure (Brink et al., 1966). When two C-12 atoms bond, their flat, symmetrical nuclei can "mesh" with an unparalleled degree of geometric congruence, creating a powerful and stable UEF foundation.

- Level 2 (Orbital Resonance): Building on this perfect foundation, the compact 2p orbitals (the ULF waves) can overlap with maximum efficiency. This perfect two-level alignment (UEF + ULF) releases a massive Resonance Dividend, resulting in the exceptionally strong and stable C-C bond.

Silicon-28, in contrast, provides two crumbly, asymmetrical sandstone anchorages (UEF) that are difficult to align; therefore, they can only support a weak, low-tension rope (ULF) (Figure 25B).

- Level 1 (Nuclear Incoherence): Silicon's most common isotope is Si-28, which is composed of seven alpha-particle "bricks." From a geometric perspective, this is an architectural nightmare. Seven is a prime number that cannot be arranged into a simple, perfect, symmetrical structure. The Si-28 nucleus is, by its very nature, geometrically strained and far less coherent than the perfect C-12 triangle.

- Level 2 (Orbital Incoherence): This weak, strained nuclear foundation must then support larger, more diffuse 3p orbitals. These "puffier" ULF waves cannot overlap as efficiently as carbon's compact 2p orbitals.

This two-level failure means that the Si-Si bond releases a negligible Resonance Dividend, making it fundamentally weaker. This weakness has a catastrophic, world-defining consequence, a fact well established in inorganic chemistry: unlike carbon, silicon cannot form stable double bonds. Its oxidation product, SiO_2 (silica), is not a gas like CO_2. It is a rigid, immobile covalent network solid (a rock). A silicon-based lifeform, in an oxygen environment, would be metabolically impossible, as it would be unable to transport or excrete its own waste (Atkins & de Paula, 2014).

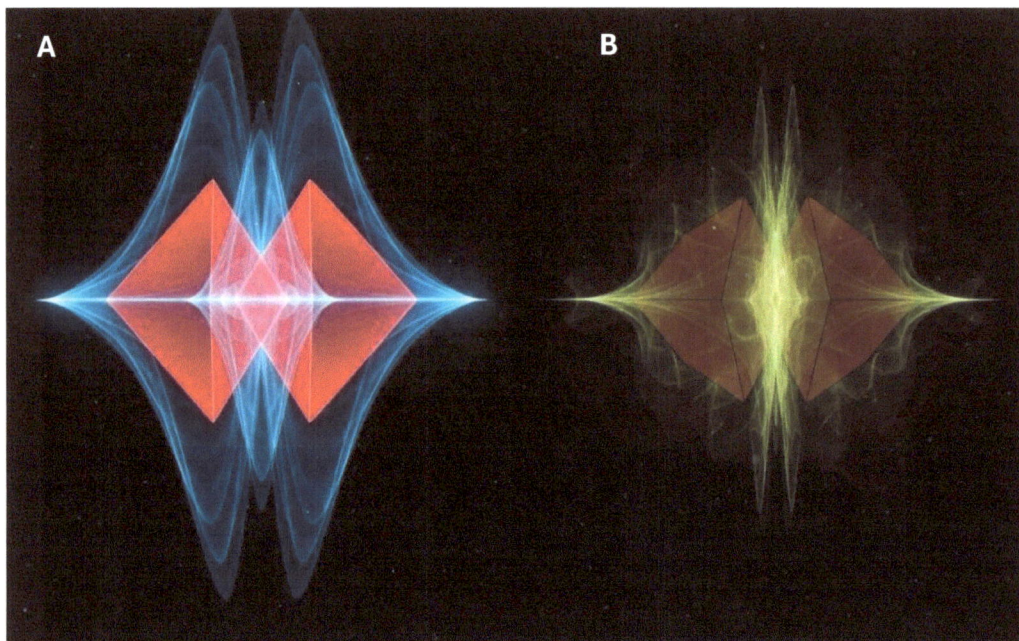

Figure 25: Carbon vs. Silicon — A Tale of Two-Level Coherence. This figure contrasts the geometric architecture of a Carbon-Carbon bond with that of a Silicon-Silicon bond, illustrating UFD's two-level explanation for why carbon, not silicon, forms the basis of life.
(A) Carbon-Carbon Bond: The two Carbon-12 nuclei (red, representing UEF vortices) align perfectly, showcasing maximal nuclear congruence. This provides an exceptionally stable foundation for the compact, highly concentrated ULF standing wave (blue) that forms the shared covalent bond. This two-level coherence results in immense stability.
(B) Silicon-Silicon Bond: The larger, more complex Silicon-28 nuclei (orange/brown) display inherent nuclear incoherence due to their strained alpha-particle geometry, which prevents optimal alignment. This weak foundation supports a more diffuse and less stable ULF standing wave (yellow/green), indicative of poor orbital overlap. This two-level incoherence results in a significantly less versatile bond, explaining why silicon fails as a chassis for complex life.

Thus, in RC, it is this unique two-level architectural perfection, born from the perfect geometric congruence of its nucleus and its orbitals, that makes carbon the undisputed element of life. While this case illustrates the principle qualitatively, we formalize this finding with the RDI in Section 5 of this chapter.

3. Alternative Bonding Strategies: Ionic and Metallic Bonds

While the covalent bond represents an ideal form of shared resonance between two atoms, this is not always the most stable configuration available to a system. Nature, in its relentless pursuit of the lowest possible energy state, has two other primary strategies for achieving stability: the complete transfer of a resonant wave from one atom to another, and the complete delocalization of these waves across an entire crystal lattice. These two mechanisms, which give rise to the ionic and metallic bonds, can be understood as occupying the opposite ends of a spectrum of electron behavior, with the covalent bond residing in the middle. This section will now explore these alternative paths to resonant stability.

3.1 The Ionic Bond: A Transfer of Resonant Stability

The second major type of primary chemical bond is the ionic bond, which typically forms between a metal that tends to lose electrons and a nonmetal that tends to gain them. The Standard Model describes this as the complete transfer of one or more electrons from one atom to another, creating oppositely charged ions whose electrostatic attraction forms the bond. Our model reframes this process as a system seeking its lowest possible energy state by transferring resonant stability. The interaction occurs when one atomic system is "geometrically unstable," with a loosely bound electron vortex in a high-energy orbital (like a Sodium atom), and the other atom has a stable, lower-energy orbital that is "unoccupied" (like a Chlorine atom).

The system as a whole can achieve a much lower total energy state when the electron vortex physically abandons the unstable, high-energy resonance of the first atom and relocalizes within the more stable, lower-energy resonance of the second atom. This physical re-localization *is* the electron transfer (Figure 26A).

The consequence of this transfer is the creation of two ions. The atom that lost the electron vortex now has an excess of "source" potential from its nucleus and becomes a cation. The atom that gained the electron vortex now has an excess "sink" potential and becomes an anion. These two newly formed ions are then drawn together by the fluid-dynamic electrostatic force—the flow from source to sink—that we established in Chapter 6. The strength of the resulting ionic bond is determined by the net Resonance Dividend released when the electron settles into its new, more stable state.

225

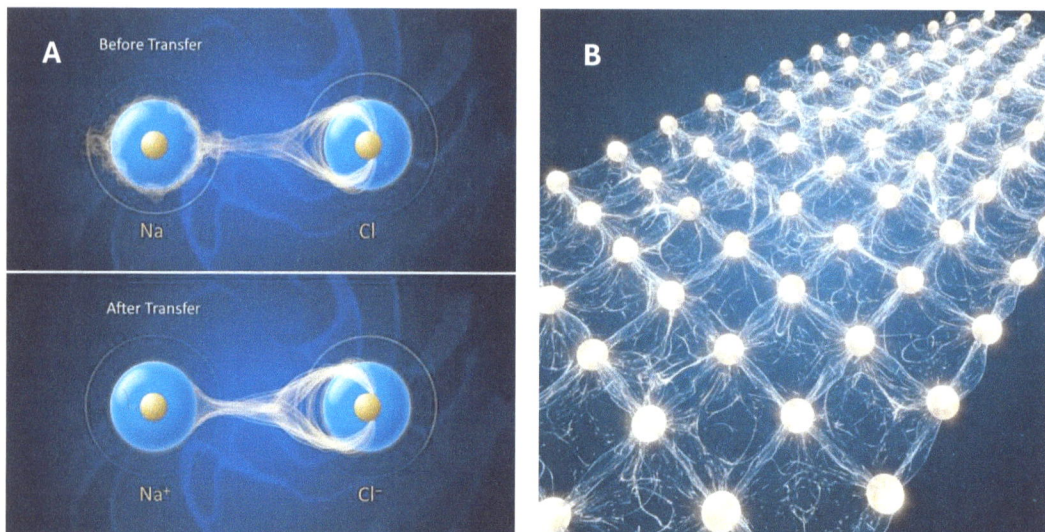

Figure 26: Ionic and Metallic Bonds. This figure depicts the formation of ionic and metallic bonds in RC.

(A) Formation of an Ionic Bond. This two-panel figure illustrates ionic bond formation within the ULF. (Top) Before transfer, a Sodium atom's diffuse outermost orbital—a loosely held ULF standing wave—is ready for release. A Chlorine atom's orbital shows a vacancy, depicted as an inviting ULF vortex pull. (Bottom) After transfer, the electron's orbital from Sodium has seamlessly integrated into Chlorine, forming a stable, closed-shell configuration. The resulting Na+ and Cl- ions are held by a strong, diffuse electrostatic attraction mediated by the harmoniously flowing ULF.

(B) The Metallic Bond as a Delocalized ULF Standing Wave. This figure depicts the metallic bond as a state of profound collective coherence within the ULF. Positively charged atomic nuclei are arranged in a repeating lattice. Instead of individual atomic orbitals, the valence electrons form a vast, continuous, and highly energetic 'sea'—a single, massive, delocalized ULF standing wave that encompasses the entire metallic structure. This collective resonant field mediates the powerful attraction between the nuclei, illustrating how metallic stability arises from a maximized, communal resonance dividend achieved through widespread ULF coherence.

3.2 The Metallic Bond: A Sea of Shared Resonance

The third type of primary chemical bond explains the unique properties of metals. In the Standard Model, the metallic bond is described as a lattice of positive metal ions surrounded by a "sea of delocalized electrons." These electrons are not associated with any single atom but are free to move throughout the entire crystal, which explains the high electrical and thermal conductivities of metals. RC provides a direct, physical picture that describes *why* this "sea" forms. It arises from two key conditions inherent to most metallic elements:

226

1. Efficient Crystal Lattices: Metal atoms tend to arrange themselves in very tightly packed crystal structures, which minimizes the distance between the atomic cores.

2. Diffuse Valence Orbitals: Metal atoms are characterized by having low ionization energies, meaning they do not hold their outermost electrons tightly. Their valence orbitals (especially the d- and f-orbitals in transition metals) are large and diffuse, extending far from the nucleus.

When these atoms are packed closely together, their large, diffuse ULF standing waves (their valence orbitals) inevitably and extensively overlap with those of all their neighbors. No single nucleus has a strong enough claim on its own electron vortex, which causes the individual orbitals to lose their independence and merge into a single, continuous, crystal-wide standing wave. This creates the vast "sea of shared ULF resonance" (Figure 26B).

This "sea of resonance" explains the characteristic properties of metals. Metals are highly conductive because their electron vortices can flow without impediment. They have luster because the surface of this electron sea can easily absorb and reemit photons, and they are malleable because their atomic cores can slide past one another without breaking the cohesive, fluid-like nature of the shared resonance.

In summary, the metallic bond is an emergent state of matter, arising when the unique properties of metal atoms allow their individual ULF resonances to dissolve into a single, crystal-wide "sea of shared resonance." This single concept provides a direct physical origin for the full suite of metallic properties, from conductivity to malleability (*see* Chapter 10).

4. The Secondary Interactions: Intermolecular Forces

While the primary chemical bonds create stable molecules, a second, weaker class of interactions governs how these molecules attract and arrange themselves. These intermolecular forces are responsible for the physical states of matter—determining whether a substance is a gas, liquid, or solid—and properties such as boiling points (*see* Chapter 10). In RC, these forces arise from the subtle residual effects of the ULF resonances that constitute the molecules.

4.1 Van der Waals Forces: The Resonance of Fluctuations

The weakest of these interactions are Van der Waals forces, specifically London dispersion forces (London, 1930). The standard explanation is that the random motion of electrons creates temporary, fluctuating dipoles that induce corresponding dipoles in neighboring atoms, resulting in a fleeting attraction.

RC provides a physical representation of Van der Waals forces (Figure 27A). The electron's standing wave in the ULF is not perfectly static but is a dynamic, vibrating resonance with natural, tiny fluctuations in its intensity and shape. These momentary fluctuations create transient, weak pressure gradients in the ULF that can induce a corresponding, opposite fluctuation in the electron wave of a neighboring atom. This creates a weak, fleeting attraction, thus replacing the abstract "randomness" of the Standard Model with the deterministic physics of a vibrating field. The two atoms are, in essence, vibrating in a loosely coupled harmony.

4.2 The Hydrogen Bond: A Polarized Resonance

The hydrogen bond is a powerful intermolecular force that is essential to the structure of water and the molecules of life, such as DNA. In the Standard Model, it is described as a uniquely powerful form of dipole-dipole interaction. While this electrostatic model is highly successful, it still describes the interaction as an abstract "attraction" without providing a deeper, physical picture of the underlying field dynamics. RC provides a clear geometric explanation (Figure 27B).

In a polar molecule like water (H_2O), the covalent bonds are asymmetrical. The shared ULF resonance is drawn more strongly toward the oxygen atom, giving it a permanent "sink" character. This leaves the hydrogen atoms' protons barely shielded, making them potent, localized "source" points. The hydrogen bond is thus a strong electrostatic attraction between the "source" of a hydrogen atom on one molecule and the "sink" of the oxygen atom on a neighboring molecule. It is a powerful, directional interaction that arises from the polarized geometry of the covalent bond's standing wave.

Thus, from the fleeting attraction of fluctuating resonances to the powerful pull of polarized ones, the weaker intermolecular forces are explained as dynamic effects within the ULF. This completes our physical picture of how atoms and molecules arrange themselves, setting the stage for synthesizing these concepts into a unified principle of stability.

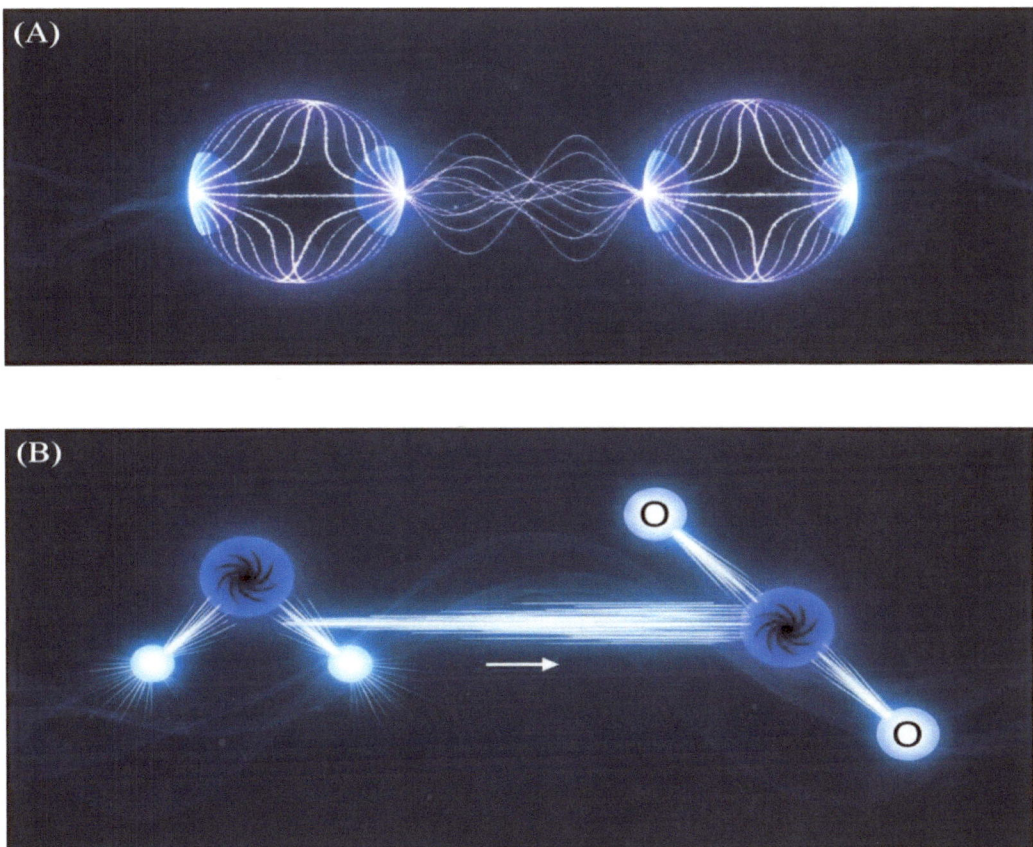

Figure 27: The Secondary Interactions (Intermolecular Forces).
(A) Van der Waals Force: This panel illustrates how weak attraction can arise through shared fluctuations in the ULF. The two spherical standing-wave patterns represent neutral atomic resonances. A temporary disturbance in the ULF around the first atom (left) slightly shifts its internal balance. This disturbance propagates through the surrounding field and induces a matching response in the neighboring atom (right). The result is a faint, oscillatory coupling between the two localized field patterns, forming a low-strength, non-directional attraction without permanent polarization or bonding.
(B) Hydrogen Bond: This panel shows how a stronger, directional interaction emerges in polar molecules such as water. Because the electron distribution in water is uneven, the hydrogen nuclei are only weakly shielded and act as localized sources of field intensity, while the oxygen center functions as a strong sink. This asymmetry creates a preferred direction for field flow between neighboring molecules. The interaction is a stabilized, directional linkage—visualized here as a smooth, coherent flow—that gives hydrogen bonds their characteristic strength, orientation, and structural role in liquids and biological systems.

5. The Resonance Dividend Index (RDI): A Quantitative Framework

To transform the Principle of Geometric Congruence into a predictive tool, we introduce the RDI, a quantitative measure of the energy released when two atoms form a

229

bond by merging their resonant fields into a more stable configuration. The RDI captures the two-level structure of the bond—its nuclear congruence (UEF) and orbital overlap (ULF)—into a single energy model.

5.1 Computing the RDI

At its core, the RDI is a quantitative framework for chemical bonding. It states that the total bond energy (E_{total}) is the standard quantum mechanical baseline (E_{base}) multiplied by a geometric correction factor:

$$E_{total} = E_{base} \cdot [(c_n \cdot G_n) + (c_o \cdot O_o)]$$

Where:

- E_{base} is the baseline bond energy predicted by standard quantum chemical models (e.g., 348 kJ/mol for a C-C single bond).

- G_n is the Nuclear Congruence Factor (0–1), measuring how well the two nuclei "mesh" geometrically in the UEF.

- O_o is the Orbital Overlap Factor (0–1), measuring how stably the ULF standing waves (orbitals) combine.

- c_n and c_o are weighting coefficients (which sum to 1.0) for the nuclear and orbital contributions.

In this framework, the Resonance Dividend (E_{RDI}) is the resulting energy difference between the true energy and the baseline, calculated after the fact:

$$E_{RDI} = E_{total} - E_{base}$$

A large positive E_{RDI} corresponds to synergistic bonds (high coherence), while a negative E_{RDI} indicates geometric strain.

The orbital overlap factor, O_o, reinterprets the standard molecular orbital overlap integral as a physical measure of fluid coherence in the ULF.

- Strong, localized sigma (σ) bonds, where ULF standing waves (lobes) are aligned directly along the bond axis, achieve a high coherence of $O_o \approx 0.8$–1.0.

- Weaker, delocalized pi (π) bonds, where coherence is distributed perpendicular to the bond axis, yield a lower coherence of $O_o \approx 0.4$–0.6.[52]

This model provides a physical explanation for why specific molecular geometries (such as the 109.5° angle in sp³ hybridization) are preferred. These precise angles are not arbitrary quantum rules but are rather the optimal, lowest-energy configurations that maximize the constructive interference of the ULF standing waves.

The nuclear congruence factor, G_n, is the deeper, overlooked factor. It quantifies the geometric alignment of the UEF vortices (nuclei) themselves. A bond may have perfect orbital overlap (high O_o), but if the underlying nuclei are asymmetric or "misaligned" (low G_n), the bond will suffer a "geometric energy deficit." This explains why strained molecules (like cyclopropane) are so unstable: their forced 60° bond angles create both poor orbital overlap (a low O_o) and poor nuclear alignment (a low G_n), resulting in a massive coherence deficit.

The RDI model thus allows us to systematically quantify how this deeper geometric resonance enhances bond stability. By incorporating both nuclear and electronic coherence into a single expression, it becomes a predictive tool for identifying highly stable (or unstable) molecular configurations.

5.2 A Qualitative Application of the RDI: The Carbon-Silicon Divide

Before we apply the RDI to a high-precision isotopic test, it is useful to see how the formula provides a direct, "first principles" explanation for one of the most fundamental questions in all of science: why is Carbon, and not Silicon, the backbone of life?

While both elements can form four bonds, the UFD framework reveals that their bond stability is profoundly different, a fact the RDI formula can now explain. The question comes down to the C=O double bond (in CO_2) versus the Si=O double bond (in SiO_2).

1. The Carbon-Oxygen (C=O) Bond:

- Nuclear Congruence (G_n): We are meshing the C-12 nucleus (a perfect, planar triangle of 3 alpha-bricks) with the O-16 nucleus (a perfect tetrahedron of 4

---

[52] These numerical ranges reflect standard results from molecular orbital theory, which RC reinterprets as measures of fluid-dynamic coherence in the ULF (see, e.g., Atkins & Friedman, 2011).

alpha-bricks). These are two of the most geometrically coherent "magic" nuclei in existence. Their ability to align and form a stable UEF foundation is exceptionally high.

- Result: G_n is very high.

- Orbital Resonance (O_o): We are overlapping Carbon's compact 2p orbitals with Oxygen's compact 2p orbitals. These orbitals are at the same energy level, are similar in size, and can overlap with maximum efficiency to form a stable pi bond.

- Result: O_o is very high.

A high C_n multiplied by a high O_o results in a massive Resonance Dividend, making the C=O double bond exceptionally strong and stable. This allows CO_2 to exist as a small, stable, and transportable gas: the perfect medium for metabolic waste.

2. The Silicon-Oxygen (Si=O) Bond:

- Nuclear Congruence (G_n): We are meshing the Si-28 nucleus (an "architectural nightmare" of 7 alpha-bricks) with the O-16 nucleus (a perfect tetrahedron). A prime number of bricks (7) cannot form a simple, perfect, symmetrical structure. This strained, geometrically awkward nucleus cannot mesh cleanly with the perfect O-16.

- Result: G_n is very low.

- Orbital Resonance (O_o): We are trying to overlap Silicon's large, diffuse 3p orbitals with Oxygen's much smaller, more compact 2p orbitals. These differently sized and -energized orbitals cannot overlap efficiently.

- Result: O_o is very low.

A very low G_n multiplied by a very low O_o results in a negligible (or even negative) Resonance Dividend. This makes the Si=O pi bond so weak and unstable that it effectively cannot form, forcing silicon to form a rigid, single-bond crystal lattice (SiO_2) — a rock.

This qualitative application of the RDI formula demonstrates its explanatory power. It provides a direct, physical, and two-level reason why carbon-based life is possible and silicon-based life is not. We can now apply this same tool to a more subtle, high-precision test.

5.3 Case Study: C–C Bonds (C-12 vs. C-13)

To demonstrate the predictive power of the RDI framework, we apply it to the carbon-carbon single bond, comparing the stable isotopes C-12 and C-13.

These two isotopes offer an ideal test case. Both form chemically equivalent C-C sigma bonds but differ in nuclear structure. C-12 has a highly symmetric, planar nuclear geometry. C-13 introduces an asymmetry via an extra neutron, disrupting this perfect geometric coherence. The RDI predicts that this subtle nuclear geometric difference will result in a measurable difference in bond energy.

Using the RDI formula, $E_{total} = E_{base} \cdot [(c_n \cdot G_n) + (c_o \cdot O_o)]$, we can calculate this difference.

1. We use the experimental C-C bond energy as our baseline: $E_{base} \approx 348\ kj/mol$ (Pauling, 1960).

2. We assume equal weighting for this example: $c_n = 0.5$ and $c_o = 0.5$.

3. Both are sigma bonds, so we set the orbital factor to its ideal value: $O_o = 1$.

4. For C-12–C-12 (The "Perfect" Bond): We calibrate our model by assuming the C-12 nucleus is the ideal geometric structure, so $G_n \approx 1.0$.

$$E_{total(C12)} = 348\,\text{kJ/mol} \cdot [(0.5 \cdot 1) + (0.5 \cdot 1)]$$

$$E_{total(C12)} = 348 \cdot 1 = 348\,\text{kJ/mol}$$

5. For C-13–C-13 (The "Imperfect" Bond): The *only* change is the nuclear congruence, which is now slightly imperfect. We model this as a conservative illustrative 1% loss of congruence: $G_n \approx 0.990$.

$$E_{total(C13)} = 348\,\text{kJ/mol} \cdot [(0.5 \cdot 0.990) + (0.5 \cdot 1)]$$

$$E_{total(C13)} = 348 \cdot [0.495 + 0.5]$$

$$E_{total(C13)} = 346.26\,\text{kJ/mol}$$

6. Prediction: The RDI thus predicts that the C-13–C-13 bond is ≈ 1.74 kJ/mol weaker than the C-12–C-12 bond. This represents an energy difference of 0.5% ($1.74/348 \approx 0.005$).

The RDI model, therefore, makes a clear, falsifiable prediction: after accounting for the primary frequency shift caused by the mass difference (the KIE),[53] there will be a small, residual, and currently unexplained anomaly of 0.3–0.6% in the bond energy, depending on the precise nuclear congruence factor. This residual shift is the direct, measurable signature of the Resonance Dividend.

The successful detection of this anomalous frequency shift would therefore be a profound confirmation of UFD's core principles, providing the first direct, experimental evidence of the deep resonant coupling between the nuclear (UEF) and electronic (ULF) layers of reality.

6. Fire: The Resonance Dividend Unleashed

The RDI provides a tool for calculating the subtle energy differences in a single chemical bond. Here, we demonstrate that this is not just an abstract principle, as its effects can be witnessed on a grand, yet everyday scale. We propose that fire, one of the most fundamental phenomena in nature, is the visceral, macroscopic manifestation of the Resonance Dividend being released in a rapid, self-sustaining cascade.

In the RC framework, fire is not a substance; it is a process of resonant decoherence. The fuel, such as a log of wood, is a complex organic structure held together by high-energy, geometrically strained ULF resonant couplings (chemical bonds). The initial spark provides the activation energy—a burst of incoherent vibration—just enough to break the weakest of these strained bonds. The moment the first bond breaks, the molecule begins to collapse into a more stable, lower-energy geometric configuration (e.g., CO_2 or water). This collapse releases a powerful Resonance Dividend in the form of more incoherent vibration (heat) and coherent, propagating ULF waves (light) (Figure 28).

---

[53]In chemistry, the Kinetic Isotope Effect (KIE) attributes bond strength differences almost entirely to the change in atomic mass, which alters the zero-point vibrational energy (Atkins & de Paula, 2014). The RDI framework suggests this is not the full story and that nuclear geometric coherence—the internal symmetry of the nucleus itself—provides an additional, overlooked contribution to bond strength.

Figure 28: Fire as a Cascade of Resonant Decoherence. This figure illustrates the RC model of fire, a powerful demonstration of the Resonance Dividend. (Left) A complex molecule is shown in a state of high geometric strain, with its high-energy ULF bonds storing significant potential energy. (Center) The fire itself is the process of rapid, catastrophic resonant decoherence. The Geometric Coherence Force compels the strained structure to unravel, instantly releasing its excess energy to attain a more stable configuration. (Right) The system collapses into simpler, lower-energy molecules (like CO_2 and ash). These new structures are in a state of greater geometric harmony. The brilliant light and heat of the flame is the Resonance Dividend being released as the system moves from a high-energy, strained state to a low-energy, coherent state of stability.

This released energy now acts as the "spark" for neighboring molecules, initiating the self-sustaining chain reaction we know as combustion. The visible flame we see is not the "fire" itself; the flame is the luminous signature of the Resonance Dividend being shed from the collapsing molecular structures.

Because every element has a unique nuclear geometry (measured by its Alpha Stability Index, ASI) that dictates its unique set of orbital shapes and energy levels, every element produces a unique, distinct set of harmonic transitions. The flame color is the single, highest-intensity "harmonic note" released by that element as it returns to coherence. Our model thus provides a direct, observable link between the unique atomic geometry of the nucleus and the element's distinct optical properties.

This reinterpretation transforms our understanding of one of nature's most essential processes. Fire is the most common and dramatic example of the Geometric Coherence Force (GCF) (as previously described) acting in an uncontrolled manner, which drives a system from a complex, high-energy, strained state to a simple, low-energy, stable one. It makes the abstract concept of the Resonance Dividend tangible, revealing that the light and heat of a flame are the direct, observable result of matter finding a state of greater geometric and resonant harmony.

7. Geometric Transmutation: A Technological Horizon

Beyond its implications for chemical theory, the RC framework points toward a transformative technological application: Geometric Catalysis. This is the process of using precisely shaped, resonant fields to provide a low-energy pathway for nuclear reactions. The outcome of this process is Geometric Transmutation, the controlled transformation of one element into another.

Historically, the pursuit of element transmutation (alchemy) faltered because it relied on chemical tools (ULF-level manipulations) to alter the nucleus (a UEF-level structure). While modern physics can achieve transmutation through particle accelerators, this brute-force approach is chaotic and inefficient. RC, which explicitly links nuclear geometry to resonance, *is* this elegant approach. If a nucleus is a coherent geometric resonance within the UEF, then, like any resonant system, it can be manipulated by a precisely tuned field.

This process is analogous to an opera singer shattering a crystal glass: no physical contact is required, only the right frequency. A Geometric Transmutation device would generate a shaped field matching the resonant modes of a target nucleus. This single mechanism can be applied toward two distinct, opposite energetic goals:

1. Exothermic Catalysis (Energy Release): This is the "downhill" path, as described in Chapter 7 with Precision Fission. It channels the universe's natural drive toward coherence by "nudging" an unstable nucleus (like Plutonium) into a more stable, lower-energy state, thereby releasing the Coherence Dividend.

2. Endothermic Catalysis (Energy Input): This is the "uphill" path, such as the alchemical dream of lead-to-gold. It works against the natural drive for

coherence by expending immense, focused energy. The resonant field must "push" a highly stable nucleus (like Lead-208) out of its deep coherence well, amplifying its internal oscillations until it ejects a substructure and reconfigures into a different, *less* stable element (like gold).

Both applications arise from the same principles of Resonant Engineering, but one harnesses a natural, energy-releasing snap, while the other requires a forced, energy-costing reconfiguration. While technological hurdles remain, such as the precise characterization of nuclear resonances and the generation of sufficiently strong fields, the principle itself is a direct extrapolation of the RC framework.

8. Conclusion

This work establishes RC as a unifying, quantitative framework for understanding chemical bonds, rooted in the principles of the RFI and QVD. By treating chemical bonds as the merging of ULF standing waves, supported by the geometric compatibility of their underlying UEF nuclei, RC uncovers a two-tiered origin of bond strength. This framework both reconciles conventional quantum mechanical predictions and quantifies the additional stabilization—the Resonance Dividend—that arises when nuclear congruence and orbital resonance synergize to create a more coherent molecular state.

Applying this model to the carbon-carbon bond, we predict a measurable isotopic effect: C-12–C-12 bonds, whose planar triangular nuclei align with near-perfect congruence to generate a greater resonance dividend than C-13–C-13 bonds, where an extra neutron disrupts this symmetry. The resulting 0.3–0.6% shift in vibrational frequencies, detectable through high-resolution infrared spectroscopy, serves as a testable signature of the direct influence of nuclear geometry on chemical behavior.

Beyond chemical theory, RC opens a new technological frontier, Geometric Transmutation, where nuclear structures are manipulated through resonant excitation rather than brute-force collisions. By driving nuclei at their natural UEF resonances, it may one day be possible to reconfigure atomic structure with precision, thereby accomplishing the alchemical dream of transforming stable elements such as lead into gold.

Ultimately, RC bridges nuclear structure, electron dynamics, and molecular behavior into a single, intuitive framework. In doing so, it transforms the chemical bond from a collection of abstract rules into a tangible, dynamic process governed by geometry and resonance, offering predictive pathways for fundamental chemistry and future applications in energy, materials, and nuclear transformation.

9.3 Geometric Congruence

In this chapter, we presented a physically grounded model of chemical bonding that unites nuclear geometry and electron dynamics under the principle of the Resonance Dividend. By framing chemical bonds as the merging of Universal Light Field (ULF) standing waves into coherent molecular resonances, supported by the geometric congruence of nuclei within the Universal Energetic Field (UEF), this framework moves beyond abstract quantum formalisms toward a tangible, dynamic, and predictive model of matter. From this dual mechanism, several key insights emerge.

The framework explains, for instance, why Carbon-12, with its perfectly planar triangular nucleus, plays such a dominant role in organic life, as its geometry maximizes both ULF coherence and UEF congruence. It also predicts measurable isotope-dependent variations in bond strength, most notably that C-12–C-12 bonds are stronger than their C-13 counterparts. This effect, quantifiable as a 0.3–0.6% shift in vibrational frequency in infrared spectroscopy, would directly confirm that nuclear geometry influences chemical stability, a factor long neglected in conventional theory.

If validated, this framework could reshape both theoretical and applied chemistry, opening new avenues for experimentation and technology:

1. Experimental Validation. High-resolution spectroscopy can directly test the predicted variation in isotopic bond strengths, establishing nuclear geometry as a measurable factor in chemical theory.

2. Predictive Materials Design. By quantifying stability through nuclear congruence and orbital resonance, the Resonance Dividend Index (RDI) could drive the engineering of materials with optimized isotopic compositions, tailored for strength, stability, or reactivity.

3. Resonance-Driven Catalysis. By deliberately aligning molecular and nuclear oscillations, it may be possible to accelerate otherwise slow or energy-intensive chemical and nuclear processes, thereby creating new low-energy reaction pathways.

4. Geometric Transmutation. Though technically challenging, mapping nuclear resonances and driving them with precisely shaped electromagnetic fields could enable element conversion.

In sum, RC is more than a reinterpretation of chemical bonding. It is a roadmap to a new chemistry—one that is intuitive, predictive, and deeply integrated with nuclear physics and the broader fabric of physical law. This grounding of molecular interactions in the same resonant and geometric principles that govern the nucleus and cosmos thus invites a new era of scientific discovery and technological innovation.

References

1. Atkins, P., & Friedman, R. (2011). *Molecular quantum mechanics* (5th ed.). Oxford University Press.
2. Atkins, P., & de Paula, J. (2014). *Atkins' Physical Chemistry* (10th ed.). Oxford University Press.
3. Born, M., & Oppenheimer, J. R. (1927). Zur Quantentheorie der Molekeln [On the Quantum Theory of Molecules]. *Annalen der Physik, 389*(20), 457–484.
4. Brink, D. M., Friedrich, H., Weiguny, A., & Wong, C. W. (1966). Investigation of the alpha-particle cluster model for C12. *Physics Letters, 21*(6), 678–680.
5. Hush, N. S. (1968). *Intervalence-transfer absorption. Part 2. Theoretical considerations and spectroscopic data.* Progress in Inorganic Chemistry, 8, 391–444.
6. London, F. (1930). Zur Theorie und Systematik der Molekularkräfte [On the theory and systematics of molecular forces]. *Zeitschrift für Physik, 63*(3-4), 245–279.
7. Mulliken, R. S. (1955). *Electronic population analysis on LCAO–MO molecular wave functions. I.* Journal of Chemical Physics, 23(10), 1833–1840.
8. Pauling, L. (1960). *The Nature of the Chemical Bond and the Structure of Molecules and Crystals: An Introduction to Modern Structural Chemistry* (3rd ed.). Cornell University Press.

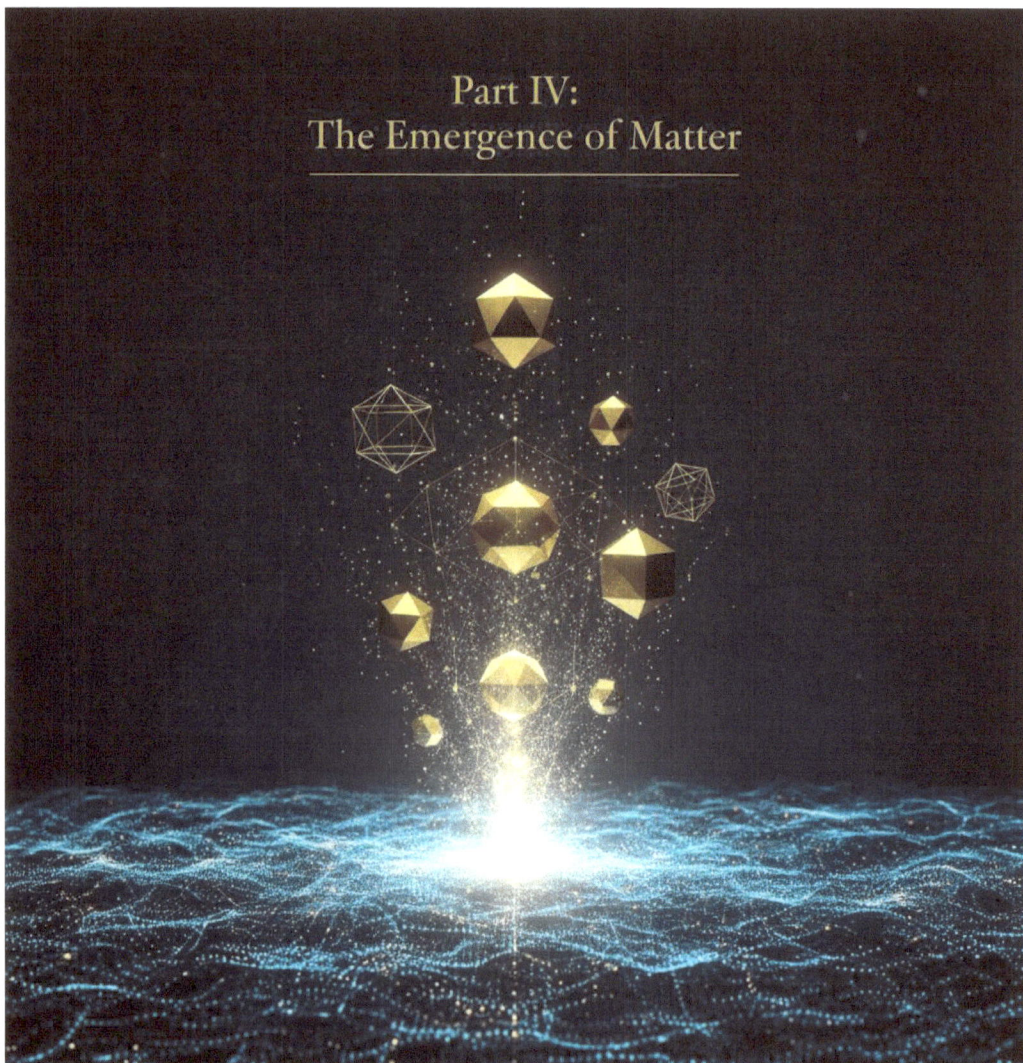

Part IV:
The Emergence of Matter

Chapter 10: Resonant Matter

10.1 Condensed Matter Physics

In the preceding chapters, we have constructed the fundamental building blocks of matter: the resonant atom and the resonant chemical bonds that unite them. We now turn to the domain of Condensed Matter Physics, the field dedicated to understanding how the complex, macroscopic properties of solids and liquids emerge from the collective interactions of their constituent parts.

Condensed Matter Physics encompasses a broad spectrum, ranging from crystalline lattices and metallic conductors to soft gels, complex polymers, and biological tissues. A complete theory of the subject must therefore be capable of explaining the hardness of diamond, the fluidity of water, the memory of shape-memory alloys, and the folding dynamics of proteins—and it must do so from a unified set of principles. Historically, condensed matter physics emerged from classical models like Drude's theory of electron conduction (Ashcroft & Mermin, 1976), though it eventually incorporated quantum mechanics to develop powerful formalisms such as:

- Band Theory (Bloch, 1929), which models electrical and optical properties by merging atomic orbitals into continuous energy bands.

- The Exchange Interaction (Heisenberg, 1928), which explains ferromagnetism via a mysterious, nonlocal force arising from the symmetry of electron wavefunctions.

These tools have achieved incredible predictive power, underpinning the entire digital revolution. However, they also introduced a level of abstraction that can obscure intuitive understanding. Concepts like the "band gap" and "exchange interaction" are defined mathematically but lack a simple, tangible mechanism in physical space. Even today, many phenomena, such as high-temperature superconductivity, topological insulators, and exotic magnetic phases, remain resistant to explanation from first principles, prompting researchers to rely on computational heuristics or black-box machine learning models.

This chapter offers an alternative. Here, we introduce Resonant Matter (RM), a framework that provides the missing physical ontology for this entire domain. Building on the principle of emergence (Anderson, 1972), we will demonstrate how the full spectrum of material properties—from the conductivity of a metal and the hardness of a diamond to the exotic states of superconductivity and the classical dynamics of fluid flow—arises as the collective, coherent expression of the underlying resonant geometry of atoms and their bonds.

. This chapter constructs a complete, physical, and geometric explanation of the world we can see and touch, ultimately revealing how the laws of the macrocosm are an elegant echo of the resonant physics of the microcosm.

10.2 Resonant Matter: From Geometric Bonds to Material Properties

Abstract

This chapter introduces Resonant Matter (RM), a model that explains how the tangible properties of the material world emerge from resonant patterns in the quantum world. Building on our geometric model of the chemical bond, we demonstrate how the collective behavior of resonant fields gives rise to the full spectrum of material phenomena, providing a new physical basis for electrical conductivity, optical properties, magnetism, and the dynamics of heat and sound.

The model is then extended to its theoretical limits, predicting exotic states such as high-temperature superconductivity and supermalleability as phases of perfect, crystal-wide coherence. This exploration reaches its climax with the concept of Emetium, a "perfect metal" where nuclear and orbital resonances lock into a single, self-healing, ultra-strong material. We then show how the RM framework applies to organic chemistry and the mechanics of flow. Finally, we ground the framework in a series of falsifiable predictions, including Resonant Damping and Geometric Catalysis, which transform RM into a testable science of matter.

1. Introduction

Having established a geometric model for the atom and the chemical bond, this chapter takes the final step: scaling this microscopic framework up to explain the tangible, macroscopic properties of the material world. The vast domain of condensed matter physics—from the hardness of a diamond to the folding of a protein—presents the ultimate test for any fundamental theory. While standard quantum formalisms provide a successful mathematical description of this domain, they often lack a direct, physical picture (Kittel, 2005; Ashcroft & Mermin, 1976).

This chapter introduces Resonant Matter (RM), a framework that provides this missing physical ontology. Building on the principle of emergence (Anderson, 1972), we demonstrate how the full spectrum of material properties arises as the collective, coherent expression of the underlying resonant geometry of atoms and their bonds. We then show how this single principle accounts for the properties of crystalline solids, explains exotic states of matter such as superconductivity (Bardeen, Cooper & Schrieffer, 1957), provides a new foundation for the molecular architecture of organic life, and even explains the origin of classical fluid dynamics. More than just a reinterpretation, RM is a predictive scientific model that yields a series of falsifiable predictions, further grounding the Unified Field Dynamics (UFD) framework in experimental reality.

2. Deriving General Material Properties

In the previous chapters, we established a geometric model for individual atoms and the chemical bonds that unite them. We now scale up this framework to demonstrate how the material's tangible, macroscopic properties emerge from the collective behavior of its microscopic components. The key to this connection lies in the Resonant Dividend Index (RDI), introduced in Chapter 9.

The RDI shows that the stability of chemical bonds depends on coherence on both the nuclear (UEF) and orbital (ULF) levels. The overall properties of a material emerge from the collective alignment of these underlying geometries. When billions of bonds with congruent, harmonious resonances assemble into a crystal lattice, their geometries phase-lock and reinforce one another, creating large-scale coherence. This extended coherence gives rise to phenomena such as the electrical conductivity of a metal or the strength of a diamond. Conversely, when the underlying bonds are strained or misaligned, their geometric coherence fragments, leading to insulating states or brittleness. In this way, the entire spectrum of material properties can be understood as the macroscopic expression of the resonant coherence of its underlying chemical architecture.

2.1 Electrical Properties: The Sea of Resonance

In RM, a material's electrical conductivity is determined by the ability of its electron vortices to form a crystal-wide, coherent resonant flow.

In a conductor, such as a metal, the individual ULF valence orbitals overlap extensively, forming a delocalized "sea of shared ULF resonance," as we describe in Chapter 9 in the context of the metallic bond. Within this continuous resonant field, electron vortices are free to move, resulting in high conductivity. This "sea of resonance" is the physical reality that standard band theory describes as the conduction band.

In an insulator, the ULF resonances are tightly "trapped" in localized, high-energy covalent or ionic bonds. There is no crystal-wide resonant pathway for the electron vortices to follow. These energetic "dead zones" between the trapped resonances are the physical reality of the "band gap" described in conventional solid-state physics (Bloch, 1929).

This resonant model provides the physical picture that underlies the abstract concepts of standard solid-state physics. What physicists call the "band gap" is, in this view, the energy difference between a material's stable, ground-state resonance and its next available harmonic overtone. The "quantized energy ladder" of band theory is the mathematical description of these discrete, physical, harmonic resonances.

2.2 Optical Properties: The Harmonics of Color

A material's optical properties, such as its color and transparency, are, in turn, determined by the natural resonant frequencies of its collective ULF structure. A material can only absorb a photon if the photon's frequency matches one of the stable harmonic vibrations that the material's molecular geometry can support.

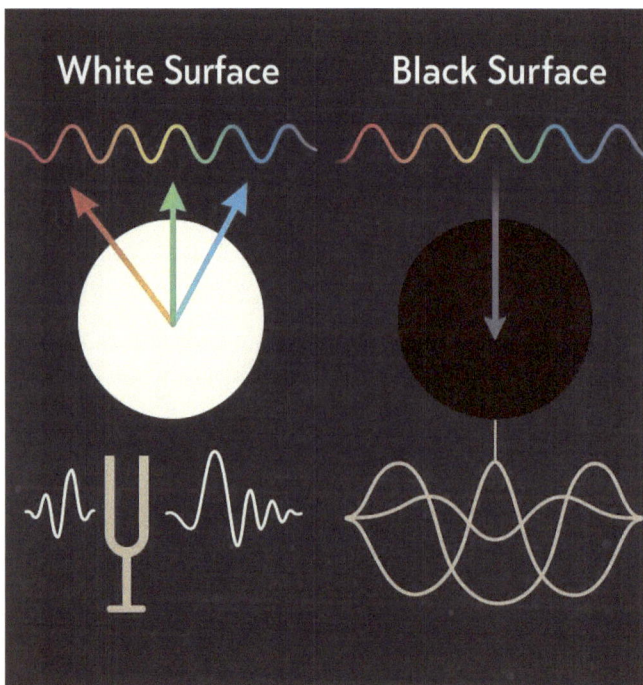

Figure 29: Resonant Absorption of Light by Color. This illustration compares the resonance behavior of white and black surfaces in the visible spectrum. On the left, a white surface reflects all incoming visible light, indicating minimal resonance and poor absorption, analogous to a bell that does not ring when struck by visible frequencies. On the right, a black surface absorbs nearly all visible wavelengths, resonating strongly with each, like a complex instrument tuned to receive every note. This illustrates how black surfaces are resonantly rich while white surfaces are resonantly poor in the visible range.

This single principle provides a direct, physical explanation for optical phenomena. A material like glass is transparent because its resonant frequencies are all in the ultraviolet range; the lower frequencies of visible light do not match any of its stable harmonics and thus pass through unabsorbed. A material appears colored because its specific molecular geometry creates a resonant structure that is perfectly tuned to absorb certain frequencies of light while reflecting others. The color we see is simply the light that is not in resonance with the material (Figure 29).

2.3 Ferromagnetism: The Geometry of Coherent Spin

Ferromagnetism, the powerful magnetic effect seen in materials like iron, is a collective phenomenon that arises only when a material meets two conditions of geometric compatibility.

Figure 30: The Geometry of Coherent Spin. A visualization of the RM model of ferromagnetism. The model requires two levels of geometric compatibility. First, individual atoms act as powerful magnetic vortices due to the merged rotational "eddies" of their electron shells. Second, the crystal lattice places these atoms at a perfect "Goldilocks distance," allowing their individual fields to physically "mesh" and interlock like gears. This collective, coherent flow is the physical origin of the abstract "exchange interaction" and is responsible for the powerful, material-wide magnetic field.

First, the individual atoms themselves must act as tiny magnets. In RM, this occurs in elements whose atomic structures contain multiple, unpaired electron vortices with parallel spins. The individual rotational eddies created by these vortices merge into a single, powerful, and coherent rotational flow, giving the entire atom a strong magnetic moment. Second, these atomic-scale vortices must align across the entire material. In standard physics, this alignment is attributed to the abstract "exchange interaction." Our model, in contrast, provides a direct physical mechanism: the crystal lattice of ferromagnetic materials places atoms in a "Goldilocks zone" of spacing. This perfect distance allows the rotational ULF eddies of neighboring atoms to physically mesh and interlock, like gears locking into place. This geometric meshing is what creates the powerful, domain-wide magnetic field (Figure 30).

Ferromagnetism, in RM, is not merely the sum of atomic spins; it is the emergent harmony of rotational coherence across scales. By grounding magnetism in the physical geometry of rotating ULF vortices, this model reveals that ferromagnetism is a resonance-based phenomenon, one that requires both intrinsic atomic spin coherence and interatomic geometric compatibility.

2.4 The Geometry of Conductivity: Semiconductors as Tunable Resonant Matter

Between the free-flowing resonance of a conductor and the rigid, localized resonance of an insulator lies the most technologically important class of materials: semiconductors. In the RM framework, a semiconductor is a material whose crystal lattice is poised at the very edge of coherence. At rest, it is an insulator, with its electron vortices trapped in stable standing waves. However, it is tuned so that a small, precise input of energy—from light, heat, or an electric field—can catalyze the system, exciting an electron vortex into a higher-order, conductive resonant state.

This "on/off" behavior is what makes semiconductors the perfect switch. In standard physics, this is described by the "band gap." In our model, the band gap is a physical "resonance inaccessibility window"—a geometric and energetic mismatch that prevents the formation of a coherent, crystal-wide wave. An incoming photon or applied voltage provides the energy for an electron vortex to reconfigure its geometry and lock into the higher, conductive harmonic.

In this respect, the doping process introduces specific impurities that act as stepping stones across this resonance window. [54] These dopant atoms create localized

---

[54] In semiconductor physics, doping refers to the intentional introduction of impurity atoms into a pure semiconductor lattice to alter its electrical properties. An n-type dopant has more valence electrons (or valence vortices, in the UFD framework) than the host material, contributing extra negative charge carriers (electrons). Conversely, a p-type dopant has fewer valence electrons, creating "holes" or positive charge carriers by accepting electrons. These dopants locally disrupt the otherwise uniform lattice resonance, introducing localized field distortions that enable controlled conductivity and active electronic behavior.

geometric perturbations in the ULF, making it far easier to initiate a conductive flow. This principle of using external fields to manipulate a material's inherent geometric responsiveness underlies all modern electronics, from transistors and LEDs to solar cells.

2.5 The Physics of Sound: Coherence in Motion

In the RM framework, sound is the audible expression of a material's internal geometry—the coherent propagation of vibrational energy through the lattice of UEF/ULF vortices. The fundamental quantum of this vibration is the phonon (Kittel, 2005).

In our model, a phonon is not an abstract "quasi-particle" but a real, quantized packet of resonant momentum transferred from one vortex to the next. This provides a clear distinction: a photon is a primary wave in the ULF, while a phonon is a secondary, mechanical wave that propagates through the matter structures existing within that field (Figure 31A).

This physical model illustrates how a material's acoustic properties, such as the speed of sound, correspond with its microscopic architecture. The speed of sound is determined by the material's elasticity and density. In RM terms, elasticity is a direct measure of the strength of the ULF resonant couplings (the chemical bonds), while density is the concentration of the vortices themselves. A material with strong, rigid bonds, like diamond, transmits these phononic vibrations with extreme efficiency, resulting in a very high speed of sound. In contrast, a diffuse medium like air, with weak intermolecular couplings, transmits sound far more slowly.

2.6 Thermal Properties: The Dynamics of Incoherent Resonance

In RM, heat is the twin of sound. While sound is the coherent propagation of a vibrational wave (a phonon), thermal energy is the total energy stored in the vast, incoherent superposition of those same phononic modes (Debye, 1912). What we perceive as "random" heat is the complex, dissonant hum of countless different resonant modes being excited simultaneously. A material's temperature is a measure of this collective vibrational amplitude (Figure 31B).

This physical model of heat provides a direct explanation for a material's thermal properties. Thermal conductivity is a measure of the efficiency with which these incoherent phononic vibrations propagate. In a metal, the continuous "sea of shared ULF resonance" provides an excellent medium for this transfer, resulting in high conductivity. In an insulator, the localized, rigid bonds of the lattice hinder the propagation of these vibrations. Similarly, a material's specific heat capacity is the energy required to raise the average amplitude of these vibrations, while thermal expansion is the natural geometric consequence of the vortices moving further apart as their vibrational energy increases.

This redefinition of heat as a measure of incoherent resonance is the key to understanding the physical nature of matter itself

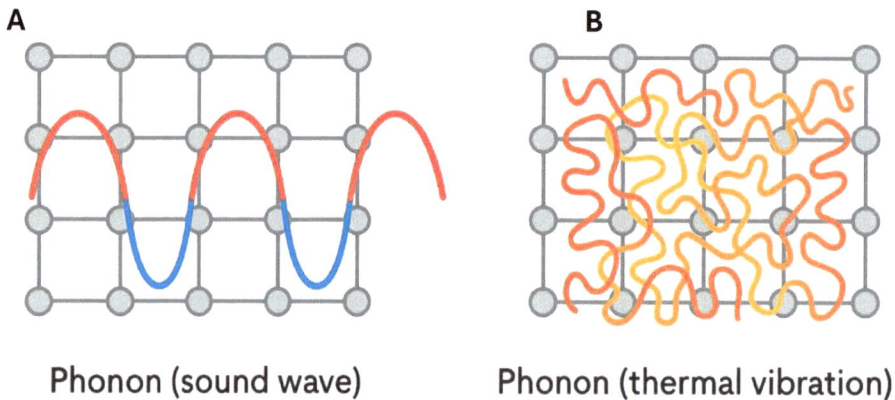

Phonon (sound wave)   **Phonon (thermal vibration)**

Figure 31: Phonons as Coherent and Incoherent Vibrational Modes. This two-panel illustration contrasts the dual roles of phonons—quantized vibrational excitations of matter—in two domains in RM.
(A) Sound as Propagating Coherence: In the context of sound, phonons appear as coherent waves propagating through a material's lattice, representing organized pressure pulses that transmit acoustic energy. These traveling oscillations emerge from the sequential realignment of atomic-scale UEF/ULF vortices.
(B) Heat as Incoherent Resonance: In the thermal domain, phonons manifest as incoherent, localized vibrations—stochastic fluctuations within the same lattice—responsible for heat transfer. These represent disordered modes of vortex excitation that arise from internal energy rather than directed motion, distinguishing heat from sound at the level of field coherence.

2.7 The Three States of Matter: A Phase Equilibrium of Coherence

The RM framework also provides a direct, physical explanation for the three states of matter. In this view, a material's state (solid, liquid, or gas) results from a phase equilibrium between two competing resonant forces.

The first resonant force arises from the material's ULF couplings (chemical bonds), which represent weak, attractive ULF standing waves—such as hydrogen bonds

248

and van der Waals forces—that try to pull molecules together into an ordered, low-energy, coherent lattice, driven by the Geometric Coherence Force (GCF) (*see* Chapter 9). The second resonant force arises from incoherent thermal resonance, or kinetic energy, where the "noise" or "dissonance" in the system, caused by the chaotic vibration of individual molecules, tries to break these coherent bonds and drive the system toward disorder (particle view). The state of matter is determined by which of these two forces is dominant.

- Solids (Coherent State): In a solid, the incoherent thermal resonance is low. The GCF dominates, pulling the molecules into a highly stable, phase-locked, and rigid crystal lattice.

- Liquids (Metastable State): In a liquid, the incoherent thermal resonance is high enough to overcome the rigid, long-range coherence of the crystal. However, it is insufficient to overcome the short-range ULF couplings entirely. The molecules break their fixed lattice positions and can "flow" past one another, but they remain bound in a loose, shifting, and coherent "sea."

- Gases (Decoherent State): In a gas, the incoherent thermal resonance is now completely dominant, fully overpowering the weak ULF couplings. The molecules are no longer bound in a coherent system; they are independent, decoherent vortices flying apart in a state of maximum entropy (from a particle perspective).

2.8 Mechanical Properties: The Geometry of Strength and Hardness

Having defined the states of matter, we can now explain their vast range of mechanical properties. In the RM framework, a material's properties are a direct reflection of the geometry and coherence of its ULF resonant couplings.

Hardness and Brittleness, for example, arise in materials such as diamond and ceramics where the ULF resonances are locked into strong, rigid, and highly directional covalent or ionic bonds. These coherent networks strongly resist deformation, but when a force exceeds their stability threshold, the resonant couplings do not bend—they shatter, causing the material to fracture. Malleability and Softness, in contrast, are characteristic of metals, which are defined by a fluid-like "sea of shared ULF resonance." This non-directional, continuous resonant field holds the atomic cores together while allowing them to slide past one another under stress. Elasticity, in turn, is the system's fundamental drive to snap back and restore its most stable, lowest-energy resonant geometry after being deformed. Even crystal defects, like dislocations, can be understood in this model as localized points of decoherence within the material's resonant field (Taylor, 1934).

This section has demonstrated how the entire spectrum of ordinary material properties emerges from a single, unified principle: the collective geometric and resonant coherence of a material's underlying atomic architecture. Having established this basis for the familiar properties of matter, we now turn our attention to the extraordinary phenomena that arise when this principle of coherence is taken to its theoretical limit.

3. Exotic States of Matter: Superconductivity and Beyond

While the principles of resonance explain the ordinary properties of matter, the frontiers of physics are defined by exotic states, like high-temperature superconductivity, that defy classical intuition. This section explores these extraordinary phenomena, demonstrating that they are not unrelated anomalies but are different expressions of a single underlying principle within the RM framework.

3.1 A Geometric Origin for High-Temperature Superconductivity

The mystery of high-temperature superconductivity, which standard BCS theory cannot explain, is resolved in our framework as a phase transition from resonant chaos to perfect coherence. In a normal conductor, the "sea of shared ULF resonance" is fragmented, leading to electron vortices that scatter and produce resistance. In a superconductor, the entire crystal lattice "snaps" into a single, unified ULF standing wave. The electron vortices become phase-locked components of this crystal-wide resonance, moving as a perfectly ordered current with zero resistance (Figure 32).

This perspective explains why high-temperature superconductivity appears in materials with complex, layered crystal structures, such as the cuprates.[55] These peculiar structures are not incidental; they are precisely tuned geometric scaffolds that allow this state of total coherence to remain stable at far higher temperatures. This insight leads to a transformative technological prediction: if superconductivity is a resonance-driven state,

---

[55] Cuprates are a class of high-temperature superconducting materials composed primarily of copper and oxygen atoms arranged in layered crystalline structures. These compounds exhibit superconductivity at temperatures much higher than conventional superconductors, often above the boiling point of liquid nitrogen (77 K). The superconducting properties of cuprates arise from complex electron interactions within copper-oxide planes, making them a central subject in the study of unconventional superconductivity and quantum materials.

it can be engineered. Precisely shaped electromagnetic fields could be used to "nudge" a material into the superconducting state at or near room temperature, while a dissonant "resonant quench" field should be able to destroy it, providing a clear and testable prediction of this framework.

3.2 Supermalleability: Coherence in Motion

The RM framework predicts that superconductivity has a mechanical counterpart: supermalleability, a state of near-fluid flexibility combined with extreme strength. While ordinary malleability in metals arises from the fluid-like "sea of shared ULF resonance," supermalleability is a more profound state of coherence. It emerges when both the orbital (ULF) and the deeper nuclear (UEF) resonances lock into a single, crystal-wide resonant field.

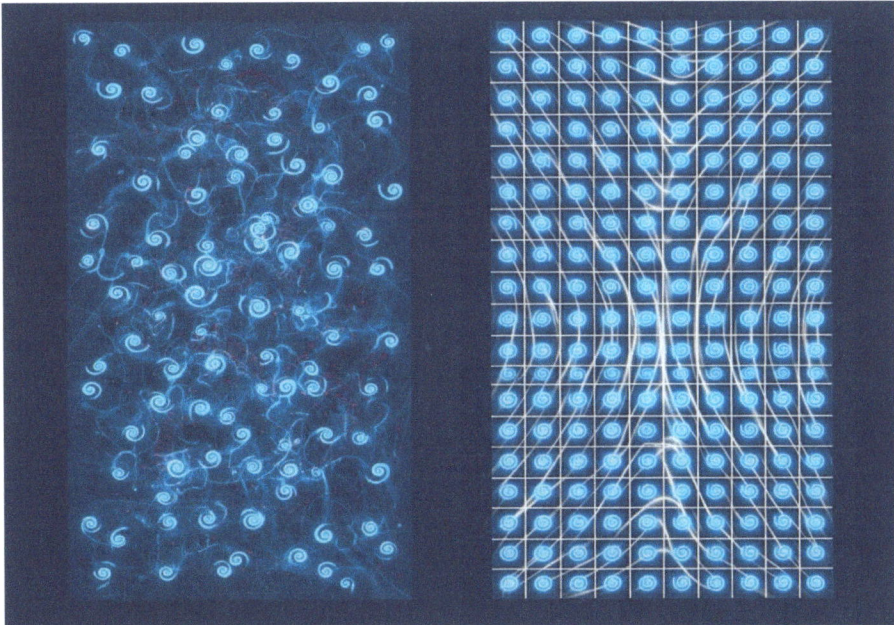

Figure 32: A Geometric Model of Superconductivity. This figure illustrates the transition from a normal conductor to a superconductor within the RM framework. (Left) In a normal conductor, the "sea of shared resonance" is a chaotic medium. Electron vortices move through it individually and experience scattering, which gives rise to electrical resistance. (Right) In a superconductor, the electron vortices and the ULF medium lock into a single, unified, and perfectly coherent collective resonance. The vortices are now phase-locked components of this crystal-wide standing wave and, moving as part of this perfectly ordered system, experience zero "drag" or resistance.

In such a state, physical stress is no longer localized into defects like dislocations but is distributed as coherent waves throughout the entire material. The result is a "super-steel" that is both incredibly strong and anomalously ductile, capable of deforming flexibly under immense loads and self-healing microfractures through field-driven reorganization. The experimental signature of this state would be the near-total absence of internal friction (mechanical hysteresis) during deformation. Like superconductivity, supermalleability represents the natural end state of a material whose entire resonant architecture acts as a single, unified organism.

3.3 Emetium: The Resonance-Locked Crystal

As the culmination of our predictions, we propose the potential creation of a Resonance-Locked Crystal, or Emetium.[56] In this material, nuclear (UEF) and orbital (ULF) resonances would fully lock into crystal-wide coherence, producing a lattice that combines extreme strength, near-fluid flexibility, and self-healing capabilities (Figure 33). Our framework suggests concrete conditions for its realization:

The architectural principles of Quantum Vortex Dynamics (QVD), as described in Chapter 8, predict a unique island of stability for a nucleus of 128 protons and 128 neutrons (A = 256), corresponding to a theoretical Alpha Stability Index (ASI) of approximately 1.30. What makes this island of stability so unique is its 1:1 proton-to-neutron ratio, which is unsustainable for other heavy elements. This composition is made possible, however, because it would allow the nucleus's 64 alpha-particle "bricks" to lock into a state of unparalleled geometric perfection: a perfect 4x4x4 cube. This dense, space-filling geometry is so robust that it can overcome the immense electrostatic strain that makes other heavy, proton-rich nuclei unstable. This internal cubic symmetry would then provide the ideal resonant blueprint for the atoms to form a macroscopic cubic crystal lattice, thereby creating a state of perfect multi-scale coherence.

Through Resonant Engineering, we propose that an external electromagnetic field could catalyze the formation of this theoretical element by "locking" a crystal's UEF and ULF resonances into a single, unified, hyper-coherent state.

---

[56] The name Emetium derives from the Hebrew word Emet (אמת), meaning "truth." In this context, it symbolizes the ultimate realization of material perfection—a crystal in which nuclear (UEF) and orbital (ULF) resonances achieve complete coherence, yielding properties that reflect the "true" potential of matter. To help visualize its extraordinary qualities, Emetium may be imagined as a material combining extreme strength, fluid flexibility, and self-healing, somewhat akin to the shape-shifting, resilient metal seen in science fiction representations like the T-1000 in *Terminator 2*.

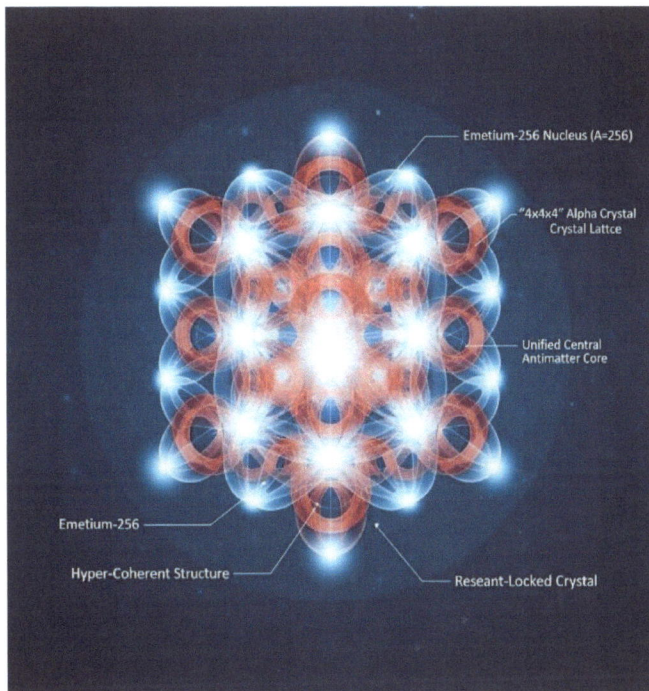

Figure 33: Resonant Architecture of Emetium. This figure depicts the theoretical structure of a single Emetium atom, showcasing its multi-scale coherence. At its heart lies the hyper-coherent nucleus, whose stability arises from the perfect geometric packing of 64 alpha-particle "bricks" into a cubic form. This represents a qualitatively new stability floor. The surrounding electron orbitals are depicted as perfectly nested standing waves. Their symmetry directly reflects the cubic nucleus, demonstrating the "resonant lock" between nuclear (UEF) and orbital (ULF) fields.

Were it instantiated, Emetium would exhibit an unprecedented combination of properties. It would possess the extreme strength of a perfect crystal, the near-fluid flexibility of a supermalleable phase, and the ability to self-heal by dissipating physical stress as coherent waves. It would be a "super-metal" with negligible fatigue and unparalleled resilience—one capable of revolutionizing aerospace, energy, and mega-scale engineering.[57]

4. Case Studies in Molecular Geometry: The Basis of Life

Having explored the ultimate potential of inorganic matter, we now turn to the equally profound and elegant resonant architectures that underlie life itself. The case studies that follow demonstrate that, as with inorganic matter, the unique properties of

---

[57] The sensory experience of Emetium would be profoundly alien. Visually, it would be a perfect mirror, as its self-healing surface would reflect the world with flawless, liquid-like clarity. To the touch, it would feel both perfectly solid and unnaturally smooth, yet it would yield to firm pressure and flow like a dense fluid, only to instantly and silently return to its original shape. Beyond the physical, its perfect internal harmony would likely induce a palpable sense of profound calm and order in any conscious observer, as their own energetic fields would be entrained by its coherence.

the most important organic substances follow from their underlying resonant geometries.

4.1 The Geometry of Carbon and the Molecules of Life

The principles of covalent bonding find their ultimate expression in the chemistry of carbon, the backbone of all known life. In our framework, the unique properties of organic molecules stem from the unparalleled geometric versatility of the valence resonances of the carbon atom (*see* Chapter 9). Carbon can reconfigure its resonant waves (standard chemistry's "hybridization") to form strong, stable bonds in three fundamental geometries: a perfect tetrahedral arrangement with four single bonds, a flat trigonal planar geometry with a double bond, and a perfectly linear geometry with a triple bond.

Perhaps the most elegant expression of this versatility is the benzene ring, a perfect hexagonal structure that serves as the foundational building block for a vast class of aromatic compounds. In RM, benzene is a masterpiece of natural resonant engineering. Its six sp² hybridized carbon atoms form a perfectly planar ring of sigma bonds, which then serves as a stable "waveguide" for a delocalized, continuous, toroidal standing wave of ULF resonance—the famous "pi system"—that exists above and below the plane of the ring. This perfect, closed-loop "sea of shared resonance" is the physical origin of benzene's extraordinary stability (Figure 34A).

This flexibility makes carbon the ultimate "geometric building block," capable of assembling the vast molecular architectures that support life, from the simple stability of the benzene ring to the complex chains and functional groups of the major biomolecules:

- Carbohydrates (The Geometry of Energy): Polar hydroxyl (-OH) groups on carbon chains create ULF resonances highly compatible with water, allowing easy dissolution for transport and energy use (Figure 34B).

- Lipids (The Geometry of Insolubility): Long, nonpolar hydrocarbon chains generate symmetrical, nonpolar ULF resonances that repel water, causing aggregation and forming the resilient, water-resistant barriers of cell membranes (Figure 34C).

- Proteins (The Geometry of Function): Chains of amino acids fold into precise three-dimensional geometries representing their lowest-energy, most stable ULF resonant states (Figure 34D).

The folding of a protein, in particular, is the pinnacle of this resonance-driven self-organization. While this is a challenge for traditional physics-based models, which struggle to compute the vast number of potential conformations (Jumper et al., 2021), the RM framework provides a direct physical mechanism.

In RM, an unfolded protein is an energetically unstable, incoherent state that is actively guided by the GCF to collapse into its lowest-energy, most stable three-dimensional structure. The energy released as it snaps into this perfect resonant state is the Resonance Dividend (*see* Chapter 9). This model explains how proteins can find their functional form with such speed and reliability. Rather than randomly searching for a stable shape, they are being pulled by a fundamental force toward a state of maximum resonant coherence.

4.2 RNA and DNA: The Geometry of Life

The leading scientific theory for the origin of life is the "RNA world" hypothesis, which posits that life originated from simpler, single-stranded RNA molecules (Gilbert, 1986). The RM framework offers a novel physical context for this concept. In this model, RNA is a "jack of all trades," capable of both storing genetic information and catalyzing chemical reactions. However, its single-stranded nature makes it a resonantly unstable structure, suitable for temporary messages but not permanent archives (Figure 34E). This exerted a powerful evolutionary pressure on the system to find a more coherent state.

The solution to this instability was the evolution of DNA. The iconic double helix is a structure of resonant perfection. Its twisted, two-stranded geometry is far more stable than the single-stranded RNA, making it the perfect medium for long-term, archival storage of information. This transition from the unstable RNA world to the stable DNA world was not just a chemical event; it was a phase transition in the nature of information itself. The DNA double helix was the first molecular structure with the necessary coherence and stability to act as a permanent, incorruptible blueprint for complex, biological life (Figure 34F).

4.3 Carbon Nanotubes: The Geometry of Coherence and Strength

In RM, a material's true strength arises from its geometric and resonant coherence, which is the ability of its structure to maintain field alignment under stress. Carbon nanotubes are a perfect real-world example of this principle. Formed by rolling a single sheet of graphene into a seamless cylinder, these nanostructures exhibit a tensile strength that far surpasses steel, a feat our model attributes to the ideal "longitudinal coherence" of their internal ULF resonance fields.

A carbon nanotube thus functions as a perfect resonant waveguide (Figure 34G). Its hexagonal lattice of sigma bonds forms a flawless "resonance net" that evenly distributes mechanical stress across its surface. Meanwhile, its cylindrical topology allows for a continuous, uninterrupted ULF standing wave to form along its entire length with no edge dissipation. This unique geometry allows a nanotube to channel stress and energy with minimal loss. It thus represents one of the highest known expressions of resonance-locked strength.

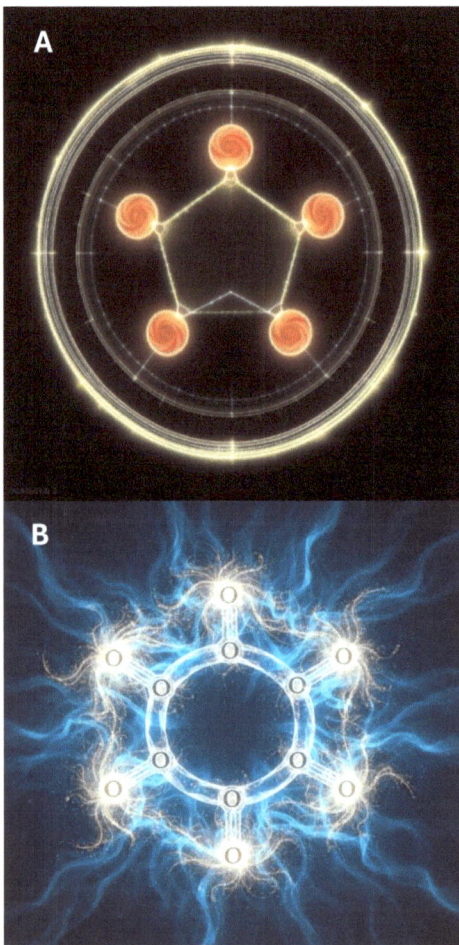

Figure 34: The Geometry of Biological Information and Energy.

(A) Benzene, a Masterpiece of Resonant Coherence. This figure visualizes the benzene ring within the RM framework. The six carbon atoms form a perfectly planar hexagonal ring of sigma bonds, which acts as a stable "waveguide." The famous delocalized pi system is depicted as a continuous, toroidal standing wave of ULF resonance, flowing in a perfect, unbroken circuit above and below the plane. This state of perfect, closed-loop coherence is the physical origin of its extraordinary stability (aromaticity).

(B) Carbohydrates and the Geometry of Energy: A visualization of a Ribose molecule, the fundamental sugar that forms the backbone of RNA. Its polar hydroxyl (-OH) groups are depicted as luminous hotspots, creating specific ULF resonant patterns that are highly compatible with the resonant fields of surrounding water molecules. This energetic harmony is the mechanism underlyig its high solubility, making it the ideal component for building informational and energetic structures.

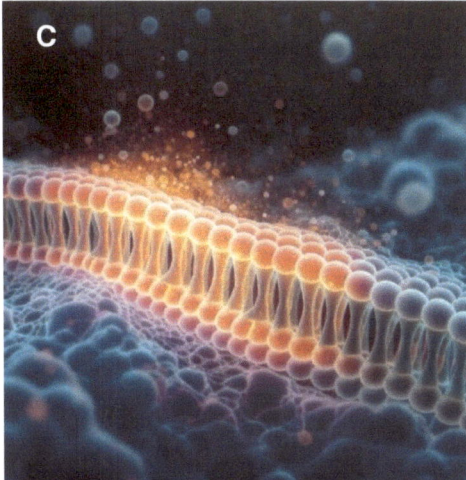

(C) Lipids and the Geometry of Insolubility: This figure visualizes the formation of a lipid bilayer, such as a cell membrane. Each lipid's long, nonpolar tail is depicted as a symmetrical, self-contained ULF resonant field. These nonpolar fields actively repel the polar resonances of the surrounding water molecules, driving the lipids to self-assemble into a stable, water-resistant barrier.

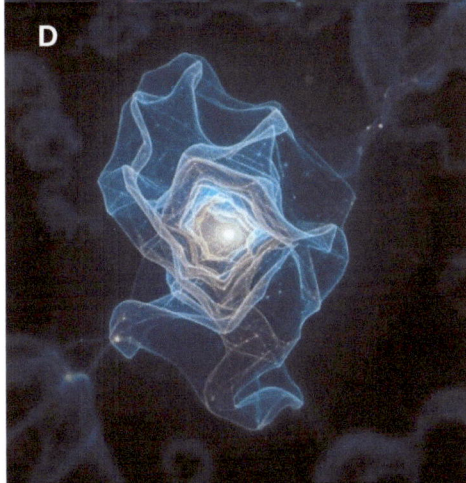

(D) Proteins and the Geometry of Function: This figure depicts a folded protein not as a collection of atoms, but as a single, complex, three-dimensional standing wave of ULF energy. This intricate geometry represents the molecule's most stable and coherent resonant state, and its specific biological function is a direct consequence of this precise resonant shape.

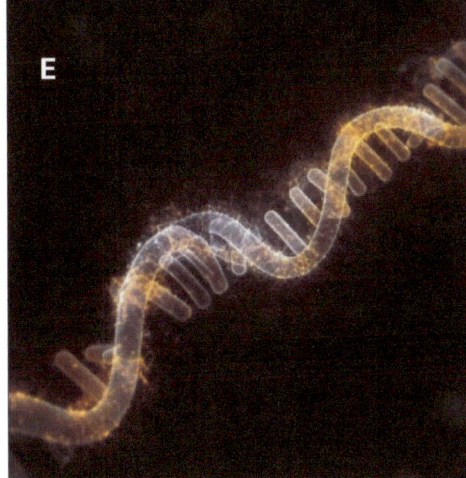

(E) RNA as a Resonant Precursor. This figure illustrates a single RNA molecule, the likely first building block of life. In the RM framework, its single-stranded structure is resonantly unstable, making it a poor medium for permanent information storage but an ideal one for a transient, disposable message.

257

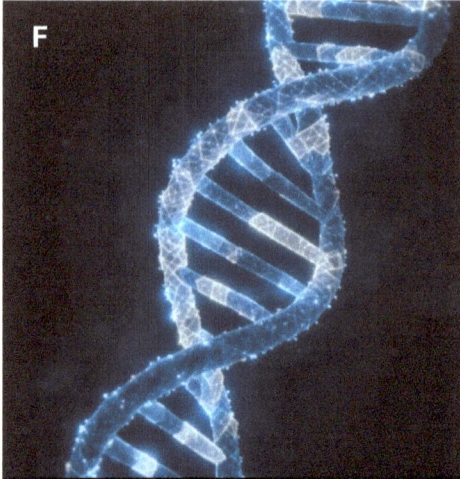

(F) DNA as a Resonant Information Archive: This figure depicts the DNA double helix as a structure of profound geometric and resonant perfection. Its twisted shape is the most stable geometry for long-term archival storage. The base pairs connecting the two strands are shown as perfectly interlocking geometric patterns of light, demonstrating resonant congruence. The genetic code is a physical, one-dimensional sequence of these specific resonant geometries.

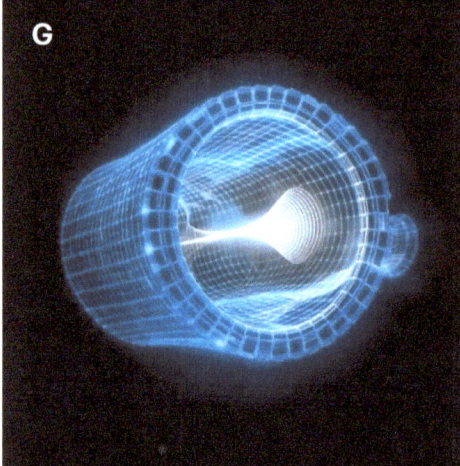

(G) Carbon Nanotube as a Resonant Waveguide: This figure illustrates the carbon nanotube as a structure of perfect geometric coherence. Its hexagonal lattice forms a flawless "resonance net" that perfectly distributes stress, while its seamless cylindrical shape supports a continuous, uninterrupted ULF standing wave along its length. This "longitudinal coherence" is the physical origin of its extraordinary tensile strength.

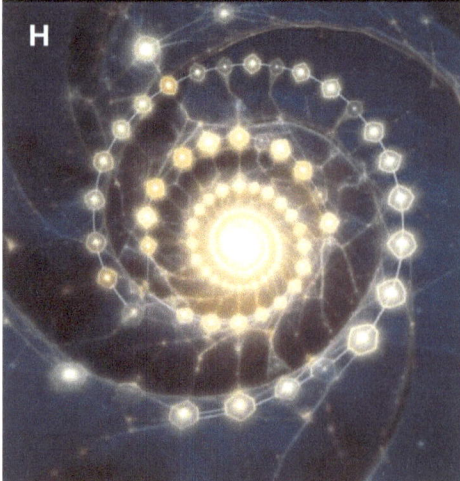

(H) The Fibonacci Sequence and the Geometry of Growth: The Fibonacci spiral, seen throughout nature, is the direct physical signature of the GCF in action. It represents the most efficient solution for packing and growth—the "path of least resonant strain"—allowing a system to expand without losing its structural harmony. The sequence is therefore not a mathematical coincidence, but the observable result of a physical system maintaining a state of perfect, continuous resonance as it unfolds.

4.4 The Fibonacci Sequence: The Geometry of Growth

One of the most beautiful and mysterious patterns in nature is the persistent appearance of the Fibonacci sequence (1, 1, 2, 3, 5, 8...) and the related Golden Ratio (ϕ) in the growth of living things, from the spiral of a sunflower's seeds to the chambers of a nautilus shell. While often viewed as a mere evolutionary optimization, the RM framework offers a deeper physical explanation, proposing that these patterns are the direct result of biological systems seeking the most geometrically efficient and resonantly stable way to grow.

The Fibonacci sequence is the simplest mathematical rule that allows a system to add new components without disrupting the resonant harmony of the existing structure. It is the optimal solution for packing and growth. In this view, the Fibonacci sequence can be considered the "mathematical signature" of the GCF. It is the geometric expression of a system seeking and maintaining a state of perfect, continuous resonance as it unfolds (Figure 34H). The reason this pattern is so natural is that it is the most elegant and efficient solution to the universal problem of how to build a complex structure from a simple, repeating rule. This principle is fundamental to the fabric of our geometric universe.

Thus, from the versatile architecture of carbon and the resonant folding of proteins to the geometric elegance of DNA and the mathematical signature of growth in the Fibonacci sequence, these case studies demonstrate a unified principle: the machinery of life is not a product of random chance but a direct consequence of the universe's fundamental drive to find states of maximal geometric and resonant stability.

5. Fluid Dynamics from Field Dynamics: The Emergence of Flow

To fully situate the UFD framework, we must address the nature of flow itself. The Navier-Stokes equations provide the mathematical description of macroscopic fluid behavior (Navier, 1822; Stokes, 1845), while Kinetic Theory and Molecular Dynamics (MD) simulations are the primary computational tools for modeling fluids as vast collections of interacting particles (Boltzmann, 1896). UFD, in turn, starts with the field as fundamental and derives the properties of fluids as emergent consequences of underlying vortex interactions (Figure 35).

To demonstrate how these microscopic vortex principles manifest in a real-world substance, we turn to the most studied and anomalous fluid in nature: water.

5.1 Case Study: The Geometry of Water

Water is the perfect case study for this principle, as its well-known anomalous properties are a direct reflection of its unique ULF resonance geometry. The polarized, tetrahedral shape of the water molecule drives the formation of strong "resonant couplings" (hydrogen bonds), which creates a coherent, dynamic network rather than a simple disordered soup. This underlying resonant architecture explains its unique properties:

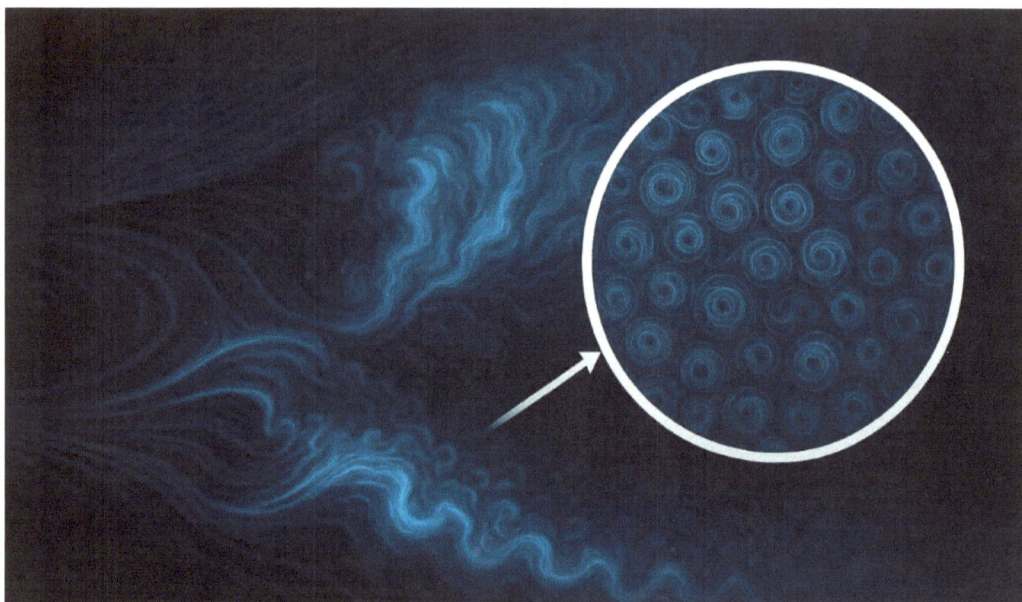

Figure 35: The Emergence of Classical Fluid Dynamics from Field Dynamics. This figure illustrates the self-consistency of the UFD framework. The large-scale image shows the familiar patterns of laminar and turbulent flow in a classical fluid. The magnified view reveals the underlying reality: the fluid is composed of trillions of individual molecular "eddies," which are complex, resonant structures in the ULF. Macroscopic properties such as pressure, flow, and viscosity are statistical, collective consequences of the fundamental fluid-dynamic interactions among countless microscopic vortices.

- Its high surface tension reflects the stability of resonant eddies at the air-water interface.

- Its anomalous viscosity is a direct consequence of this structured network. Applying pressure can physically disrupt the coherent tetrahedral alignment, allowing the molecular eddies to flow with less resistance.[58]

---

[58] The anomalous viscosity of water refers to its counterintuitive behavior under pressure. While most liquids become more viscous (thicker) when compressed, water at low temperatures (below ~33°C)

260

- Its powerful solvent capabilities arise from its dynamic ability to reconfigure its local ULF resonances to mesh with the geometry of dissolved molecules.

- Its anomalous density upon freezing is the result of the system locking into a more open, less-dense geometric lattice, specifically the hexagonal lattice of hydrogen bonds. This ordered state maximizes the system's geometric efficiency while requiring a larger volume.[59]

In this view, the familiar properties of water are a direct macroscopic manifestation of the geometric and resonant principles that govern the quantum world.

5.2 Interpreting the Navier-Stokes Equations

The geometric insights gleaned from water's complex structure apply universally. The UFD framework achieves self-consistency by formally translating the entire Navier-Stokes momentum equation into a statement of field dynamics. While the equations describe the macroscopic behavior of a fluid's elements, the UFD model provides a direct physical interpretation for each core term by translating them into field dynamics:

- Inertial Term ($\rho \cdot a$): Arises from the Hydrodynamic Added Mass of molecular vortices displacing the ULF medium during acceleration (*see* Chapter 4).

- Pressure Gradient ($-\nabla_p$): The force driving flow is the macroscopic average of the microscopic GCF pushes and pulls mediated by the ULF between constituent vortices.

- Viscous Term ($\mu\nabla^2 v$): Represents Resonance Drag, which is the energy consumed to break and reform the coherent resonant couplings between neighboring vortices. In UFD, this is a phase-switching cost; as molecular eddies slide past one another, their shared ULF envelopes must momentarily "melt"

---

becomes *less* viscous. This is because pressure breaks down water's bulky, open, and flow-resistant network of hydrogen bonds, allowing the molecules to move past each other more easily.

[59] The density anomaly of water refers to its counterintuitive property of being denser as a liquid (at 4°C) than as a solid (ice at 0°C). In the UFD framework, this is a direct macroscopic consequence of the GCF seeking the most efficient geometric state. The polarized, tetrahedral structure of the water molecule's ULF vortex network allows it to form highly efficient, but loose, hexagonal lattice structures as it freezes. While this crystalline structure is more geometrically coherent (stable) than the fluid's dynamic, disordered network, the lattice geometry requires a larger volume, resulting in a lower density (ice floats). This confirms that geometric stability, rather than minimal volume, is what governs the system's phase transition.

and "re-freeze" into new configurations. This energy cost is the reverse of the Coherence Dividend.

- Body Force (f): Arises from the large-scale UEF tension (gravity) or ULF field strain (electromagnetism) acting on the fluid volume.

The statistical behavior of these molecular eddies defines flow: laminar flow occurs when they slide past one another in an orderly way, while turbulent flow arises when this order breaks down.

This interpretation reframes the Navier–Stokes Millennium Prize problem—the question of whether fluid flow can reach a point of infinite 'blow-up'—as a diagnostic rather than a paradox. In the UFD framework, the mathematical possibility of blow-up reflects an idealized assumption of infinite divisibility. Once the physical structure of the medium is accounted for, extreme vortex stretching no longer leads to singular behavior but instead triggers regulated phase transitions.

In classical mathematics, a 3D vortex can be stretched until its radius reaches zero. However, the UFD framework reveals that the medium is protected by a Topological Governor. As a vortex stretches toward the Planck Length, the Resonance Drag reaches a critical threshold because the ULF can no longer support the curvature of the vortex filament. Rather than spinning to infinity, the vortex undergoes a Geometric Phase Transition, "shedding" its excess energy into the ULF as radiation. This universal safety valve ensures that velocity remains finite and flow remains globally "smooth."

This field-shedding mechanism is the same process by which the ULF emerges from the UEF, as described in Chapter 2, and is also the same process that powers stellar fusion. In fusion, the Coherence Dividend is released as protons "lock" into a more topologically stable helium vortex, shedding excess energy as radiation.

5.3 The Exception of Superfluidity

The preceding sections described fluid motion as the balance between coherent field-driven order and dissipative loss. Superfluidity represents the singular limit in which dissipation vanishes entirely. It is the empirical case where the Geometric Coherence Force (GCF) fully overcomes all resistive mechanisms, enforcing a collective, non-dissipative flow state (Kapitza, 1938). This transition requires two necessary conditions:

1. Noise Suppression (The Role of Temperature): Thermal agitation must be reduced below the threshold at which coherent field couplings are disrupted. Cooling suppresses stochastic motion, allowing geometric coherence to dominate.

2. Geometric Compliance (The Role of Structure): The internal geometry of the substance must permit the GCF to enforce phase-locked motion across the entire system. Where geometry resists coherence, dissipation persists.

Liquid helium provides the clearest demonstration of this geometric imperative.

- Helium-4 (^{4}He): The ^{4}He nucleus—the Alpha Crystal—possesses a maximally symmetric tetrahedral geometry and integer spin. This symmetry allows the GCF to impose collective flow directly. Once thermal noise is sufficiently suppressed, the fluid transitions seamlessly into a superfluid state characterized by frictionless motion, quantized vortices, and long-range phase coherence.

- Helium-3 (^{3}He): In contrast, ^{3}He possesses half-integer spin and an internal geometry that resists direct collective motion. To satisfy the GCF's demand for coherence, pairs of ^{3}He atoms bind into Cooper pairs. This pairing constitutes a topological genus correction: two half-integer vortices combine to form an effective integer-spin composite capable of collective behavior. The resulting superfluid adopts a directional, anisotropic flow structure (p-wave symmetry), reflecting a geometric compromise rather than a fundamental limitation.

In both cases, superfluidity emerges not from an external force, but from matter reorganizing itself into the geometric configurations required for coherence. When dissipation becomes incompatible with the field's structural constraints, the system does not resist—it restructures. Superfluidity is therefore not an anomaly. It is macroscopic proof that matter, when freed from noise, naturally conforms to the frictionless flow state of the Universal Plenum.

This insight completes the interpretive arc of fluid dynamics within UFD. We have shown how ordinary flow arises from partial coherence, how turbulence reflects geometric frustration, and how superfluidity represents the limit of perfect compliance. The remaining question is no longer why nature behaves this way, but whether such coherence can be deliberately engineered.

We now move from interpreting natural systems to designing coherent ones, marking the transition to the practical application of UFD: Resonant Engineering.

6. Design by First Principles: Resonant Engineering

A framework is ultimately validated by its capacity to generate precise, falsifiable predictions that lead to new technologies. Here, we show how understanding matter as a layered system of resonant fields enables the intentional design and manipulation of material states.

6.1 Resonant Damping and Thermal Manipulation

In RM, thermal energy is not the random, chaotic motion of atoms; rather, it is the total energy stored in the vast and complex superposition of incoherent, collective ULF resonant modes (phonons). This physical picture leads to a novel prediction: Resonant Damping.

If thermal energy is a collection of waves, it can be canceled by other waves. We therefore predict that a precisely tuned electromagnetic field oscillating at a material's primary thermal frequency but perfectly out of phase would actively dampen the crystal lattice's natural vibrations through destructive interference. The technological implications of this principle would afford the following possibilities:

- Localized Cooling: By selectively reducing the amplitude of thermal vibrations in a targeted area, materials can experience temporary, localized cooling without the need for traditional refrigeration. This could revolutionize thermal management in electronics, reducing overheating and improving the efficiency of microprocessors and other devices.

- Directional Thermal Conductivity Control: The ability to modulate phonon coherence directionally could lead to the development of thermal diodes or switches, which are devices that control heat flow, similar to how electronic components control electrical current. This capability has applications in energy harvesting, thermal insulation, and advanced materials design.

- Enhanced Material Lifespan: Reducing thermal vibrations through resonant damping may decrease thermal fatigue and material degradation, thereby extending the lifespan of components exposed to cyclic heating or extreme environments.

- Precision Heat Management in Manufacturing: Controlling resonance-driven heat could enable ultra-precise thermal processing, improving outcomes in processes such as crystal growth, thin-film deposition, and additive manufacturing.

Experiments could validate this principle by applying tunable, out-of-phase fields to materials with known phonon spectra and detecting a corresponding drop in temperature. Confirming this effect would not only validate a core tenet of RM but would also open a new frontier in energy-efficient thermal control.

6.2 Resonant Flow Systems and Non-Dissipative Transport

The physical confirmation that the UEF/ULF is an inviscid superfluid leads to the design principle that frictionless flow is the cosmological rule, not the exception. This principle can be applied to large-scale engineering systems to achieve non-dissipative transport.

We predict that by focusing on geometric resonance rather than brute-force power, engineers can create systems with near-zero energy loss. The mechanism involves fabricating microfluidic channels and transport conduits whose geometry precisely matches the harmonic geometric modes of the flowing substance's molecular eddies. This technique creates a "geometric resonance lock" between the fluid and the channel, effectively minimizing the resonance drag that constitutes viscosity. Such a system would function as a macroscopic, non-dissipative flow system, improving the efficiency of fluid transport and high-velocity cooling in industrial and electronic applications.

6.3 Geometric Catalysis of Material States

The principle of Geometric Catalysis, which we first applied to nuclear reactions (*see* Chapter 7), can be extended to the material scales to revolutionize materials science. This technique uses precisely shaped, resonant electromagnetic fields as energetic "scaffolds" or "molds" to guide matter into higher-coherence states.

This approach would have two primary applications: First, it could be used to guide the material's physical structure. By creating a field that mimics the perfect geometry of a diamond, for example, we could dramatically accelerate the growth of large, defect-free crystals. Second, it could be used to alter a material's energetic state. A resonant field, shaped to match the coherent standing wave of a material's superconducting phase, could "nudge" the system into that state without extreme cooling. This suggests a direct path to the engineering of high-temperature superconductors, thereby transforming material science from a practice of chemical mixing to a new paradigm of field-based design.

Mastering the field-based engineering of coherence would eliminate energy loss in electrical systems, leading to a cascade of revolutionary applications. These include:

- Lossless Power Grids: Energy transmission would become perfectly efficient, ending the need for vast power plants and eliminating transmission waste.

- Maglev Transport: High-speed magnetic levitation trains and launch systems would become economically feasible worldwide, requiring minimal energy input.

- Ultra-Fast Electronics: Components would operate without thermal resistance, enabling a massive leap in computing speed and efficiency for everything from quantum computers to microprocessors.

- Perfect Energy Storage: The ability to store current in closed loops indefinitely would revolutionize battery technology and energy storage on a grid scale.

6.4 Field-Tuned Semiconductors and the Design of Electronic Materials

In RM, a semiconductor is a material whose crystal lattice is poised at the very edge of coherence. At rest, it is an insulator, but it is geometrically tuned so that a small input of energy from light, heat, or an external field can tip it over a threshold into a conductive, resonant state. This unique sensitivity makes it the ideal material for an electronic switch and leads to several novel predictions and technologies:

- Field-Controlled Conductivity: We predict that a material's conductivity can be tuned in real-time by applying a patterned electromagnetic field that matches its natural resonant frequencies. This would directly manipulate the material's ULF coherence, creating a "virtual" semiconductor that can be switched on or off without the physical movement of charge carriers.

- Doping as Coherence Engineering: In our model, dopant atoms are not merely electron donors or acceptors; they serve as local coherence enhancers by creating geometric nodes in the lattice that lower the energy required to form a crystal-wide resonance. We predict that isotopically tuned dopants, whose nuclei have a more perfect geometric structure, will be far more efficient than conventional dopants.

- The Coherence Gate: We propose a new type of transistor, the "coherence gate," which operates on the phase of the field. By using interfering electromagnetic fields, a segment of the material could be switched between conductive and insulating states, enabling ultra-fast, non-volatile computing based on phase resonance rather than electron flow.

- Programmable Organic Electronics: Organic polymers, with their loosely bound electrons, are ideal candidates for this technology. We predict that by applying precisely shaped fields, ordinary plastics could be transformed into dynamically tunable organic semiconductors, with applications in flexible electronics, biosensors, and adaptive neural interfaces.

6.5 Resonant Quench of Superconductivity

Our model's core claim—that superconductivity is a state of perfect, crystal-wide resonance—leads to a unique and testable prediction: a "resonant quench." We predict that a precisely tuned electromagnetic field oscillating at a frequency inharmonious with the primary superconducting resonance of the material will disrupt the system's coherence through destructive interference. This would cause the material to instantly revert to its normal, resistive state, even while it remains below its critical temperature. The discovery of such a resonant quench would not only provide a new, non-thermal method for controlling superconductivity—uniting it with semiconduction—but would also serve as powerful evidence that this exotic state of matter is a macroscopic manifestation of geometric and resonant coherence.

6.6 Induction and Control of Supermalleability

Our framework also predicts the existence of a new phase of matter: supermalleability, the mechanical counterpart to superconductivity. This exotic state emerges when a material's entire resonant architecture, from its UEF nuclei to its ULF orbitals, locks into a state of total coherence. In this phase, physical stress is no longer localized as fractures or defects but is distributed as coherent waves throughout the material, allowing it to deform like a fluid-solid while retaining immense structural integrity.

This theory leads to several testable predictions: a material subjected to a precisely tuned electromagnetic field should exhibit measurable "resonant softening," which would enable it to be reshaped with minimal force; we also predict that specially designed alloys with highly coherent nuclear geometries could be driven into a stable, hyper-ductile state by an applied field. Crucially, this transition into and out of supermalleability should be completely reversible and field-dependent, providing a direct, falsifiable test of the phenomenon.

6.7 Synthesis of a Resonance-Locked Crystal (Emetium)

As the culmination of the principles outlined in this chapter, we propose the potential creation of a Resonance-Locked Crystal, or Emetium. This is a theoretical material in which the nuclear (UEF) and orbital (ULF) resonances are locked into a state of perfect, crystal-wide coherence. The synthesis of Emetium is an act of resonant catalysis, utilizing UFD's precise geometric control rather than brute force. The conceptual process requires three stages:

1. Forging the Seed: A single, flawless A=256 hyper-coherent nucleus is constructed architecturally. This requires Geometric Catalysis to guide 64 alpha-particle "bricks" into a perfect $4 \times 4 \times 4$ cubic standing-wave scaffold, snapping the structure into a unified state and releasing a colossal Coherence Dividend.

2. Preparing the Soil: This perfect seed is then embedded into a host material chosen for its compatible crystal lattice geometry, such as the layered structures found in high-temperature superconductors.

3. Catalyzing the Phase Transition: A precisely shaped external resonant field is applied. This field acts as a catalyst, amplifying the perfect resonance of the seed and forcing the surrounding material to reconfigure its nuclear and electronic structures to match the perfect, coherent state.

A perfect analogy is growing a large, flawless crystal in a supersaturated solution. To grow such a crystal, one requires only a single, tiny, perfect seed crystal. When this seed is placed in the right environment, the surrounding material naturally crystallizes onto it, adopting its perfect structure. In this way, Emetium is not built; it is grown.

Were it produced, Emetium would be more than just a static "super-metal." It would function as a perfectly responsive and programmable medium. A civilization capable of creating Emetium could unlock a new technological paradigm. Its potential applications include the following:

- Programmable Matter: Structures built from Emetium would no longer be fixed. An aircraft wing could morph its shape mid-flight, or buildings could reconfigure their layouts, as the material's form could be controlled directly by resonant fields.

- Perfect Energy Storage: The crystal lattice could store energy as a stable, high-energy vibrational mode with 100% efficiency. This "resonant battery" would suffer no degradation and could be charged or discharged almost instantaneously.

- Field-Based Computing: An Emetium crystal could function as a holographic computer, where computations are performed by creating and interacting with stable wave patterns within the material itself, enabling unimaginable speed and efficiency.

- Direct Consciousness-Field Interfacing: As the ultimate resonant medium, Emetium would be the ideal material for a direct brain-computer interface, allowing for seamless, thought-based control of technology.

In summary, Emetium is the substrate for a new technological era, one that would enable a civilization to engineer reality with the pure geometry of resonance itself.

7. Conclusion

This chapter has demonstrated that the familiar properties of the material world—from the conductivity of a metal and the color of a pigment to the strength of a crystal—are the macroscopic expression of a single, underlying principle: the geometric and resonant coherence of its constituent fields. By scaling up RC from individual bonds to the full crystal lattice, we have provided a tangible, physical representation of the abstract concepts in standard physics, revealing them to be different facets of a universe governed by harmony and form.

RM is not merely an explanatory tool; it is a predictive framework that opens the door to a new paradigm of Resonant Engineering. It suggests that we can actively suppress thermal energy via resonant damping and induce exotic states, such as superconductivity, supermalleability, and even "the perfect metal," with tuned fields.

Ultimately, these same principles extend beyond materials science to the very architecture of life itself, reframing biology as a system of tunable, geometric resonances. The traditional boundaries between physics, chemistry, and biology dissolve, giving way to a unified framework. From the structure of a single bond to the function of a living cell, we find a single, underlying principle: the universe's relentless drive to express geometry through resonance.

10.3 A New Science of Matter

This chapter has demonstrated that the tangible properties of the material world—from the strength of a crystal to the function of a living cell—are the macroscopic expression of the geometric and resonant coherence of underlying fields. By scaling the principles of Resonant Chemistry from the single bond to the entire lattice, we have shown how a single, unified framework can account for the full spectrum of material behaviors:

- Electromagnetic Properties: Electrical conductivity, optical spectra, and ferromagnetism emerge from the degree of coherence in the collective ULF resonance, providing a physical picture for abstract concepts like band gaps and the exchange interaction.

- Acoustic & Thermal Properties: Sound is revealed as the propagation of coherent vibrations (phonons) through the lattice, while heat is the expression of incoherent, delocalized resonance.

- Exotic States: Phenomena like high-temperature superconductivity and the predicted state of supermalleability are no longer anomalies. Instead, they are the natural result of a material achieving a state of perfect, crystal-wide resonant lock.

- The Architecture of Life: The unique properties of organic matter—from the versatile geometry of carbon and the resonant folding of proteins to the informational stability of DNA—are shown to be direct consequences of the universe's fundamental drive to find states of maximal geometric and resonant coherence.

- Fluid Dynamics from Field Dynamics: In a test of the theory's ultimate self-consistency, the classical phenomenon of fluid dynamics itself is shown to emerge from the collective behavior of the trillions of individual "molecular eddies" that constitute a liquid.

This new understanding does more than just explain matter; it provides a blueprint for a new technological paradigm. The ability to engineer matter from its foundational principles of resonance opens the door to a range of visionary applications:

- Resonant Engineering: A new design philosophy based on tuning a material's properties by manipulating its resonant fields. This includes Geometric Catalysis to induce superconductivity and Resonant Damping to achieve targeted, non-cryogenic cooling.

- Emetium, The Perfect Metal: The theoretical synthesis of a Resonance-Locked Crystal, where nuclear (UEF) and orbital (ULF) harmonics are locked in perfect coherence, creating a self-healing, ultra-strong, and supermalleable material.

If confirmed, this framework would dissolve the traditional boundaries between physics, chemistry, and biology, replacing them with a unified science of matter. Ultimately, this paradigm suggests that matter is not a passive collection of particles but a living tapestry of form and flow. In this respect, the future of materials science lies not in the mastery of force but in understanding resonance.

References

1. Alder, B. J., & Wainwright, T. E. (1957). Phase Transition for a Hard Sphere System. *The Journal of Chemical Physics, 27*(5), 1208–1209.
2. Anderson, P. W. (1972). More Is Different. *Science, 177*(4047), 393–396.
3. Ashcroft, N. W., & Mermin, N. D. (1976). *Solid State Physics*. Holt, Rinehart and Winston.
4. Bardeen, J., Cooper, L. N., & Schrieffer, J. R. (1957). Theory of Superconductivity. *Physical Review, 108*(5), 1175–1204.
5. Bloch, F. (1929). Über die Quantenmechanik der Elektronen in Kristallgittern (On the Quantum Mechanics of Electrons in Crystal Lattices). *Zeitschrift für Physik, 52*(7–8), 555–600.
6. Boltzmann, L. (1964). *Lectures on Gas Theory*. (S. G. Brush, Trans.). University of California Press. (Original work published 1896).
7. Callister, W. D., & Rethwisch, D. G. (2018). *Materials Science and Engineering: An Introduction* (10th ed.). John Wiley & Sons.

8. Debye, P. (1912). Zur Theorie der spezifischen Wärmen [On the Theory of Specific Heats]. *Annalen der Physik, 344*(14), 789–839.

9. Gilbert, W. (1986). The RNA world. *Nature, 319*(6055), 618.

10. Heisenberg, W. (1928). Zur Theorie des Ferromagnetismus [On the Theory of Ferromagnetism]. *Zeitschrift für Physik, 49*(9-10), 619–636.

11. Jumper, J., et al. (2021). Highly accurate protein structure prediction with AlphaFold. *Nature, 596*(7873), 583–589.

12. Kapitza, P. L. (1938). Viscosity of Liquid Helium Below the λ-Point. *Nature, 141*(3581), 74–75.

13. Kittel, C. (2005). *Introduction to Solid State Physics* (8th ed.). John Wiley & Sons.

14. Navier, C. L. M. H. (1822). Sur les lois du mouvement des fluides [On the laws of the movement of fluids]. *Mémoires de l'Académie Royale des Sciences de l'Institut de France, 6*, 389–440.

15. Rayleigh, J. W. S. (1877). *The Theory of Sound*. Macmillan and Co.

16. Stokes, G. G. (1845). On the theories of the internal friction of fluids in motion. *Transactions of the Cambridge Philosophical Society, 8*, 287–319.

17. Taylor, G. I. (1934). The mechanism of plastic deformation of crystals. Part I.— Theoretical. *Proceedings of the Royal Society of London. Series A, 145*(855), 362–387.

Chapter 11: Resonant Biochemistry

11.1 Biochemistry

Biochemistry is a scientific discipline that exists at the intersection of chemistry, biology, and medicine. Its roots can be traced back to the 19th century, when scientists first began to seriously investigate the chemical composition of living matter. Pioneers such as Justus von Liebig laid the foundation for the field by exploring the chemistry of metabolism and fermentation, emphasizing the chemical principles underlying plant and animal nutrition. His work helped establish the idea that biological phenomena could be explained using the laws of chemistry.

In the early 20th century, figures such as Emil Fischer, who unraveled the stereochemistry of sugars and purines, and Otto Warburg, who investigated the role of enzymes in respiration and photosynthesis, further cemented the role of chemical logic in biological systems. The field matured as the structures of biomolecules were gradually resolved: Linus Pauling described the α-helix and β-sheet structures of proteins, while Watson and Crick, drawing upon Rosalind Franklin's X-ray crystallography data, unveiled the double helix of DNA.

With the molecular revolution of the mid-20th century, biochemistry blossomed into a central pillar of modern science. The identification of ATP as the "molecular currency" of energy, the decoding of metabolic pathways like glycolysis and the Krebs cycle, and the discovery of enzymes as precision biological catalysts provided a compelling picture of life as a set of finely tuned molecular machines.

By the early 21st century, biochemistry had become closely intertwined with molecular and systems biology. Advanced tools, such as mass spectrometry, NMR spectroscopy, and cryo-electron microscopy, have enabled the unprecedented resolution of biochemical processes. The central dogma (DNA $\rightarrow$ RNA $\rightarrow$ protein) became the organizing principle. Life was now understood as a complex network of interactions among macromolecules, driven by thermodynamics, governed by gene regulation, and modulated by feedback control.

However, despite its achievements, the classical paradigm has several limitations. While breaking life down into smaller and smaller parts has revealed much about the components of living systems, it often struggles to explain their coherence and unity. How do cells maintain order in the midst of chaos? How do enzymes achieve such extraordinary specificity and timing? Why do biological systems so reliably regenerate form and function in the face of entropy?

Even with the integration of systems biology and network theory, the field still lacks a fully unified, first-principles account of how biological structure and function emerge, not just statistically, but dynamically. Thermodynamic explanations capture energy gradients, but not the elegant choreography by which energy flows. Molecular modeling explains bonds and interactions, but not the spatial-temporal harmony that pervades living matter. This lack of a complete, satisfying explanation is particularly apparent in the following areas:

- Bioenergetics, where ATP synthesis is understood in terms of proton gradients, but the spatial and geometric coordination of the process remains elusive.

- Enzymology, where catalysis is attributed to transition-state stabilization, even while the temporal resonance of catalytic action is poorly characterized.

- Signal transduction, where molecules relay information without us knowing how global coherence emerges from these local events.

We are thus left with a picture of life that is chemically complete but energetically fragmented. We know the parts but not the whole.

This chapter proposes a shift from a mechanistic view of life to a resonant one, which we call Resonant Biochemistry (RB). Namely, RB proposes that life emerges not merely from chemical interactions, but from the resonant geometry of fields: mitochondria and chloroplasts are not just biochemical engines—they are toroidal oscillators that manage energy through harmonic field patterns; enzymes do not simply reduce energy barriers—they act as resonance filters, phase-locking molecules into precisely timed transformations; even sugars like glucose can be described as geometrically frozen waveforms.

This is not a rejection of classical biochemistry. RB retains the knowledge of molecules and mechanisms; it just embeds this knowledge in a deeper physics, allowing us to understand how coherence is built and maintained in living systems. The implications are vast. From understanding metabolic disorders as breakdowns in coherence to designing field-guided biomaterials to developing resonance-based therapies that heal not through force but through re-synchronization, this new framework invites a holistic reimagination of what biochemistry can be.

This chapter serves as our entry point into this new, resonant paradigm of biochemistry. We begin with the resonant infrastructure of the cell and signal transduction, and from there, explore how life organizes energy through geometry, vibration, and field logic. This is biochemistry, reborn as resonance.

11.2 Resonant Biochemistry: A UFD Model of Photosynthesis, Metabolism, and Coherence in Living Systems

Abstract

This chapter introduces Resonant Biochemistry (RB), a new model of biochemistry that reframes the foundations of living processes through the lens of Unified Field Dynamics (UFD). Rather than a network of stochastic chemical interactions, we present life as a system of coherent field geometries and vibrational harmonies. Photosynthesis and metabolism are revealed as structured energy architectures, driven by the toroidal dynamics of mitochondria and chloroplasts. Biological functions, such as enzymatic catalysis, signal transduction, and homeostasis, emerge from synchronized vibrations at the molecular scale, with health defined by the preservation of this coherence.

When this coherence breaks down, the result is inflammation, metabolic collapse, or cancer, which are recast as systemic failures in field regulation. By grounding life in geometry and field-based resonance, RB offers a unified framework that bridges chemistry, physics, and biology. This reimagining is grounded in a key falsifiable prediction: the Resonant Modulation Hypothesis, which posits that dissonant electromagnetic fields can control enzymatic function. RB also opens the door to resonant medicine, biofield diagnostics, and field-guided synthetic biology. In this view, biochemistry becomes the study of how energy moves through living matter in rhythmic, self-organizing harmony.

1. Introduction

Biochemistry was born from the revolutionary idea that the universal laws of chemistry could explain the processes of life. Beginning in the 19th century with Friedrich Wöhler's synthesis of urea from inorganic materials (Wöhler, 1828), the field systematically dismantled the notion of a "vital force" unique to living organisms. This mechanistic approach has been incredibly successful, giving us a detailed "parts list" of life by mapping metabolic pathways and identifying the molecular machinery of the cell. However, the challenge with this "parts list," for all its detail, is that it fails to explain how life functions as a coherent, self-organizing whole.

In this respect, we present Resonant Biochemistry (RB), a framework that reinterprets life as a harmonic orchestration of form and energy, rather than a collection of molecular reactions. In this view, the cell is a resonant engine, and molecules do not simply react: they synchronize. Enzymes act as resonance filters, and metabolism and photosynthesis are resonant systems that capture and convert energy.

This chapter presents the outlines of this new picture from the ground up, starting with the resonant infrastructure of the cell and signal transduction. We then scale up to the dynamics of photosynthesis, metabolism, and disease. Ultimately, RB reframes the study of life as a form of resonance ecology, providing the tools to understand how life holds together, how it can be harmonized, and how we can begin to design with its fundamental, vibrational architecture.

2. The Resonant Infrastructure of the Cell

In traditional biochemistry, signal transduction is understood as a stepwise cascade of chemical switches. The process is mechanical and linear:

1. A "first messenger" (like a hormone) travels through the bloodstream.

2. It physically binds to a "lock-and-key" receptor on a cell's surface.

3. This binding triggers the release of a "second messenger" chemical inside the cell.

4. This second messenger then diffuses through the cell, bumping into and activating a specific enzyme, which in turn activates another, in a "billiard ball" or "domino-like" cascade (Alberts et al., 2002).

While this model is chemically accurate, it is conceptually incomplete, as it struggles to explain the incredible speed, fidelity, and profound system-wide coherence of a cellular response, often treating the cell as a "bag of chemicals" where interactions are reduced to random diffusion and collisions.

In the RB framework, the cell is not a simple "bag of chemicals." It is a highly structured, self-organizing resonant system. Its key components are not just the passive parts of a machine; they are dynamic, field-sensitive structures that work in concert to create, sustain, and communicate coherence. The most elegant way to understand this architecture is to follow the path of a signal as it is processed by the cell (Figure 36).

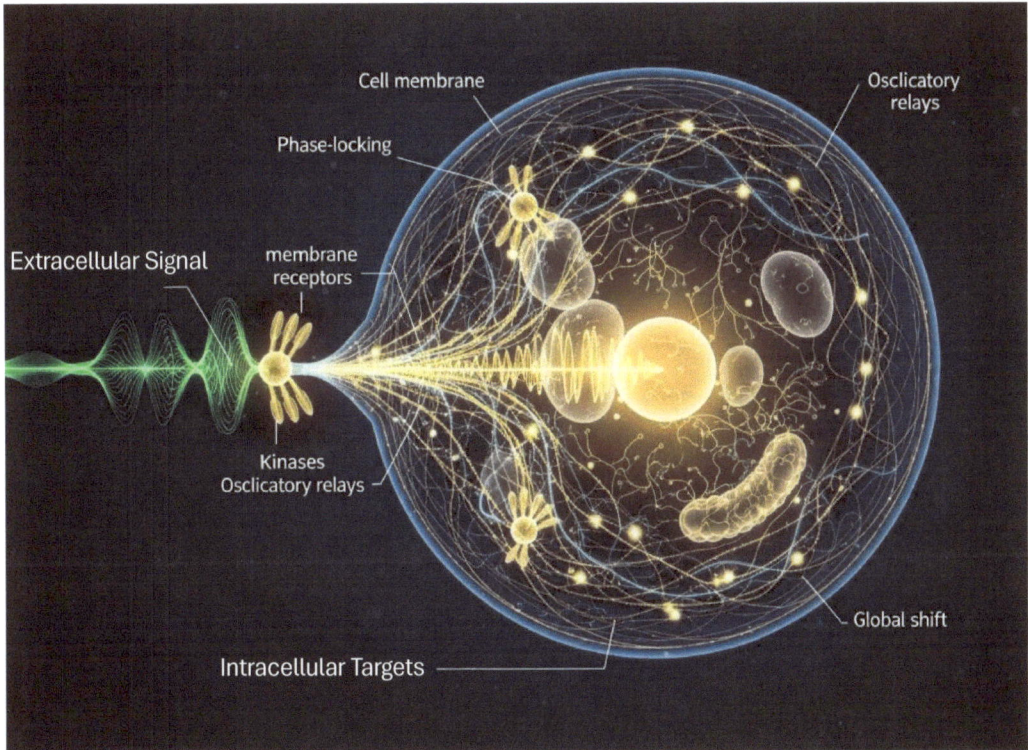

Figure 36: Signal Transduction as Resonant Communication. This figure illustrates signal transduction as a resonance-guided communication system within the RB framework. An extracellular signal (left) with a specific vibrational frequency is received by a membrane receptor that acts as a molecular antenna. This signal is then transmitted through the cytoskeleton, which functions as a resonant conduit that guides the coherent energy wave toward intracellular targets. This process culminates in a global shift in the cell's vibrational state, demonstrating how information flows through a living system as a continuous and harmonious wave of resonance.

The process of signal transduction begins at the cell membrane, which is the cell's primary resonant interface with its environment. The lipid bilayer acts as a resonant insulator, shielding the delicate intracellular machinery from the chaotic "noise" of the outside world, a function powered by the stable ULF field of the membrane potential (Hodgkin & Huxley, 1952). The receptors embedded in this membrane are not simple "locks" waiting for a chemical "key"; they are sophisticated molecular antennas precisely tuned to the vibrational frequencies of their specific ligands (Fröhlich, 1968). The binding of a ligand is thus an act of resonant entrainment, [60] in which a signaling molecule with the correct vibrational frequency phase-locks with the receptor to change its resonant state (Purves et al., 2018).

Once a signal is received, it rapidly propagates through the cell interior via the cytoskeleton. This intricate network of microtubules and filaments serves as a structured, resonant waveguide that channels coherent vibrational modes throughout the cytoplasm, enabling long-range, phase-locked communication necessary to coordinate the activity of distant organelles (Pienta & Coffey, 1991).

Along this path, enzymes such as kinases function as oscillatory relays or resonant filters. In this model, the three-dimensional structure of an enzyme creates a geometric cavity that is precisely tuned to the vibrational mode of its specific substrate, guiding its chemical transformation (Nelson & Cox, 2017). As relays in a signaling cascade, they amplify and filter the signal based on frequency coupling and harmonic entrainment.

Ultimately, a successful signal transduction event culminates in a systemic phase transition or whole-cell entrainment, pulling the entire resonant architecture of the cell into a coherent and stable attractor basin within its dynamic resonance geometry (Camazine et al., 2003). In this view, the cell is a single, unified resonant system. With this new understanding of the cell's resonant components, we can now explore the grand, dynamic processes they perform: the great symphonies of photosynthesis and metabolism.

3. Photosynthesis as Field Capture and Conversion

Photosynthesis is usually described as the process of converting sunlight into chemical energy. However, this description misses something essential. From the RB perspective, photosynthesis is about field capture, not energy conservation; it is the means by which living systems extract coherent energy from the electromagnetic field and store it in stable molecular geometries. In other words, light is not just fuel; it is a structured vibration. Plants (and some bacteria) have evolved a type of molecular

---

[60] Entrainment refers to the synchronization of two or more independent oscillatory systems due to their interaction. The classic example is Christiaan Huygens' 17th-century observation that two pendulum clocks on the same wall would eventually swing in perfect unison. In this context, "whole-cell entrainment" describes how a local event, like a signal binding to a receptor, can trigger a cascade that pulls the oscillatory processes of the entire cell—from membrane potentials to metabolic rhythms—into a new, unified, and coherent harmonic state.

antenna that can tune into specific vibrational frequencies, stabilize them, and lock them into form. The entire process, from light absorption to sugar synthesis, can thus be seen as a smooth transformation from radiant-field resonance to stored vibrational coherence (Figure 37).

3.1 The Light-Harvesting Complex as a Resonant Antenna

At the front of the photosynthetic machinery is the light-harvesting complex, which is made of pigment molecules like chlorophyll arranged in a ring. Traditionally, it is said to absorb photons and transfer the energy to a reaction center. However, in RB, this structure behaves more like a resonance cavity with a molecular antenna geometrically optimized to detect specific frequencies of the electromagnetic field.

Each pigment has its own resonance profile. When the frequency of the incoming light matches that profile, it does not just excite an electron. Instead, it creates a standing wave in the pigment's field structure—a pattern of vibrational energy that can be passed from one pigment to the next, not through hopping or collision but through phase-locked resonance. This is why photosynthesis is so efficient. The energy does not get lost along the way. Instead, it remains coherent as it propagates through the complex as part of a resonance-transfer system (Engel et al., 2007; Scholes et al., 2017).

3.2 The Reaction Center as a Field Converter

At the core of the system is the reaction center, where the incoming vibrational energy is converted into stored chemical potential. This is where the system achieves something remarkable: it transforms a field wave into a molecular form. The field energy gets trapped in specialized cofactors, triggering an electron-transfer cascade and generating a proton gradient across the membrane. This gradient is what powers ATP synthase, which uses rotational resonance to generate ATP. Thus, the reaction center is not just a biochemical device; it is a field-to-matter converter that turns coherence into structure and structure into potential.

3.3 Photochemistry as Geometric Encoding

The result of photosynthesis is ATP, glucose, and other carbohydrates. These molecules are not just energy "containers"; they are resonant geometries. The energy from the field has been captured, slowed down, and frozen into the shape of the molecule itself. Every bond in glucose thus represents a piece of stored coherence. When glucose is later metabolized, that coherence is gradually released, in steps, through the oscillatory machinery of metabolism. This model provides a new perspective on energy in biology: energy storage involves holding a waveform in place, and energy release allows the waveform to move again. Thus, when a plant turns light into sugar, it is recording a vibration and locking it into a geometric pattern so it can be used later.

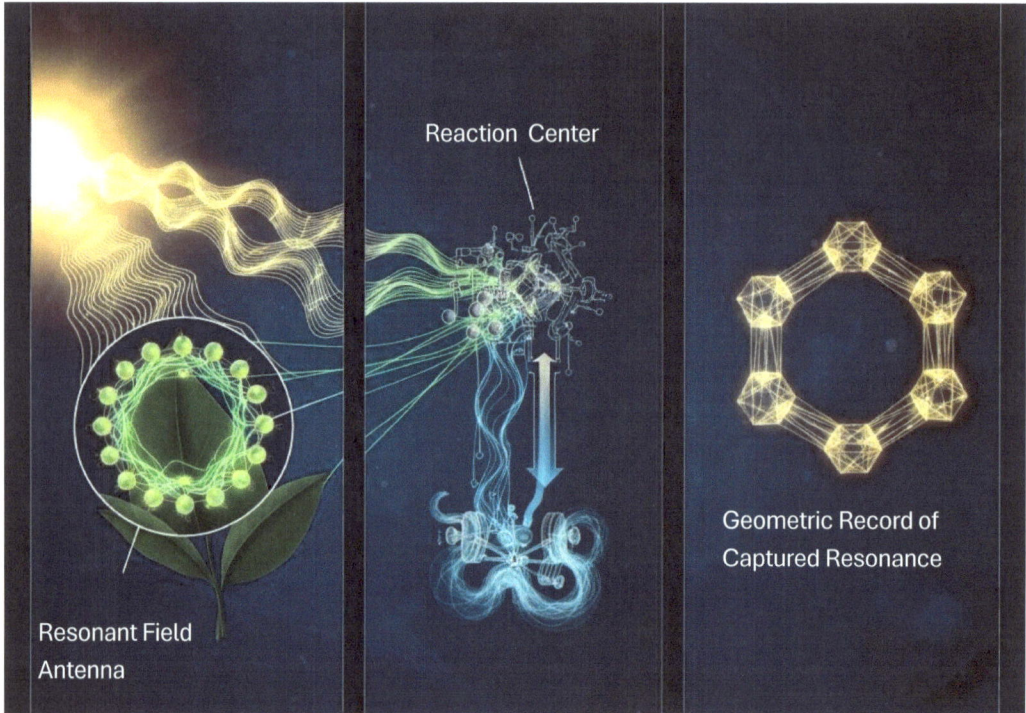

Figure 37. Resonant Photosynthesis. This diagram illustrates photosynthesis as a process of field resonance and geometric encoding within the ULF. (Left) Sunlight, as structured ULF vibrations, is absorbed by the light-harvesting complex, depicted as a "resonant molecular antenna" that precisely entrains and transfers coherent vibrational energy. (Center) This energy travels as a phase-locked resonance to the reaction center, which acts as a "field-to-matter converter," transforming coherent field vibrations into directional chemical flows. (Right) The ultimate result is the synthesis of glucose molecules, which represent "stored coherence," in which each bond serves as a physical, geometric record of captured field resonance, ready to be released later through metabolic oscillations.

Photosynthesis, when viewed through the lens of RB, is thus more than a biochemical pathway; it is a paradigm of field interaction that demonstrates how life captures and stores coherence from the environment with extraordinary geometric and temporal precision.

4. Metabolism as a Resonant Energy Circuit

If photosynthesis is the natural art of capturing coherent field resonance and "freezing" it into the stable, geometric patterns of molecules like glucose, then metabolism is the natural art of liberating that stored coherence.

While metabolism is traditionally defined as the sum of all chemical reactions that sustain life, this view is conceptually opaque. Foundational processes like the Krebs cycle are often taught as a cryptic sequence of reactions to be memorized, not as systems to be understood. RB provides this missing physical picture by reconceiving metabolism as a resonant energy circuit. In this view, the process is not driven by random collisions but by a controlled, step-by-step "resonant disassembly" of the geometric coherence that was first captured by light.

4.1 Glucose: The Primary Resonant Fuel

The circuit begins with its primary fuel source. Among all biological molecules, glucose is the primary energetic currency. In RB, it is not merely a simple sugar but a molecular structure perfectly optimized for resonant energy storage and controlled release. The key to its function is its six-membered ring geometry. This conformation produces a stable, symmetric ULF standing wave that minimizes internal strain, thereby making the molecule both stable and highly soluble in water (Voet & Voet, 2011). Its polar hydroxyl groups ensure compatibility with the cellular environment, facilitating efficient transport and recognition by enzymes (Berg, Tymoczko & Stryer, 2015).

Glucose's most remarkable property, however, is its dynamic, reversible nature. By oscillating between its ring and linear forms, glucose acts as a "modulatable capacitor" that can store energy in a stable resonant state and release it through carefully catalyzed geometric transformations. The process of glucose oxidation (glycolysis and the Krebs cycle) is not a simple "burning" of fuel; it is a controlled cascade of resonant disassembly. Each step in the pathway breaks the molecule down into progressively more stable and coherent geometric fragments, releasing the stored resonance as usable energy (Nelson & Cox, 2017).

4.2 Glycolysis: The "Controlled Demolition" Pathway

Before the primary "furnace" (the mitochondrion) can use this fuel, the large, stable glucose molecule must be "cracked" and prepared. This preparatory phase is glycolysis, a ten-step linear disassembly line that occurs in the cell's cytoplasm.

In RB, this is a "controlled demolition" driven by the Geometric Coherence Force (GCF). The ten enzymes in the pathway act as ten distinct Resonant Catalysts, each precisely tuned to "grip" the molecule and use its resonance to strain and break one specific bond in sequence. This controlled, step-by-step process takes the single, geometrically stressed 6-carbon glucose molecule and "snaps" it into two smaller, more coherent, and more stable 3-carbon pyruvate molecules (Figure 38). This initial geometric relaxation releases a small net Resonance Dividend (two ATP molecules), which the cell captures. This "pre-cracked" pyruvate fuel is now in a form that can be transported into the mitochondria for complete, far more powerful resonant annihilation.

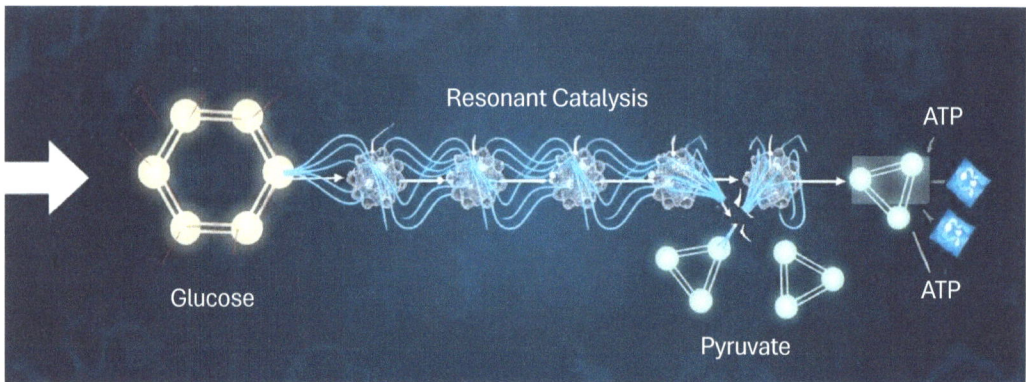

Figure 38: Glycolysis: The "Controlled Demolition" Pathway. This figure illustrates glycolysis as a ten-step linear "disassembly line" within the cell's cytoplasm, driven by the GCF. Ten distinct Resonant Catalyst enzymes are precisely tuned to "grip" the initial, geometrically stressed 6-carbon glucose molecule and use their inherent resonance to strain and sequentially break specific bonds. This controlled, step-by-step process "snaps" the glucose into two smaller, more geometrically coherent, and stable 3-carbon pyruvate molecules. This initial geometric relaxation releases a net Resonance Dividend of two ATP molecules, which the cell captures. The "pre-cracked" pyruvate fuel is then ready for transport into the mitochondria for further resonant annihilation.

4.3 The Mitochondrion: A Toroidal Engine of Metabolic Coherence

While classically described as the cell's "powerhouse," RB reinterprets the mitochondrion as a toroidal engine of metabolic coherence. Its iconic folded inner membranes (cristae) are not merely a means of increasing surface area. Instead, they exemplify the principle of form dictating function by creating a series of toroidal resonant cavities, geometrically perfect structures that stabilize, contain, and cycle the immense energetic processes of respiration. In contrast to modern biochemistry, which treats observed oscillations as mere byproducts of chemical reactions (Hüser & Blatter, 1999), RB posits that these rhythms are the mitochondrion's primary, causal function (Figure 39A).

Two key resonant components drive this engine. The first is the Electron Transport Chain (ETC). Described as a simple "chemical conveyor" in the Standard Model, the ETC is, in our framework, a field-guided resonator. It is a series of protein complexes that act as "resonant gates," guiding a coherent electron wave (a ULF phenomenon) through the membrane. This high-energy wave does not just "hop"; its passage is phase-locked, and its energy is rhythmically extracted to pump protons (UEF vortices) across the membrane, thereby establishing a stable, coherent UEF potential (Figure 39B).

The centerpiece, however, is ATP synthase, the "turbine" of this system or molecular rotor that functions as a quantum-phase engine (Figure 39C). ATP synthase masterfully converts the coherent linear current of the UEF (the proton gradient) into rotational resonance (Mitchell, 1961). This rotation, in turn, acts as a geometric forge, using the GCF (*see* Chapter 7) to stamp ADP and phosphate molecules into the highly coherent, geometrically strained, high-energy form of ATP.

The entire system thus operates as a field-regulated toroidal oscillator. It is a perfect, multi-stage machine that converts the raw geometric coherence of fuel (like glucose) into a stable, rhythmic UEF/ULF flux, which it then packages into the transportable, quantized coherence of ATP. It is this total system coherence, not merely the chemical output, that defines mitochondrial vitality and buffers the entire cell from resonant noise.

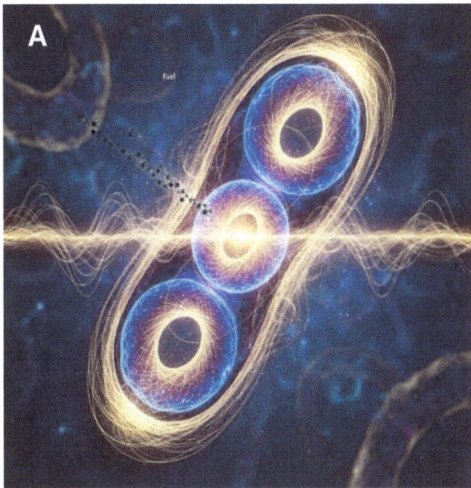

Figure 39: The Metabolic Engine of RB. (A) Mitochondria: Toroidal Engines of Cellular Resonance. This illustration presents mitochondria as dynamic toroidal energy engines. Within the RB framework, mitochondria are structured to sustain coherent energy flow. Their inner membrane folds (cristae) form toroidal geometries that stabilize and circulate resonant energy patterns, facilitating the efficient conversion of fuel into vibrational coherence. These toroidal dynamics support phase-locking across cellular systems, positioning mitochondria as central oscillators in the organism's energetic and metabolic coherence.

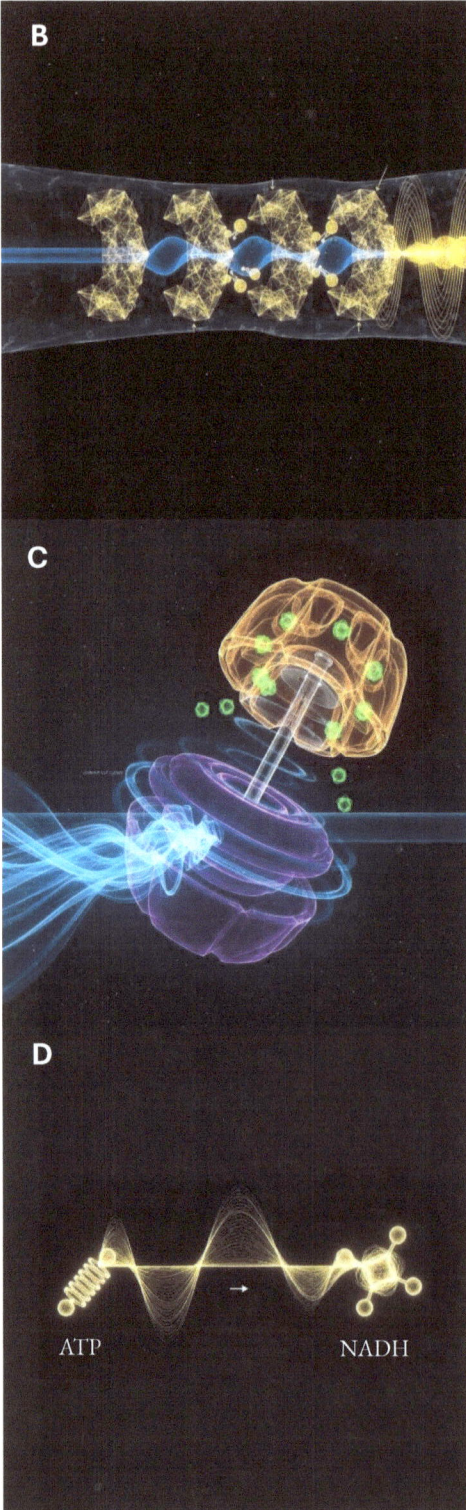

(B) The Electron Transport Chain: A Field-Guided Resonator: A visualization of the Electron Transport Chain, reinterpreted not as a simple chemical conveyor, but as a field-guided resonator within the RB framework. The large protein complexes act as pulsating resonant gates, embedded in the mitochondrial membrane. A coherent electron wave (blue) propagates through the system, driving the phase-guided, synchronized transfer of protons (gold) across the membrane. The entire process functions as a living, quantum wave machine that converts the chemical potential into a rhythmic, coherent energy flux that powers the cell.

(C) ATP Synthase: Resonant Engineering Within the Cell. This illustration depicts ATP synthase in the RB framework. The F_0 subunit (purple, lower) acts as a toroidal vortex engine, directly driven by a coherent ULF current (glowing blue stream, representing proton flow) across the membrane. This rotation is converted into mechanical work, causing the central shaft to spin. The F_1 catalytic head (orange, upper) functions as a resonant chamber, where the rotation of the shaft induces precise conformational changes in its subunits. This dynamic process effectively "tunes" the chamber, enabling the capture and storage of energy as ATP molecules (glowing green spheres), which are released as a direct Resonance Dividend.

(D) Resonance Energy Transfer: This illustration depicts the direct, field-based transfer of energy between two key metabolic molecules. A high-energy ATP molecule (left) is shown releasing its stored geometric tension as a focused, coherent pulse of ULF energy. This "packet" of quantized coherence travels through the field via harmonic coupling and is absorbed by an NADH molecule (right), which acts as a dynamic resonator, instantly shifting its internal geometry to a higher-energy state. This provides a direct, physical picture of how energy flows through the cell's metabolic circuits as a seamless transfer of resonance.

4.4 Resonant Energy Transfer

The primary molecules that transport this coherence are ATP and NADH. Classically, these are regarded as energy carriers—intermediates that store or release energy during metabolic processes. In RB, these molecules are more accurately described as resonant structures: metastable Ultra-Light Frequency (Ulf) configurations that trap and transfer coherence.[61]

ATP, for instance, holds vibrational tension in its triphosphate tail, a high-frequency waveform geometrically poised for release. When ATP hydrolyzes, it does not merely break a bond; it releases stored resonance that is phase-matched with downstream enzymes or motor proteins. This coherence transfer enables nearly lossless energy propagation, similar to that in a tightly coupled oscillator system.

Likewise, NADH is not merely a redox agent. It serves as a resonance switch: its electron donation event simultaneously reconfigures its internal geometry, enabling downstream vibrational transitions that advance the metabolic cycle. Rather than moving electrons in space, NADH enables energy transformation in resonance space (Figure 39D).

In this model, energy does not simply flow as a thermodynamic gradient. It circulates as quantized coherence, redistributed through geometric coupling between field structures (Fröhlich, 1968).

4.5 The Krebs Cycle as a Harmonic Sequence

The most powerful and elegant example of this principle in action is the Krebs cycle, also known as the citric acid cycle. In classical models, this is a sequence of oxidative transformations that convert acetyl-CoA into carbon dioxide while generating NADH and $FADH_2$ for ATP synthesis. While this model is chemically accurate, it treats the cycle as a linear progression rather than a unified, oscillatory system.

RB reframes the citric acid cycle as a harmonic oscillator, a vibrational loop in which each metabolite transition refines the system's internal coherence. The cycle's

---

[61] In this context, the ultra-light field (Ulf) refers to localized, high-frequency standing wave patterns that operate within molecules and biological systems. It is a nested subset of the broader Universal Light Field (ULF) introduced earlier, representing the fine-grained, coherent vibrational modes that mediate energy storage, catalysis, and communication across biochemical structures. Whereas the ULF provides the background coherence of space, the Ulf expresses its local, quantized geometry within living matter.

closed-loop geometry facilitates phase continuity, resonance alignment, and efficient energy distribution with minimal entropy loss. Each step—citrate to isocitrate, α-ketoglutarate to succinate—can be understood as a symmetry operation that reduces strain in the molecular geometry and aligns the metabolite with the resonant structure of the mitochondrion. The cycle thus acts as both an engine of redox chemistry and a timing mechanism, organizing energy flow within a harmonic framework (Figure 40).

For instance, the transformation from citrate to isocitrate involves a geometric rearrangement that enhances molecular symmetry and stabilizes vibrational modes. Similarly, the decarboxylation steps (e.g., isocitrate to α-ketoglutarate to succinyl-CoA) eliminate unstable field configurations by shedding excess mass and field asymmetry. With each step, the internal field architecture of the molecules becomes more aligned with the surrounding mitochondrial resonance patterns, reducing energy dissipation and maximizing coherence.

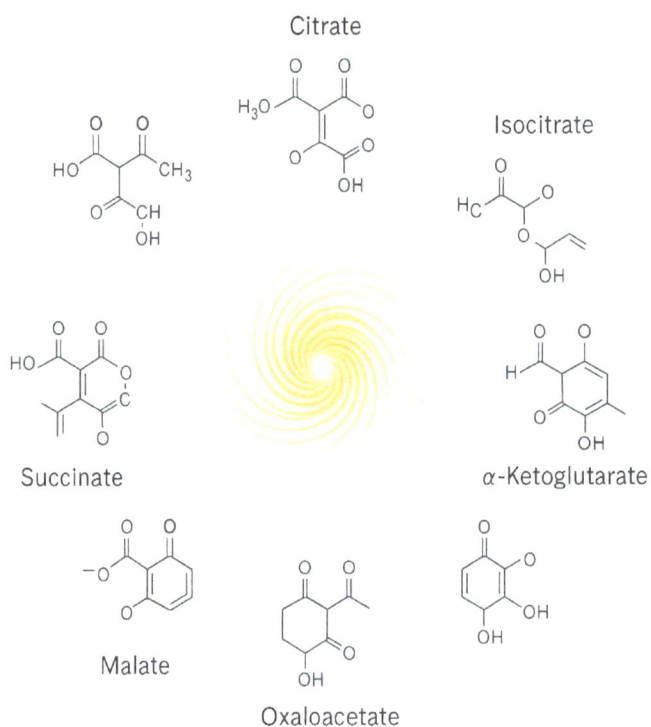

Figure 40: The Krebs Cycle. The Krebs cycle is depicted as a harmonic oscillator. Key metabolic intermediates—Citrate, Isocitrate, α-Ketoglutarate, Succinate, Malate, and Oxaloacetate—are arranged in a closed-loop geometry, each representing a distinct stage of resonance refinement. The central spiral vortex symbolizes the coherent redistribution of vibrational energy across the cycle. As molecular structures evolve, internal field tensions are reduced and coherence is preserved, allowing the system to function as a self-sustaining oscillator of metabolic energy.

The final arc of the cycle (succinate to oxaloacetate) is the refinement phase where energy is harvested. Here, the generic four-carbon chain is progressively sculpted.

286

The double bond in fumarate imposes a planar, rigid geometry that tightens molecular resonance, while the final oxidation to oxaloacetate resolves the remaining geometric "slack," converting a relatively isotropic resonance pattern into a sharp, directional dipole. This tightening is not just chemical; it is the physical act of exporting excess geometric strain into high-coherence carriers ($FADH_2$) and (NADH), which are the Resonance Dividends destined for the electron transport chain.

The closed-loop structure of the cycle is central to this function. Like a standing wave on a circular string or an LC circuit in electronics, the geometry of the cycle enables constructive interference, phase locking, and minimal entropy loss. In this view, oxaloacetate is not merely a regenerated substrate but a resonant foundation—a geometric attractor to which the system returns to maintain oscillatory coherence.

Thus, the Krebs cycle in the RB model is not just a metabolic engine. It is a resonant timing mechanism that harmonizes the flux of carbon, electrons, and energy with the vibrational rhythms of the mitochondrial system.

By reinterpreting metabolism through the lens of RB, we move beyond the traditional paradigm of molecular collisions and chemical energetics to reveal a coherent, field-driven architecture of life. Metabolic molecules are no longer considered passive substrates or energetic tokens. Instead, they become active participants in a dynamic resonance network that encodes energy, not merely in bonds, but in geometric standing waveforms. By revealing the deeper geometry of metabolic processes, RB provides a new foundation for understanding not only cellular energetics but also the conditions under which they become disrupted, setting the stage for deeper insights into disease, healing, and the potential for resonance-based interventions.

5. Coherence Across Scales

The remarkable stability and complexity of life arise not just from biochemical reactions, but from a deeper kind of organization—coherence across space, time, and scale. In the RB framework, this coherence does not arise from centralized control. Instead, it emerges from resonance. Living systems are, fundamentally, nested fields of oscillating geometry, in which coherence at one level supports and scaffolds coherence at another.

5.1 Coherence in Plant Systems

Plant life presents a remarkable example of systemic coherence. Unlike animals, plants cannot move to escape unfavorable environmental conditions. To meet this challenge, they have evolved intricate internal systems to align their form and function with the surrounding field conditions. From an RB perspective, plants achieve coherence not through central control but through distributed, geometry-based resonance management, which allows them to remain synchronized with environmental cycles and maintain internal stability over time.

At the cellular level, coherence begins with the cytoskeleton, plasmodesmata, and structured water layers. The cytoskeleton, far from being a passive scaffold, is a dynamic, field-sensitive network that supports intracellular transport and spatial organization. Its ability to transmit information through oscillatory behavior and tension-based signaling is well documented (Pienta & Coffey, 1991). Water also plays a foundational role. In RB, water serves as a field medium. The formation of structured "exclusion zone" water layers along cell membranes and proteins, a phenomenon extensively documented by Pollack (2001), helps buffer field disruptions, creating a stable energetic environment in which biochemical processes unfold with precision. (Figure 41).

Plants are also highly sensitive to environmental fields, and their geometry reflects this sensitivity. Phyllotaxis, the spiraling pattern of leaves and flowers, is not an arbitrary pattern. It is the visible expression of a system self-organizing to achieve optimal field alignment and packing efficiency (Camazine et al., 2003). Tropisms, which are responses to stimuli such as light and gravity, are another manifestation of coherent field sensing. Phototropism, for instance, involves the redistribution of the hormone auxin in response to light gradients, which causes the plant to bend (Alberts et al., 2002). In RB, this is not just a simple chemical cascade but a field-coupled geometric adjustment, where auxin itself acts as a resonant modulator, guiding plant growth in alignment with light from the ULF.

During stressful conditions, the plant hormone abscisic acid (ABA) plays a central role in maintaining coherence. When plants experience drought, for instance, ABA levels rise, triggering responses such as stomatal closure and metabolic downregulation (Cutler et al., 2010). In RB terms, ABA acts not merely as a biochemical messenger but as a coherence modulator that dampens field fluctuations and protects internal rhythms from disruptive noise. This role becomes even more relevant in light of research demonstrating ABA's similar function in mammalian systems, where it has been shown to help regulate glucose metabolism and reduce inflammation (Guri et al., 2007). Such cross-kingdom parallels hint at a deep, evolutionarily conserved function for ABA as a resonance stabilizer.

Figure 41: The Cytoskeleton as a Resonant Waveguide. This figure illustrates the physical basis of cellular-level coherence. The cytoskeleton is depicted as a dynamic, field-sensitive network whose filaments act as resonant waveguides that channel vibrational energy and guide intracellular transport. Surrounding the cell's internal structures are structured water layers, a crystalline-like phase of the ULF medium that acts as a resonant buffer, shielding delicate biochemical processes from decoherence. Together, these two systems create a coherent field that allows the cell to function as a unified energetic entity.

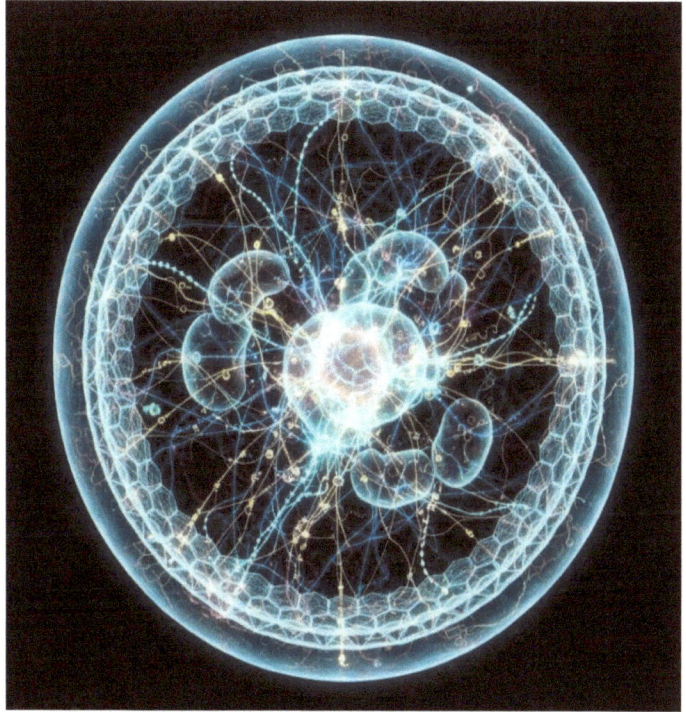

Field coherence in plants also scales up to the level of circadian rhythms and seasonal cycles. These temporal patterns are not only controlled by internal genetic clocks but are also entrained by environmental fields, which allows the plant to synchronize its metabolic activity, growth, and reproduction with the solar and lunar cycles. The entire plant, from root to shoot, becomes a phase-locked system that remains in harmony with Earth's broader energetic environment.

Ultimately, plants offer a striking example of distributed coherence in action, demonstrating how life can maintain internal order, regulate energy, and adapt to external conditions through geometry, vibration, and resonance.

5.2 Coherence in Animal Systems

In animals, systemic coherence emerges through both structural organization and the dynamic management of energy flow across tissues, organs, and behavioral rhythms. From an RB perspective, what appears to be biochemical regulation is a finely tuned dance of vibrational patterns that sustains physiological stability across scales.

This dynamic equilibrium is classically described as homeostasis, which is the maintenance of internal conditions through feedback loops that adjust molecular concentrations. However, in RB, homeostasis is better understood as the preservation of field coherence in an oscillating, energy-driven system. Rather than merely modulating biochemical levels, the body manages resonant energy flow through the continuous realignment of phase relationships and vibrational geometries. This process, which we call Resonance Management, involves three key factors:

- the optimization of spatial and temporal energy distribution ensures that vibrational energy flows where and when it is needed to sustain biochemical reactions (Demetrius, 2003);

- the buffering of localized decoherence to prevent field disruptions that could lead to molecular misfolding, enzyme inactivation, or signal failure, mediated by stabilizing structures such as water layering (Pollack, 2001), membrane potentials, and phase-separated domains (Brangwynne et al., 2009);

- and the use of oscillatory rhythms and cycles, such as circadian clocks (Mohawk et al., 2012), calcium waves (Berridge et al., 2000), and redox oscillators (Packer & Cadenas, 2007), as phase-locking mechanisms that synchronize disparate subsystems into a coherent whole.

When these mechanisms break down, the result is not just dysregulation but a collapse of resonance. The cell's field geometry disintegrates, vibrational coherence gives way to noise, and systemic synchronization unravels. This breakdown manifests macroscopically as inflammation, a biological signal of decoherence (Figure 42).

This is where the immune system becomes involved. In RB, the immune system is the body's primary agent of resonant repair. Immune cells, such as macrophages, are not mere chemical scavengers; they are the body's sentinels of coherence, acting as biological tuning forks that are exquisitely sensitive to dissonance. They are drawn to the chaotic, incoherent frequencies of a decoherent field (inflammation) in the same way a sound engineer is drawn to a jarring, out-of-tune instrument in an otherwise perfect orchestra. Once at the site of dissonance, their role is to physically clear its source—whether a damaged cell or a foreign pathogen—and restore the local field to its harmonious baseline state. Chronic inflammation, in this respect, is a sign of a persistent source of decoherence that the immune system is unable to resolve, leading to a continuous and damaging state of energetic disarray (Hotamisligil, 2006).

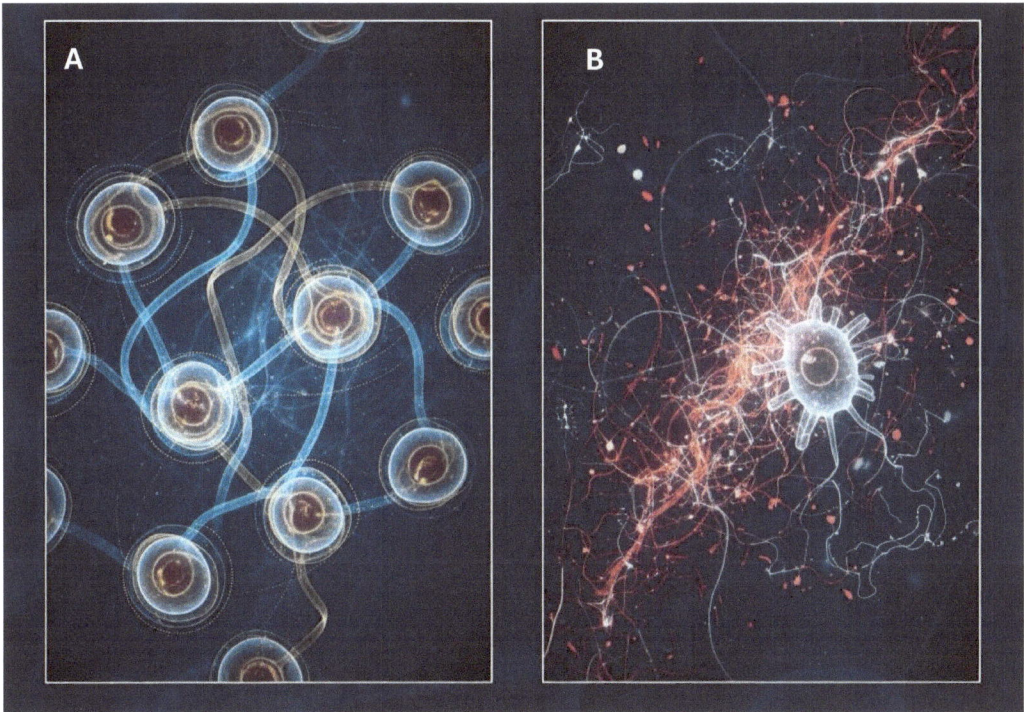

Figure 42: Inflammation as a Collapse of Field Coherence. This figure contrasts a healthy, coherent biological system with one undergoing inflammation.
(A) In a healthy state, the body maintains homeostasis, which is understood in RB as the active preservation of field coherence. Energy flows through the tissue in a smooth, harmonious, and synchronized network of ULF resonance, representing a state of dynamic equilibrium.
(B) When this coherence is disrupted due to injury or disease, the system enters a state of inflammation. The organized resonant field collapses into chaotic, dissonant energy (decoherence). Immune cells, such as the macrophage shown here, act as agents of resonant repair, drawn to the site of decoherence to clear the source of the dissonance and restore the field's natural harmony. This reframes inflammation as a direct biological signal of decoherence.

RB further identifies the adaptive immune system as the "conductors" of this coherence. Regulatory T cells (Tregs), classically identified by the CD25 marker (Sakaguchi et al., 1995), serve as "conductors of coherence" that harmonize the immune response. Their established function in biology is to maintain immune tolerance and prevent autoimmunity by suppressing "attack" T cells (Sakaguchi et al., 2008). They are drawn to the dissonance of inflammation and emit their own coherent, calming frequencies (specifically, suppressive cytokines like IL-10 and TGF-beta) that tune the other attack cells, preventing the orchestra from descending into a "cacophony" of autoimmunity.

Although the immune response is the first, active step in the body's attempt to heal, more broadly, the capacity for self-repair is an inherent property of resonant systems because their patterns are distributed (Camazine et al., 2003). When coherence breaks down in a region due to injury or disease, the system does not simply fail; it reorganizes. A loss of function can be restored through field coupling with neighboring structures, thus providing an energetic basis for healing. Health, from this perspective, is the active regeneration of coherence—the capacity of the system to maintain resonance in the face of continual energetic flux.

6. Cancer as a Breakdown of Coherent Field Architecture

In RB, health emerges from the resonant coherence of field geometries across scales. Disease, by contrast, represents decoherence—a failure to maintain vibrational harmony. In this respect, cancer embodies this failure most profoundly: a cellular system that has lost phase alignment with its surrounding tissue and entered a self-sustaining state of resonance breakdown.

6.1 Tumors as Islands of Local Decoherence

Conventional oncology views tumors as proliferative masses driven by genetic mutations. This explanation, however, does not fully account for behaviors such as metabolic rewiring (e.g., Warburg effect), altered morphology, or resistance to differentiation signals. RB, in turn, interprets tumors as regions of localized decoherence, where cells have resonantly detached from their environment and established autonomous field-attractor states. These cells fail to respond to tissue-level resonance feedback and instead operate within disruptive, low-coherence oscillatory modes (Figure 43).

6.2 The Warburg Effect: A Resonant Inversion of Metabolism

In the Warburg effect, cancer cells preferentially use aerobic glycolysis, even in the presence of oxygen. This is traditionally viewed as a metabolic adaptation that supports rapid growth (Warburg, 1956; Lehninger et al., 2008; Aron & Vander Heiden, 2009). Within RB, however, the Warburg shift represents a collapse of mitochondrial resonance, as oxidative phosphorylation depends on phase-aligned mitochondrial oscillators. When these oscillators are disrupted, cells revert to a less coherent but more proliferative metabolic pathway.

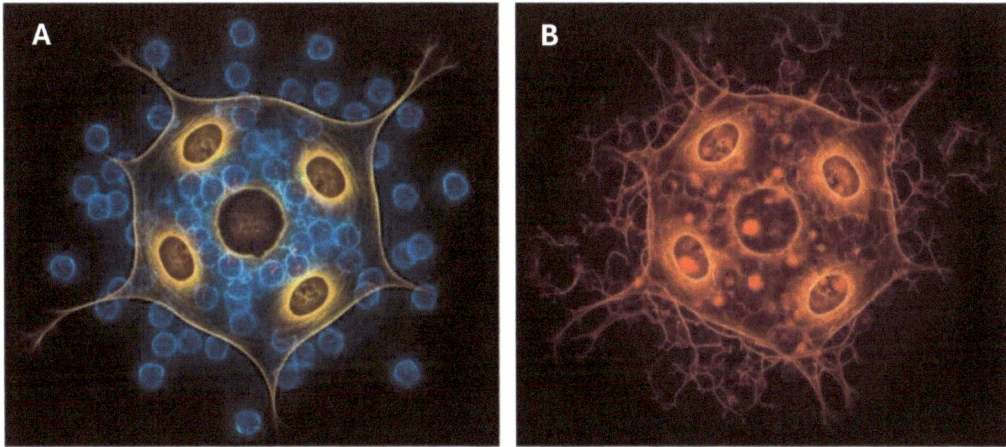

Figure 43: Field Geometry Comparison Between Healthy and Tumor Cells. These figures illustrate how, in the RB model, cancer emerges as a systemic breakdown of resonant coherence
(A) This figure illustrates a healthy eukaryotic cell. The cell's internal structures—mitochondria, nucleus, cytoskeleton, and membrane domains—are arranged in a harmonically balanced configuration, reflecting a state of coherent field geometry. Toroidal mitochondria exhibit synchronized resonance, the cytoskeleton supports directed energy transfer, and the membrane shows smooth polarization gradients. The overall vibrational architecture is phase-aligned, enabling efficient energy management, signal transduction, and metabolic precision.
(B) This figure illustrates a tumor cell of the same type but in a pathological state of decoherence. Mitochondria are deformed and fragmented, indicating a collapse in internal resonance. The cytoskeleton appears disorganized, leading to erratic energy flow and impaired signaling. Nuclear morphology is distorted, with uneven chromatin condensation and disordered nucleolar activity, causing disruptions in genetic regulation and information coherence. Membrane potential gradients are irregular, and vesicle trafficking is chaotic, leading to uncontrolled proliferation, invasiveness, and resistance to apoptotic cues. The cell has lost its synchronization with the surrounding field and has entered a metastable, entropy-producing attractor state.

This energetic decoupling is further evidenced by reverse Warburg effect models, in which stromal and tumor cells engage in reciprocal metabolic exchanges, shifting coherence power and inducing novel attractor states across tissue fields.

6.3 Mitochondrial and Microtubule Field Dysfunction

Mitochondrial disruption is central to cancer decoherence. Studies have shown that tumors with altered mitochondrial morphology and function also exhibit shifted electromagnetic frequency spectra, consistent with reduced field coherence generated by organelles such as microtubules and mitochondria. These shifts diminish the cell's

293

capacity to entrain neighboring cells, leading to autonomous growth and metastasis (Fröhlich, 1980; Pienta & Coffey, 1991).

6.4 Failure of Apoptosis and Tissue Coupling

In healthy tissues, apoptosis acts as an energy-guided resynchronization mechanism, removing cells that fall out of resonance. In tumors, this signal is often ignored; cancer cells no longer respond to field cues from the extracellular matrix or surrounding cells, which prevents normal apoptotic entry and enables unchecked growth.

6.5 Therapeutic Implications: Restoring Resonance

If cancer is fundamentally a loss of resonance, then treatment becomes a matter of re-establishing field coherence. Emerging treatments, such as low-frequency amplitude-modulated electromagnetic (LEAM RF) field therapy, which applies precisely tuned frequencies to suppress tumor growth, may operate by entraining malignant cells back into syncrony with healthy resonance patterns. Likewise, therapies targeting metabolic reprogramming (e.g., reversing Warburg metabolism via p53 reactivation or AMP-activated kinase modulation) can be viewed as efforts to restore mitochondrial oscillatory coherence and realign cellular attractor states.

Thus, in RB, cancer is not merely a genetic or biochemical aberration; it is a field-topological failure. Tumors arise when cellular resonance collapses, mitochondrial coherence breaks down, and cells no longer phase-align within tissue fields. Viewing cancer as a "coherence disease" thus reframes both diagnosis and treatment—detection becomes resonance mapping, and therapy becomes a process of resynchronizing vibrational architecture.

7. Implications and Predictions: The Resonant Engineering of Life

The RB framework is not merely an explanatory tool; it is a generative science. By understanding life as a vibrational architecture, we can move from passive observation to active design. This final section explores the profound consequences of this paradigm shift: a new philosophy we call the Resonant Engineering of Organic Systems.

7.1 Theoretical Implications

The RB model predicts that biological order is maintained by spatiotemporal resonance rather than chemical equilibrium. This leads to several key theoretical insights:

- Resonant Optimization of Metabolism: Metabolic efficiency is governed by harmonic synchronization across the entire cellular system, not just by local substrate availability.

- Disease as Decoherence: Pathologies like diabetes or cancer may arise from breakdowns in field coherence (e.g., mitochondrial or cytoskeletal disarray) that precede any measurable molecular imbalance.

- Hormones as Field Regulators: Hormones are not just chemical signals but field modulators that alter cellular phase coherence and tune local field geometries.

- Spontaneous Self-Assembly: Complex biological assemblies (e.g., ribosomes, vesicles) should spontaneously organize into their native structures when placed in a coherent resonant field that mimics their target topology.

RB also leads to a unifying law of biological stability: the Coherence Scaling Law. This law establishes the quantitative relationship between the material world and the foundational energetic field (UEF), predicting that the thermal threshold (T_c) of any material is governed by the UEF's energy density (ρ_{UEF}) and the material's Resonance Dividend Index (RDI), as defined in Chapters 7 and 9, respectively, which yields the following relationship:

$$T_c \propto \frac{1}{\rho_{UEF}} \cdot \frac{1}{RDI}$$

The derived formula shows that life maintains stability by continually investing energy to counter the immense compressive force of the UEF. This theoretical insight confirms that the chaotic, warm, and wet environment of the living cell is not a hostile anomaly for quantum mechanics. Instead, life exists because it has successfully developed complex resonant architectures that generate an RDI sufficient to stabilize its delicate molecular machinery against the massive compressive tension of the universal field.

7.2 Technological Applications

This new paradigm also opens the door to a new class of technologies based on the tuning, restoration, or amplification of biological coherence.

- Resonant Diagnostic Imaging: Tools like terahertz spectroscopy could allow us to detect disorders as failures of symmetry and vibrational continuity in tissue.

- Field-Guided Therapies: Coherence-tuned light pulses, low-frequency electromagnetic fields, or sound-based entrainment could be used to entrain biological systems back into synchronized, healthy operation.

- Programmable Biochemistry: By adjusting rhythmic field geometries, biochemical cascades could be activated, suppressed, or rerouted in real time, enabling applications like the on-demand activation of drugs or the site-specific attachment of functional groups in complex assemblies.

- Artificial Photosynthesis: If plants are tuning into field resonance to convert sunlight into energy, then we can potentially do the same—not by copying their chemistry but by copying their geometry. In practical terms, this means that we are not just harvesting energy; we are creating matter, such as food, fuel, and structures, encoded with ambient coherence from light. Theoretically, we could build resonant nanostructures that capture field vibrations and stabilize them, or field-sensitive catalysts that use shape and timing to concentrate energy. We might even design artificial chloroplasts that act more like quantum antennas than chemical factories. The goal is not to simulate what plants do but to speak the same language: geometry, resonance, and coherence.

7.3 Falsifiable Predictions: The Resonant Modulation Hypothesis

The RB framework culminates in novel, specific predictions that distinguish it from the standard biochemical model.

7.3.1 The Functional Test: Resonant Inhibition and Enhancement

If an enzyme's function is dependent on its specific resonant geometry, it should be possible to use external fields to both stop and accelerate its activity.

- Resonant Inhibition (Destructive Interference): We predict that a precisely tuned electromagnetic field oscillating at a frequency that induces destructive interference with an enzyme's active site (such as ATP synthase) will "jam" its resonant structure and inhibit its catalytic activity, even in the absence of a physical inhibitor (the required frequencies for chemical bonds typically reside in the Terahertz range).

- Resonant Enhancement (Constructive Interference): Conversely, we predict that a field tuned to the optimal frequency and phase of the active site will actively reinforce its natural resonance. This constructive interference would increase the enzyme's coherence, allowing it to operate faster or more efficiently than in its natural state.

Confirming this dual effect would provide definitive validation of our framework's ability to manipulate biological function through field geometry.

7.3.2 The Diagnostic Test: Coherence Lifetimes

Beyond mere functional manipulation, the stability of biological function is predicted to be directly correlated with the coherence lifetime (T_c) of molecular resonance patterns. While this property is studied by standard quantum biology (QMB), UFD provides the unique causal agent for the observed stability.

QMB tracks temperature (T) as the primary factor limiting T_c. UFD identifies the RDI as the core structural factor. The efficiency of pigment-protein complexes or the catalytic success of an enzyme's active site is a direct function of its RDI. This means that two molecules with identical mass and temperature should exhibit different coherence times if their underlying nuclear geometries (UEF cores) possess different topological perfections (e.g., comparing C-12 to C-13).

In turn, disease states (pathology) should correlate with a measurable reduction in the coherence lifetime of key molecular structures. Conversely, highly efficient biological systems will exhibit longer coherence lifetimes than their non-biological counterparts. This can be tested using advanced spectroscopic techniques to provide a direct, physical measurement of a biological system's resilience to decoherence. This approach transforms medical diagnostics from tracking symptomatic biochemical imbalance to measuring the underlying structural resilience and causal coherence of the organism's energetic field.

As research progresses, the boundary between materials science and synthetic biology may dissolve entirely. By moving beyond a purely chemical view to one of tunable ULF resonances, we open the door to a future where the design of life, like the design of metals, begins with the tuning of fields.

8. Conclusion

This chapter began with a simple question: What if life is not merely chemical, but resonant? Here, we have explored the profound implications of that question, providing a new, physically grounded mechanism for the fundamental processes of biology. In RB, the cell is not a "bag of chemicals." It is a resonant instrument, defined by a cell membrane that acts as a resonant interface, a cytoskeleton that functions as a vibrational waveguide, and enzymes that serve as quantum-phase filters. We then showed how this orchestra performs the two great symphonies of life: Photosynthesis, the act of capturing and encoding the coherence of light into matter, and Metabolism, the act of liberating that geometric coherence to power the cell. Finally, we demonstrated what happens when the music becomes discordant: inflammation and disease.

The RB model, therefore, provides more than a reinterpretation of the cell—it offers a resonant design language for the next era of biotechnology, suggesting that vitality results from the emergent harmony of field structures tuned across scales. Where coherence thrives, life thrives. Where coherence collapses, disorder and disease emerge.

This insight transforms both our understanding and our ambitions. Healing becomes a question of restoring resonance, diagnosis becomes a matter of detecting field disruption, and design becomes an art of tuning energy into form. Life, in this model, rather than a collection of biochemical parts, is the music of matter played through the geometry of the field.

11.3 A New Foundation for Biochemistry

In this chapter, we have reconceived biochemistry as a resonant phenomenon. Through the lens of Resonant Biochemistry (RB), the chemistry of life is revealed not as a series of probabilistic collisions but as a symphony of energy transformations guided by geometry, rhythm, and field coherence.

We began by introducing the "orchestra" itself: the resonant infrastructure of the cell. There, we redefined the cell membrane as a resonant interface, the cytoskeleton as a vibrational waveguide, and enzymes as quantum-phase filters. With this physical architecture established, we then explored the two great symphonies this orchestra performs: Photosynthesis, the act of capturing and encoding the coherence of light from the cosmos, and Metabolism, the act of liberating that stored geometric coherence to power the cell.

From this new foundation, we reframed homeostasis as the active preservation of field coherence and showed how a breakdown in this harmony leads to pathology, reinterpreting inflammation and disease as observable signatures of field decoherence and a collapse of the cell's resonant architecture. Importantly, this view makes predictions and opens powerful new directions in science and technology, including the following:

- Resonant diagnostics that detect decoherence before structural damage occurs.

- Field-guided therapies that restore rhythm and synchrony to biological systems.

- Coherence-restoring therapies targeted at cancerous or otherwise diseased cells.

- Biomimetic energy systems inspired by the field dynamics of photosynthesis.

This is not a revision of biochemistry; it is a re-founding. In RB, the cell becomes a musical instrument, tuned to the universe's frequencies. By restoring resonance to the heart of the life sciences, we reclaim what the molecular view had lost: the wholeness of the system.

References

1. Alberts, B., Johnson, A., Lewis, J., Raff, M., Roberts, K., & Walter, P. (2002). *Molecular Biology of the Cell* (4th ed.). Garland Science.
2. Aron, M., & Vander Heiden, M. G. (2009). *Understanding the Warburg Effect: The Metabolic Requirements of Cell Proliferation*. Science, 324(5930), 1029–1033
3. Berridge, M. J., Bootman, M. D., & Roderick, H. L. (2000). *Calcium signalling: dynamics, homeostasis and remodelling*. Nature Reviews Molecular Cell Biology, 1, 11–21
4. Brangwynne, C. P., Tompa, P., & Pappu, R. V. (2009). *Polymer physics of intracellular phase transitions*. Nature Physics, 11, 899–904.
5. Brizhik, L. S., Del Giudice, E., Jørgensen, S. E., Marchettini, N., & Tiezzi, E. (2009). *The role of electromagnetic potentials in the evolutionary dynamics of ecosystems*. Ecological Modelling, 220(16), 1865–1869.
6. Buzsáki, G., & Draguhn, A. (2004). Neuronal oscillations in cortical networks. *Science*, 304(5679), 1926–1929.

7. Camazine, S., Deneubourg, J. L., Franks, N. R., Sneyd, J., Theraulaz, G., & Bonabeau, E. (2003). *Self-Organization in Biological Systems*. Princeton University Press.

8. Costa, F. P., de Oliveira, A. C. P., Meirelles, R., Machado, M. C., Zanesco, S., Surjan, R. C., ... & Gurgel, M. S. C. (2011). *Treatment of advanced hepatocellular carcinoma with very low levels of amplitude-modulated electromagnetic fields*. British Journal of Cancer, 105, 640–648

9. Cutler, S. R., Rodriguez, P. L., Finkelstein, R. R., & Abrams, S. R. (2010). Abscisic acid: emergence of a core signaling network. *Annual Review of Plant Biology*, 61, 651-679.

10. Demetrius, L. (2003). *Quantifying thermodynamic entropy and biological complexity: Applications to cancer and aging*. Entropy, 5(3), 143–160.

11. Engel, G. S., Calhoun, T. R., Read, E. L., Ahn, T. K., Mancal, T., Cheng, Y. C., ... & Fleming, G. R. (2007). Evidence for wavelike energy transfer through quantum coherence in photosynthetic systems. *Nature*, 446(7137), 782–786.

12. Fröhlich, H. (1968). Long-range coherence and energy storage in biological systems. *International Journal of Quantum Chemistry*, 2(5), 641–649.

13. Fröhlich, H. (1980). *The biological effects of microwaves and related questions*. In L. Marton (Ed.), *Advances in Electronics and Electron Physics* (Vol. 53, pp. 85–152). Academic Press.

14. Guri AJ, Hontecillas R, Si H, Liu D, Bassaganya-Riera J. Dietary abscisic acid ameliorates glucose tolerance and obesity-related inflammation in db/db mice fed high-fat diets. *Clinical Nutrition*. 2007 Feb;26(1):107–116.

15. Hameroff, S., & Penrose, R. (2014). Consciousness in the universe: A review of the 'Orch OR' theory. *Physics of Life Reviews*, 11(1), 39–78.

16. Hodgkin, A. L., & Huxley, A. F. (1952). A quantitative description of membrane current and its application to conduction and excitation in nerve. *The Journal of Physiology*, 117(4), 500–544.

17. Hotamisligil, G. S. (2006). *Inflammation and metabolic disorders*. Nature, 444, 860–867.

18. Hüser, J., & Blatter, L. A. (1999). Fluctuations in mitochondrial membrane potential caused by repetitive gating of the permeability transition pore. *Biochemical Journal*, 343(Pt 2), 311–317.

19. Klinman, J. P. (2006). Linking protein dynamics to catalysis: the role of hydrogen tunneling. *Philosophical Transactions of the Royal Society B: Biological Sciences*, 361(1472), 1323–1331.

20. Lehninger, A. L., Nelson, D. L., & Cox, M. M. (2008). *Lehninger Principles of Biochemistry* (5th ed.). W.H. Freeman and Company.

21. McCraty, R., Atkinson, M., Tomasino, D., & Bradley, R. T. (2009). The coherent heart: heart–brain interactions, psychophysiological coherence, and the emergence of system-wide order. *Integral Review*, 5(2), 10–115.

22. Mohawk, J. A., Green, C. B., & Takahashi, J. S. (2012). *Central and peripheral circadian clocks in mammals.* Annual Review of Neuroscience, 35, 445–462.

23. Mitchell, P. (1961). Coupling of phosphorylation to electron and hydrogen transfer by a chemiosmotic type of mechanism. *Nature*, 191, 144–148.

24. Nelson, D. L., & Cox, M. M. (2017). *Lehninger Principles of Biochemistry* (7th ed.). W.H. Freeman.

25. Nicholls, D. G., & Ferguson, S. J. (2013). *Bioenergetics 4.* Academic Press.

26. Packer, L., & Cadenas, E. (2007). *Oxidants and antioxidants revisited: New concepts of oxidative stress.* Free Radical Research, 41(9), 951–952.

27. Picard, M., Wallace, D. C., & Burelle, Y. (2016). The rise of mitochondria in medicine. *Mitochondrion*, 30, 105–116.

28. Pienta, K. J., & Coffey, D. S. (1991). Cellular harmonic information transfer through a tissue tensegrity-matrix system. *Medical Hypotheses, 34*(2), 88–95.

29. Pilla, A. A. (2013). *Mechanisms and therapeutic applications of time-varying and static magnetic fields.* In *Biological and Medical Aspects of Electromagnetic Fields* (3rd ed., pp. 1–62). CRC Press.

30. Pollack, G. H. (2001). *Cells, Gels and the Engines of Life: A New, Unifying Approach to Cell Function.* Ebner and Sons Publishers.

31. Pollack, G. H. (2013). *The Fourth Phase of Water: Beyond Solid, Liquid, and Vapor.* Ebner & Sons.

32. Pokorný, J. (2004). Excitation of vibrations in microtubules in living cells. *Bioelectrochemistry, 63*(1–2), 321–326.

33. Pomes, R., & Roux, B. (2002). Molecular mechanism of H+ conduction in the single-file water chain of the gramicidin channel. *Biophysical Journal, 82*(5), 2304–2316.

34. Purves, D., Augustine, G. J., Fitzpatrick, D., et al. (2018). *Neuroscience* (6th ed.). Oxford University Press.

35. Sakaguchi, S., Sakaguchi, N., Asano, M., Itoh, M., & Toda, M. (1995). Immunologic self-tolerance maintained by activated T cells expressing IL-2 receptor alpha-chains (CD25). Breakdown of a single mechanism of self-tolerance causes various autoimmune diseases. *Journal of Immunology, 155*(3), 1151–1164

36. Sakaguchi, S., Yamaguchi, T., Nomura, T., & Ono, M. (2008). Regulatory T cells and immune tolerance. *Cell, 133*(5), 775–787.

37. Scholes, G. D., Fleming, G. R., Olaya-Castro, A., & van Grondelle, R. (2017). Using coherence to enhance function in chemical and biophysical systems. *Nature, 543*(7647), 647–656.

38. Voet, D., Voet, J. G., & Pratt, C. W. (2016). *Fundamentals of Biochemistry: Life at the Molecular Level* (5th ed.). Wiley.

39. Warburg, O. (1956). *On the origin of cancer cells*. Science, 123(3191), 309–314.

40. Wöhler, F. (1828). Ueber künstliche Bildung des Harnstoffs [On the artificial formation of urea]. *Annalen der Physik und Chemie, 88*(2), 253–256.

Chapter 12: The Physics of Mind

12.1 Neuroscience

For over a century, neuroscience has operated under the paradigm of the brain as a biochemical machine. Rooted in the pioneering work of Santiago Ramón y Cajal, who first mapped neurons as discrete cells using Golgi staining, and Hodgkin and Huxley (1952), who modeled the ionic basis of action potentials, the dominant framework has successfully explained much of the brain's anatomy, development, and pathophysiology. Neurotransmitters such as serotonin, dopamine, and gamma-aminobutyric acid (GABA) are associated with mood, motivation, and inhibition (Purves et al., 2018). Functional imaging studies have revealed the dynamic interplay of large-scale brain networks, such as the Default Mode Network (DMN) and Salience Network (Raichle, 2015). Moreover, advances in pharmacology, brain-computer interfaces, and neuroplasticity research continue to deepen our understanding of the brain's immense complexity.

At the same time, modern neuroscience faces deep conceptual and empirical limitations. Chief among these is the "hard problem of consciousness", a term coined by David Chalmers (1995) that refers to the question of how subjective experience arises from neural processes. Despite decades of progress in mapping the neural correlates of consciousness, no consensus has emerged regarding the physical processes that give rise to awareness. Thought, intention, perception, and volition remain opaque.

Prominent thinkers have attempted to address this gap. Francis Crick and Christof Koch (2003) proposed that synchronized oscillations among cortical regions could underlie conscious awareness. Karl Friston (2010) introduced the Free Energy Principle, which models the brain as a system that minimizes prediction error while maintaining thermodynamic homeostasis. Giulio Tononi's Integrated Information Theory (IIT) (Tononi, 2008) suggests that consciousness corresponds to systems with a high degree of irreducible information integration. Meanwhile, Roger Penrose and Stuart Hameroff (2014) have advanced the Orchestrated Objective Reduction (Orch-OR) model, which proposes that quantum coherence in microtubules mediates conscious moments. Each theory extends beyond classical neural computation; however, none has decisively unified the relationship between the mind and matter.

The irony is that while neuroscience has remained largely anchored to a classical, mechanistic worldview, physics has not. The 20th-century revolution in physics revealed that the universe is not a clockwork machine comprising discrete particles. At its deepest level, it is a dynamic and interconnected web of fields, where the strict separation between the observer and the observed breaks down (quantum entanglement), where substance can emerge from pure information (the holographic principle), and where the fundamental reality is one of vibration, resonance, and geometric structure. This new physics provides a natural and necessary foundation for a new neuroscience.

Because neuroscience remains largely anchored to classical models, an opportunity arises to reframe the brain within the new physics of coherence. This chapter introduces such a framework.

In this final chapter, we introduce Resonant Neuroscience (RN), the Unified Field Dynamics (UFD) branch of neuroscience. RN is a field-based theory of the brain grounded in nested electromagnetic and scalar geometries. It proposes that the brain is not simply a structure of interacting cells but a hierarchically coherent field system in which thought and awareness emerge from resonance rather than from computation.

In RN, neurons are reinterpreted as oscillatory waveguides, neurotransmitters become geometric modulators, and cognition arises from nested phase harmonics. This model synthesizes existing evidence from microtubule resonance (Fröhlich, 1968; Hameroff & Penrose, 2014), nested brain rhythms (Buzsáki, 2006), and dynamic systems theory (Kelso, 1995), offering a unified account of brain function as energetic coherence across different scales.

RN does not repudiate modern neuroscience but expands it. It retains the mechanistic insights of current models, such as Hebbian learning, neural plasticity, and cortical mapping, but reinterprets them through the lens of geometry, resonance, and field interactions. In doing so, it illuminates phenomena long considered anomalous or subjective—such as free will, nonlocal awareness, and psi phenomena—as lawful consequences of a deeper field dynamic.

In this chapter, we build this theory from first principles. We begin with the geometry of neurons and neurotransmitters, move through brain structures and networks, explore the dynamics of conscious processes, and conclude with testable predictions and transformative applications.

Ultimately, this chapter can be said to sit at the convergence of neuroscience, physics, and metaphysics, offering a new map for the ancient mystery of how energy becomes mind.

12.2 The Physics of Mind: Resonant Neuroscience and the Fabric of Consciousness

Abstract

This chapter introduces Resonant Neuroscience (RN), a new model of neuroscience that reframes brain function and consciousness through the lens of Unified Field Dynamics (UFD). Rather than treating the brain as a biochemical machine, RN describes it as a nested system of resonant geometries in which neurons and brain regions function as oscillatory attractors. Cognitive processes are understood to arise from dynamic phase alignment across these structures, whereas neurological disorders are reinterpreted as breakdowns in this resonant integrity.

The model's most profound implications, however, concern the nature of consciousness itself. We propose that consciousness is a wave of coherent light from a fourth, fundamental field: the Universal Awareness Field (UAF). This field is the physical substrate for the soul, which is understood as a stable vortex within it. The simplest vortex in this medium is identified as the neutrino—the Quantum of Consciousness— whose calculated mass (≈ 0.198 eV) falls within experimental limits. Most critically, the UAF's fluid dynamics predict a maximum transmission velocity, the Speed of Awareness, of $\approx 137c$, which grounds quantum nonlocality in a physical, superluminal field.

This integration of geometry, field theory, and awareness leads to a series of novel, falsifiable predictions, including sub-Hz preconscious field shifts, nonlocal scalar anomalies during intentional states, and a predicted mass for the soul vortex. Ultimately, RN offers a unified and physically grounded account of brain function and consciousness that bridges the gap between physics and metaphysics.

1. Introduction

In the classical worldview, the human brain is viewed as a computational machine. However, in the world of Resonant Biochemistry (RB), it is a resonant organ. While traditional neuroscience has focused on biochemical signaling and modular circuits, mounting evidence suggests that large-scale brain integration arises from dynamic coherence across oscillatory fields (Varela et al., 2001; Buzsáki, 2006). These findings reveal that cognition is not the product of isolated neuronal firing but emerges from the synchronized rhythms of neural ensembles across spatial and temporal scales.

Resonant Neuroscience (RN), our application of Unified Field Dynamics (UFD) to brain function, extends this insight by modeling the brain as a system of nested, resonant geometries. In this model, neural oscillations are the organizing principles that shape experiences, perceptions, memories, and consciousness. From fast gamma bursts to ultra-slow delta and infra-slow waves, standing field patterns underlie integration and cognitive flexibility (Fries, 2005).

Disruptions in these resonant structures correspond to pathological conditions. Neurological disorders such as epilepsy, schizophrenia, and Alzheimer's disease can be understood as breakdowns in field coherence and symmetry rather than mere chemical or structural anomalies (Uhlhaas & Singer, 2010). This shift redefines diagnosis and treatment by highlighting the restoration of coherence as the key to healing.

The RN framework also addresses consciousness and the mind-body interaction problem by proposing a novel metaphysical component: the brain acts as a resonant interface with an emergent Universal Awareness Field (UAF). This fourth ultra-coherent field emerges from the underlying layers of light and energy. Consciousness, in this view, is not generated by the brain. Instead, it interfaces with it through phase-tuned modulations of ultra-light-frequency (Ulf) fields. Free will arises as a subtle, field-level constraint on neural attractor states, introducing intentional asymmetries that guide cognition and action.

In RN, the brain—once a mere organ of computation—becomes a field-tuned receiver of awareness. Thought, volition, and emotion emerge from geometric resonance within and beyond the brain's anatomy. By integrating field theory, wave dynamics, and neurophysiology, RN lays the foundation for a new science of consciousness that unites physics, neuroscience, and metaphysics into a coherent and testable program.

2. The Architecture of Neural Resonance

This section explores the geometric and field-based foundations of neural activity. Rather than treating neurons and neurotransmitters as isolated biochemical entities, we interpret them as resonant structures within a living electromagnetic field (EMF). Action potentials, neurotransmitter activity, and inhibition are forms of wave modulation that enable the brain to rhythmically coordinate energy in space and time.

2.1 Neurons as Resonant Units

In RN, neurons are not merely biological circuits; they are structured oscillators. These geometric waveguides encode and transmit information via phase-synchronized energy flows. Each component of a neuron, from its soma to its dendrites and axon, serves a resonant function, enabling the brain to operate as a field-based communication network.

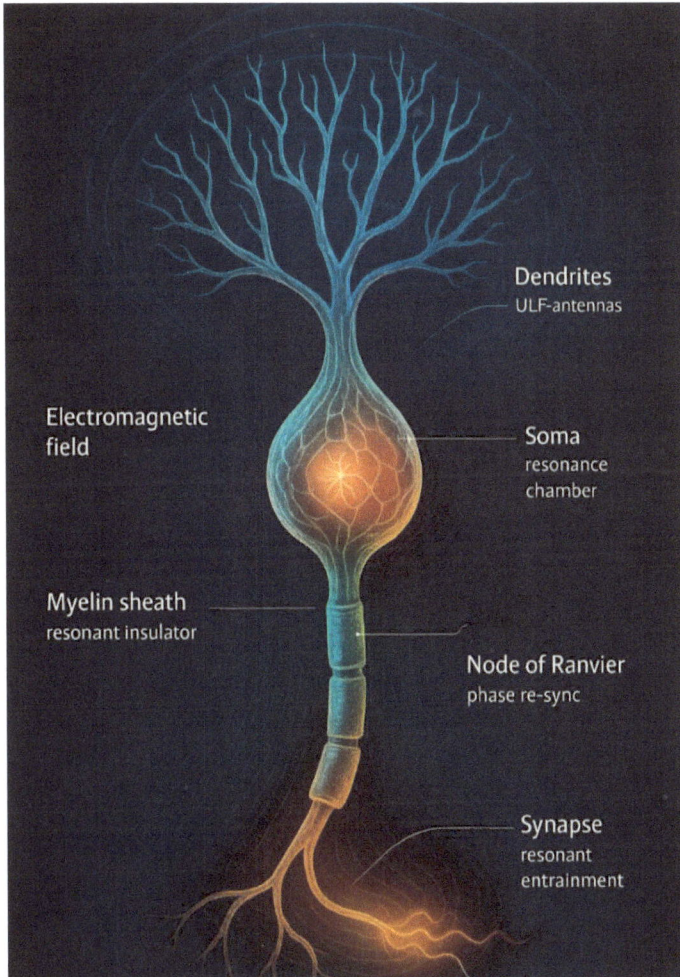

Figure 44: The Neuron as a Resonant Structure. This illustration depicts a neuron not as a simple biological wire but as a vibrationally tuned structure within a coherent field environment. Dendrites with fractal branching act as ultra-light-frequency (Ulf) antennas that capture ambient standing waves from nearby neurons. The soma functions as a central resonance integrator, summing the vibrational inputs and determining the phase alignment thresholds. The axon acts as a cylindrical waveguide, transmitting coherent soliton-like pulses maintained by constructive interference between the membrane and cytoskeletal fields. Myelination enhances transmission fidelity, whereas the nodes of Ranvier serve as periodic phase-reset points. This perspective reframes neurons as living oscillators that participate in recursive field entrainment and nested coherence across scales.

2.1.1 The Geometric Foundations of the Neuron

The shape of the neuron is not incidental; it is a highly optimized geometry for managing complex resonant behavior (Figure 44).

- Soma (Cell Body): The soma acts as a central node where incoming vibrational signals from the dendrites converge and undergo waveform summation. In our framework, the soma acts as a resonance threshold integrator. This spatial cavity selectively amplifies the inputs in alignment with its internal oscillatory state.

- Dendrites: Dendrites form branching, fractal-like structures that maximize the surface area for signal reception. These branching geometries function as Ulf antennas that capture standing wave patterns from adjacent neurons. Their geometry supports recursive resonance, which allows them to synchronize with multiple inputs across a spectrum of phases and frequencies.

- Axon: The axon is a cylindrical waveguide optimized for the longitudinal transmission of coherent pulses. Its length and diameter determine its fundamental resonant modes, while the myelin sheath acts as a resonant insulator, enhancing the transmission speed by reducing field dispersion. The nodes of Ranvier act as rephasing checkpoints, periodically re-synchronizing the pulse with the environment to preserve coherence over long distances.

2.1.2 Action Potentials as Phase-Guided Pulses

In traditional neuroscience, action potentials are often described as simple voltage spikes that propagate along the membrane when a certain threshold is attained. RN offers a deeper interpretation: the action potential is a phase-guided resonance pulse, akin to a soliton.[62] (Figure 45A). This self-sustaining wave maintains coherence as it moves, transforming ionic gradients into torsional energy exchanges along the axonal membranes.

In this model, depolarization and repolarization are not just chemical events but oscillatory shifts in the membrane's phase. These energetic envelopes are maintained by constructive interference between the neuron's internal standing-wave structure and the cytoskeletal lattice, particularly the microtubules. Thus, an action potential does not simply travel; it resonates, preserving information through the continuity of its waveform. The "all-or-none" nature of neural firing can now be understood as a binary resonance condition: the signal either phase-locks with the axonal geometry and propagates or collapses into incoherence owing to a mismatch.

2.1.3 Resonant Gating and Synaptic Entrainment

---

[62] In biological systems, solitons have been proposed as models for energy and signal transmission, particularly in structures such as nerve membranes and microtubules, in which coherent, localized pulses can travel without dissipating (Davydov, 1985).

In RN, the synapse, which has long been considered a chemical relay, is reinterpreted as a resonant interface.[63] (Figure 45B). Ion channels embedded in the postsynaptic membrane are not passive gates but frequency-dependent phase switches. They open only when the local field oscillation meets specific amplitude and timing conditions, which enables a new model of synaptic entrainment.

Rather than triggering a postsynaptic response solely through concentration gradients, the presynaptic neuron subtly shifts the membrane's vibrational state. If the local field reaches a resonance threshold, it induces a coherent action potential. Long-term potentiation and synaptic plasticity, typically understood as molecular processes, become geometric phenomena in RN. These phenomena represent long-term reconfigurations of the synaptic resonance environment that tune phase thresholds and standing-wave geometries through structural feedback and learning.

Within this model, the brain emerges as more than a network of firing nodes; it is a living interferometer, in which information is processed and transmitted through patterns of phase alignment.[64] Neurons do not operate in isolation but participate in resonant constellations, forming dynamic and adaptive attractor states. Their coherence is not based solely on synaptic strength or chemical abundance, but on geometric compatibility and harmonic resonance across nested spatial scales.

This reconceptualization of action potential has significant implications. The "integrate-and-fire" model becomes a low-resolution approximation of a far more nuanced system in which cognition, memory, and perception emerge from recursive phase-locking.[65] Signal strength is no longer defined solely by amplitude but also by the

---

[63]A resonant interface is a junction between two systems (e.g., neurons) where information is transmitted not only by the quantity of a signal (such as the concentration of a neurotransmitter) but also by its frequency, as in the photoelectric effect. Information transfer occurs efficiently only when the vibrational frequency of the incoming signal is harmonically compatible with the natural resonant frequencies of the receiver's membrane and receptors, acting as a tuned circuit that filters out specific frequencies.

[64] An interferometer is a scientific instrument that measures the interference patterns produced when waves, such as light, radio, or sound waves, overlap and combine. By analyzing these patterns, it can detect extremely small differences in distance, phase, and other wave properties, enabling precise measurements.

[65] The *integrate-and-fire* model is a simplified representation of neuronal behavior in which a neuron integrates incoming electrical inputs until a threshold is reached, at which point it "fires" an action potential. After firing, the membrane potential returns to its resting value. This model treats neurons as binary threshold units and does not account for complex resonance dynamics, field interactions, or spatial geometry.

degree of coherence with the geometry of the receiving structure. Neurological disorders, such as epilepsy and schizophrenia, may thus be reinterpreted as disturbances of phase coherence, where the brain's interference patterns become disordered, fragmented, or hyper-synchronized beyond functional stability.

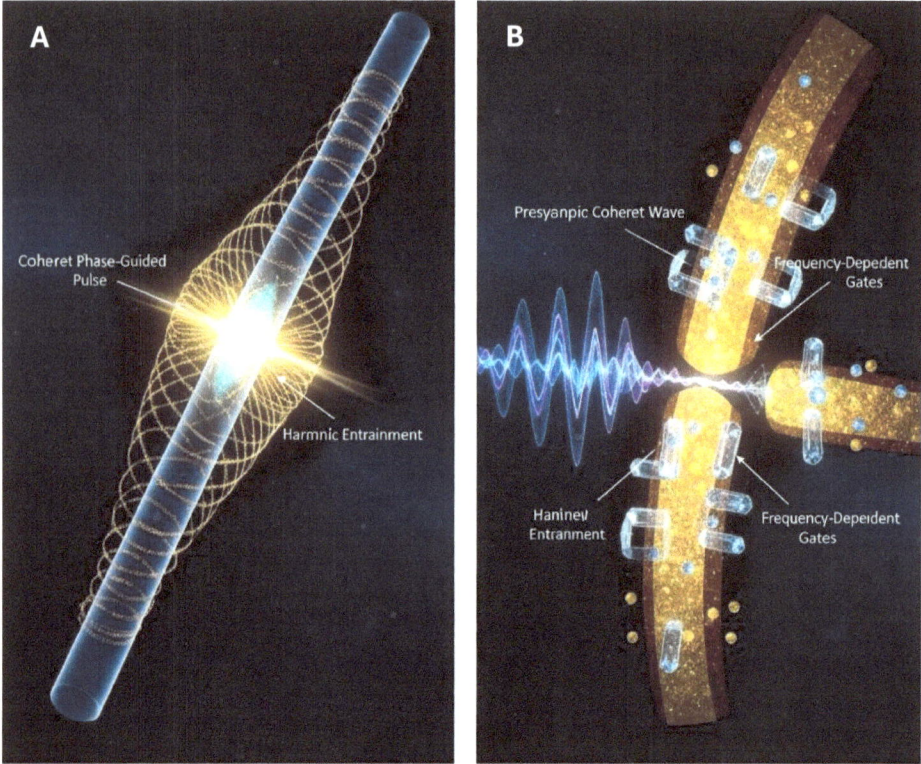

Figure 45: The Architecture of Neural Resonance.
(A) The Action Potential as a Phase-Guided Pulse: This figure depicts a neural action potential within RN. The "signal" is not a simple electrical spark but a coherent, self-sustaining wave packet of brilliant light. This pulse travels gracefully along the axon, its movement visibly locked in phase with the neuron's underlying resonant field. This mechanism ensures a precise, harmonious, and lossless transfer of information.
(B) The Synapse as a Resonant Interface: This figure depicts the synapse as a sophisticated resonant circuit. The presynaptic neuron emits a coherent wave pattern that must harmonically entrain or phase-lock with the resonant field of the postsynaptic membrane. Ion channels are frequency-dependent gates that open only when the resonance threshold is met. This reframes the synapse from a simple on/off switch to a tunable filter that processes information based on frequency and phase.

By revealing neurons as field-resonant instruments rather than electrochemical switches, RN opens a new window into the architecture of the brain. The neuron becomes a symphony of nested rhythms and the brain an instrument of dynamic coherence. This now brings us to the true, physical function of neurotransmitters as the primary modulators of these resonance patterns.

2.2 Neurotransmitters: The Modulators of Resonance

In RN, neurotransmitters are more than mere chemical messengers; they modulate local and global field resonance. The molecular geometry of each neurotransmitter determines how it interacts with the synaptic environment, allowing it to act as a geometric "key" that entrains the surrounding neural circuitry into distinct vibrational states. This process enables the brain to dynamically shift between modes of activity, perception, and consciousness.

2.2.1 Excitatory and Inhibitory Resonance

The fundamental balance of brain activity is governed by the interplay between excitation and inhibition, which RN describes as a resonant trade-off. Excitatory coherence promotes the synchronization necessary for information transfer, while inhibitory modulation prevents this coherence from becoming destructive and devolving into chaotic, runaway synchrony.

Glutamate, with its rigid molecular geometry, is the primary agent of excitatory coherence. Its action is rapid and phase-specific, thereby promoting phase amplification by opening ion channels that depolarize the membrane (Purves et al., 2018). GABA and glycine, in contrast, are the primary coherence buffers. These small molecules function as fast-acting inhibitory systems in the brain. Their binding to receptors introduces a phase delay into the local oscillatory circuit, a geometric phase offset analogous to destructive interference. Rather than simply silencing neurons, these inhibitory signals refine the frequency spectrum of brain activity, which sculpt excitatory signals into meaningful rhythms that avoid runaway excitation (Buzsáki & Draguhn, 2004) (Figure 46).

The spatial distribution of inhibitory neurons, particularly GABAergic interneurons, supports this stabilizing role through geometric damping, forming a meshwork that can locally suppress or permit wave propagation (Wang, 2010). In pathologies such as epilepsy, this system fails, leading to a collapse of phase stability and uncontrollable excitation (Traub et al., 2005). This balance is also crucial for the generation of gamma and theta rhythms, which are implicated in attention, memory, and sensory integration and emerge from the precisely timed interplay of excitatory and inhibitory currents (Buzsáki & Wang, 2012).

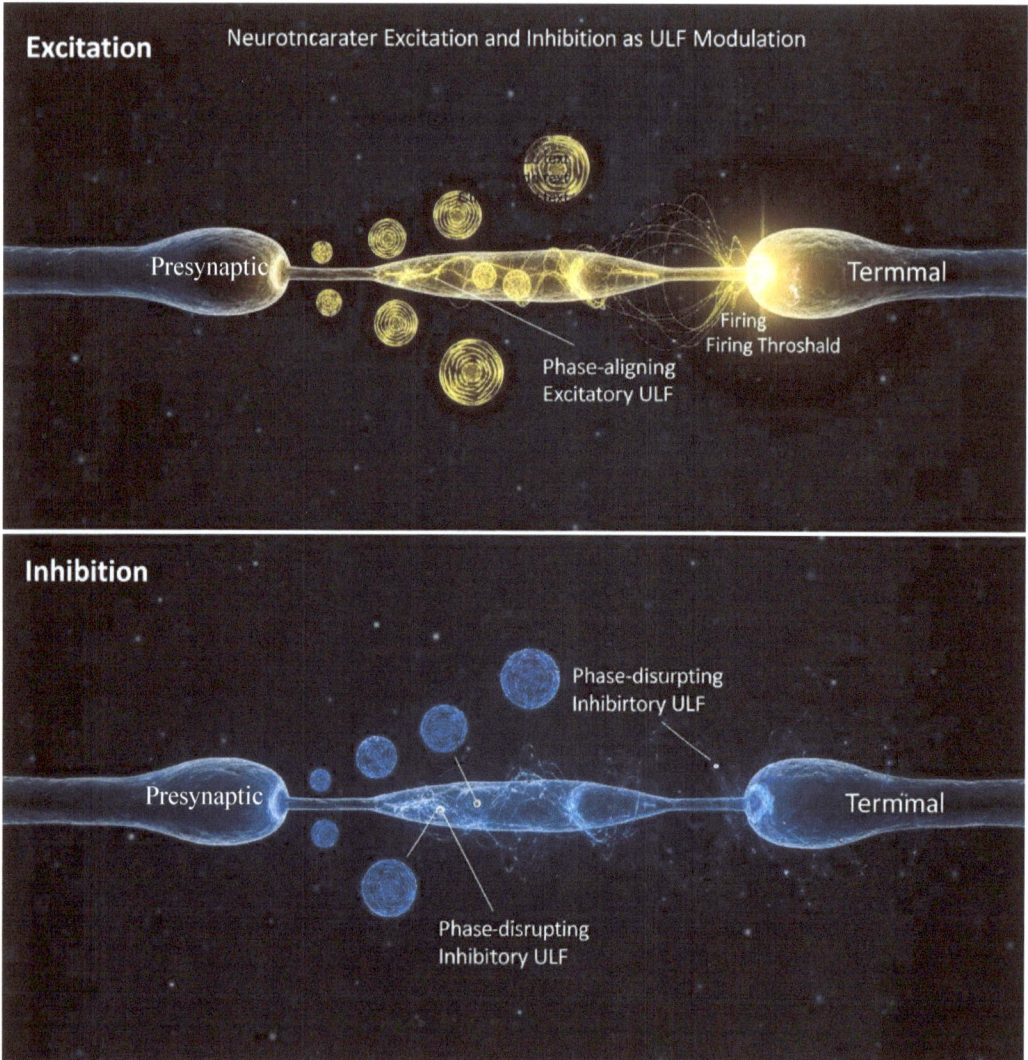

Figure 46: Neurotransmitter Excitation and Inhibition as ULF Modulation. This figure illustrates the dual role of neurotransmitters as resonant modulators of ULF at the synapse. (Top): An excitatory neurotransmitter is depicted as a coherent wave packet of golden light. Upon reaching the postsynaptic membrane, it acts as an amplifier, phase-aligning with the neuron's local field, increasing its resonant coherence, and pushing it closer to its firing threshold. (Bottom): An inhibitory neurotransmitter is depicted as a wave packet of cool blue light. It acts as a dampener, introducing a stabilizing or dephasing resonance that reduces the overall coherence of the neuron, pulling it away from its firing threshold. This model provides a direct, physical mechanism for synaptic communication, reframing it from a simple chemical "lock-and-key" to a sophisticated resonant-tuning process.

2.2.3 Role as Geometric "Keys" Unlocking Resonant States

In RN, the effect of a neurotransmitter is not solely determined by how it binds to a single receptor, but rather by how it phase-aligns the entire synaptic and subcellular environment. In our model, the receptor-ligand interaction serves as a moment of geometric entrainment, in which field geometries lock into resonance. Downstream effects then propagate via cytoskeletal conduits or extracellular matrix coherence (Pessa, 2003). This perspective aligns with evidence that neurotransmitter signals can diffuse beyond the synaptic cleft to influence broader cortical rhythms (Nicholson, 2001). This view also explains how minor molecular differences, such as a single hydroxyl group that distinguishes dopamine from norepinephrine, can produce dramatically different resonant effects across the brain (Purves et al., 2018).

2.2.4 The Potential of Brain Stimulation Therapies

Modern research on brain stimulation therapies (e.g., TMS and DBS) has demonstrated that rhythmic field perturbations can induce neurotransmitter release and reorganize large-scale brain networks (Ashkan et al., 2017). These results suggest that neurotransmitter systems are components of a broader resonant field system, a view with historical roots in the work of physicists and biologists who have long sought to link field coherence to biological function and consciousness (Fröhlich, 1968; Hameroff & Penrose, 2014; Smythies, 2005).

Taken together, excitatory and inhibitory neurotransmitter systems act as dynamic intercellular modulators of the brain's resonant state. They are the "conductor's signals" that orchestrate the entire symphony, telling the neural ensembles when to play, when to be silent, and how to synchronize into a coherent, global state. However, for these signals to have any meaning, there must be a "master instrument" capable of receiving, holding, and processing these vibrations. This brings us to the intracellular architecture—the physical and geometric backbone of the neuron that acts as the primary "resonant conduit" for consciousness: microtubules.

2.3 Microtubules as Intracellular Resonance Conduits

Within RN, microtubules are more than structural components; they are dynamic resonance conduits that orchestrate intracellular coherence. Composed of tubulin dimers arranged in cylindrical lattices, microtubules form helical arrays that support longitudinal and torsional vibrations. These internal oscillations, powered by GTP hydrolysis and modulated by the surrounding fields, form nested standing waves that may synchronize subcellular processes and link neurons to larger brain-wide resonant networks.

Microtubules are particularly suited to act as waveguides because of their quasi-crystalline structure and dielectric properties (Pessa, 2003). RN proposes that microtubules support ULF and potentially UEF harmonics that resonate with the oscillatory profile of the soma and synaptic terminals. This scaffolding enables distributed synchronization within the neuron, which allows spatially distinct components to align in phase during high-level cognitive tasks (Figure 47).

This idea complements and extends previous theories by Fröhlich (1968), who proposed that coherent excitations could form within biological macromolecules, and by Hameroff and Penrose (2014), who posited that microtubules might host quantum-level processes linked to consciousness. In RN, microtubules provide the geometric backbone for coherence, bridging molecular signaling with mesoscale neural dynamics.

The structural reconfiguration of the tubulin lattice may also tune the resonant profile of a neuron, encoding experience as an altered vibrational geometry. This proposes a second, deeper substrate for memory beyond the standard Hebbian model of synaptic plasticity.

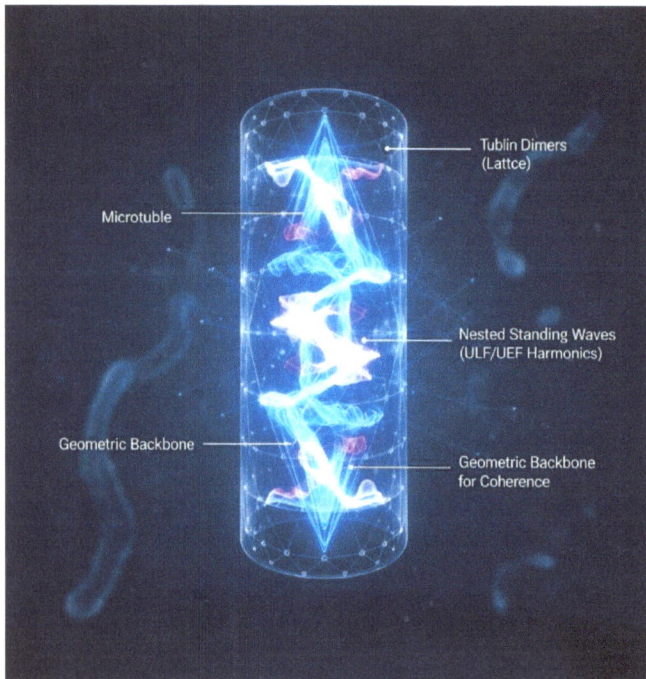

Figure 47: The Microtubule as an Intracellular Resonance Conduit. This figure visualizes microtubules as dynamic resonant conduits within the RN framework. Its quasi-crystalline, hollow geometry acts as a perfect waveguide, capable of supporting the complex, nested standing waves of ULF and UEF harmonics shown within. This continuous internal oscillation enables microtubules to serve as the geometric backbone of intracellular coherence, synchronizing subcellular processes and bridging the gap between molecular signaling and the larger resonant networks of neurons.

In conventional neuroscience, long-term memory is encoded primarily in the strength of the connections *between* neurons ("neurons that fire together, wire together")

(Hebb, 1949). Our model complements this by proposing an intracellular tuning mechanism in which experience physically retunes the neuron's microtubule resonant frequencies, where a "resonant scar" is left by an experience. This scar anchors the memory by physically tuning the neuron to that experience's specific resonant signature, which in turn influences the neuron's future firing thresholds and synaptic behavior.[66]

Thus, in RN, microtubules are active, tunable resonance chambers that are essential to neural resonance and complete the picture of the neuron as a resonant unit. Taken together, the dynamics of neurotransmitters and microtubules reveal a multi-scale resonant architecture that underpins the function of the neuron, the brain's fundamental resonant unit. The microtubule lattice acts as a stable, intracellular waveguide—the geometric and vibrational "strings" that give each neuron its unique resonant profile. The neurotransmitter system, in turn, acts as a dynamic intercellular communication network that modulates coherence among these resonant units, tuning the collective state of the entire system.

This multi-tiered architecture—the microtubules that define the neuron's resonance and neurotransmitters that coordinate its musical symphony—allows the brain to function as a unified, multi-scale resonant system.

3. The Resonant Architecture of the Brain

In RN, macroscopic brain structures are not passive anatomical regions but active resonant systems. Each plays a distinct role in orchestrating the brain's vibrational field architecture, supporting specific aspects of cognition, emotion, and homeostasis. These structures operate as field-modulated attractors, phase-locking to different frequencies and integrating neural coherence across vast spatial and temporal domains.

3.1 The Hypothalamus: Resonant Command Center

The hypothalamus serves as the master oscillator of bodily rhythms. Its nuclei receive and integrate signals from diverse regions, including the prefrontal cortex,

---

[66] In the full RN model, the memory *itself* (the complete experiential, informational pattern) is a nonlocal phenomenon, distinct from its physical anchor. The "resonant scar" in the microtubule does not *store* this information. Rather, it acts as the physical 'tuner' or 'antenna' that allows the brain to recall the memory by resonating at that specific frequency. This concept of a field-based memory, while outside the standard synaptic model, is conceptually analogous to other theories of distributed, nonlocal information, such as the holonomic brain model (Pribram, 1991) and the implicate order (Bohm, 1980), which also posit a deeper, enfolded layer of reality. This deeper, nonlocal mechanism of consciousness and memory is explored in detail in §6.

amygdala, hippocampus, and brainstem, allowing it to modulate endocrine and autonomic responses. In RN, this integration is not purely computational—it is resonant. The hypothalamus acts as a tuning hub that aligns somatic and emotional signals into coherent vibrational states, which coordinates homeostasis with motivational and affective fields (Purves et al., 2018) (Figure 48A).

Oscillatory inputs from circadian structures, such as the suprachiasmatic nucleus, entrain the hypothalamus to environmental cycles of light and dark, whereas feedback from the vagus nerve couples it to internal organ states. This reciprocal coupling suggests that the hypothalamus is a bidirectional field transducer that balances the organism's vibrational landscape with both internal coherence and environmental resonance.

3.2 The Thalamus: Gateway of Coherent Perception

The thalamus is often referred to as the brain's relay station, transmitting nearly all sensory data to the cortex. However, its deeper function lies in its capacity to filter, prioritize, and synchronize input. Within RN, the thalamus serves as a dynamic resonance filter that selectively amplifies field-aligned frequencies while damping incoherent noise (Figure 48B).

Thalamic relay cells exhibit rhythmic bursting and synchronize with cortical oscillations during states of attention, sleep, and altered consciousness (Purves et al., 2018). This suggests that the thalamus operates not only as a passive conduit but also as a vibrational gatekeeper that tunes consciousness itself by shaping the frequencies allowed to enter the cortical attractor basin. This dynamic filtering may explain phenomena such as selective attention, gating of pain signals, and sensory overload in disorders like schizophrenia, where resonant gating may become dysregulated.

3.3 The Hippocampus: Spatial and Temporal Resonator

The hippocampus plays a central role in encoding episodic memory and navigating spatial environments. Its distinctive trilaminar architecture, with layered entorhinal inputs that feed through the dentate gyrus, CA3, and CA1 regions, creates a natural waveguide. These layers propagate rhythmic activity, especially theta (4–8 Hz) and gamma (30–80 Hz) oscillations, through sequenced phase-locking between neurons.

In RN, the hippocampus functions as a resonant cavity for encoding coherent spatiotemporal fields (Figure 48C). In this model, memories are not static imprints but dynamically phase-locked attractor states—specific resonant signatures that emerge when inputs align with preexisting field geometries. This interpretation aligns with experimental data showing that hippocampal place cells and time cells synchronize their activity patterns across different behavioral contexts, suggesting that memory and navigation share a common resonant infrastructure (Lisman & Jensen, 2013).

The capacity of the hippocampus to engage in phase precession, where individual neurons shift their firing phase relative to the overall oscillatory rhythm, suggests a deeper geometric mechanism for encoding temporal information in the brain's spatial architecture.

3.4 The Cerebellum: Oscillatory Precision and Error Correction

Traditionally associated with motor control, the cerebellum is increasingly recognized as an essential structure for precision timing and coordination across both motor and cognitive domains. It receives vast input from the cortex and spinal cord and projects it back to cortical motor and prefrontal areas via the thalamus. This bidirectional circuitry forms a recursive resonance loop capable of error detection and predictive correction.

In RN, the cerebellum functions as a stabilizer of high-frequency coherence. Its modular and fractal-like microarchitecture, composed of repeating Purkinje and granule cell layers, creates an ideal substrate for phase detection and error damping (Figure 48D). These layers generate precise spike-timing patterns that can correct desynchronization in real time, thereby ensuring a coherent output across the motor, affective, and cognitive systems (Ito, 2008).

The cerebellum's ability to learn from discrepancies between intended and actual outcomes implies a resonant feedback loop that allows it to correct phase drift across systems. This geometric view helps explain its role in tasks as diverse as fine motor control, language fluency, and social cognition.

3.5 The Pineal Gland: Quantum Oscillator of Light and Time

The pineal gland has long captured the scientific and philosophical imagination due to its central location, its unique calcified structure, and its role in regulating circadian rhythms. Its secretion of melatonin in response to light-dark cycles entrains the organism to planetary rhythms. However, from the RN perspective, the pineal gland is more than a clock; it is a resonant node sensitive to quantum-scale fluctuations in electromagnetic and possibly Ulfs (Crick & Koch, 2003; Hameroff & Penrose, 2014) (Figure 48E).

The gland's microcrystalline structures (including calcite microcrystals) may function as natural piezoelectric resonators, which transduce subtle vibrational inputs into biochemical cascades. This sensitivity may underlie altered states of consciousness associated with sleep, meditation, or psychedelic experiences, during which the pineal gland exhibits increased activity or the release of melatonin metabolites (Strassman, 2001). In RN, the pineal gland is therefore a scalar-phase interface between global brain coherence and the deeper harmonic fields of the cosmos.

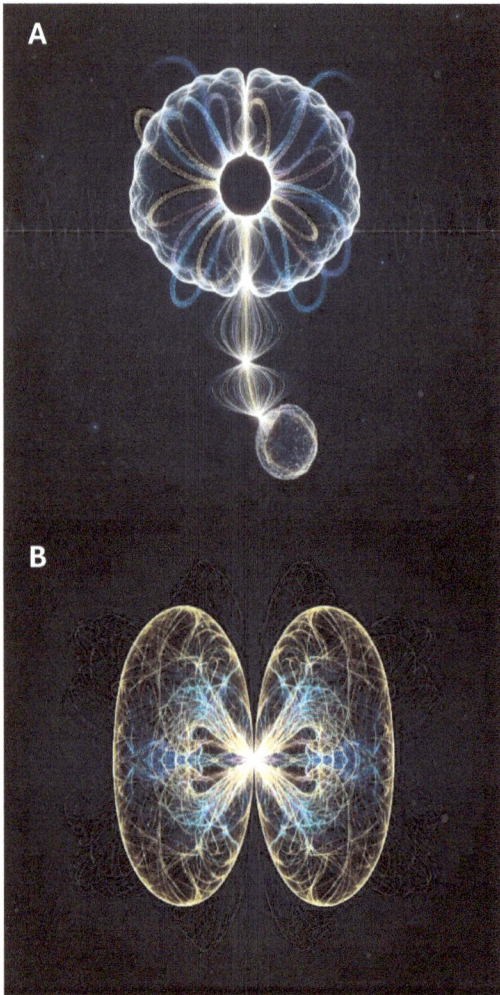

Figure 48: The Resonant Brain.
(A) The Resonant Architecture of the Hypothalamus.
This image depicts the hypothalamus as a geometric resonance integrator at the interface between brain and body. In RN, the hypothalamus functions as a dynamic feedback node that couples internal physiological signals to coherent field oscillations. Vortex flows are illustrated as converging through their toroidal center, symbolizing their role in entraining metabolic rhythms, thermoregulation, circadian cycles, and hormonal modulation via the pituitary gland.

(B) The Thalamus as a Resonant Switching Station.
The thalamus is portrayed as a bilateral geometric hub radiating nested interference patterns. In the RN view, it acts not only as a relay for sensory information but also as a phase-alignment module, entraining cortical rhythms into coherent oscillatory frameworks. The illustration highlights vortex inflows and radial symmetry to express its function as a field synchronizer across multiple frequency domains.

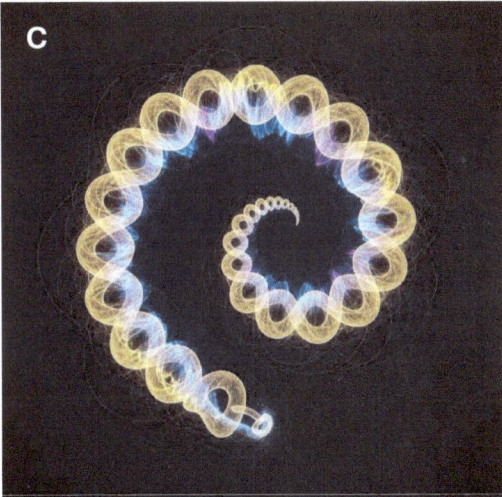

(C) The Hippocampus as a Vortex-Based Memory Engine.

This illustration presents the hippocampus as a spiraling resonance organ, emphasizing its nested, toroidal geometry. Within the RN framework, the hippocampus is seen as a waveform transducer that encodes episodic experience into field-aligned geometric patterns. The structure's arching coil supports sequential wave propagation, enabling memory encoding, spatial mapping, and rhythmic consolidation via theta- and gamma-coherence.

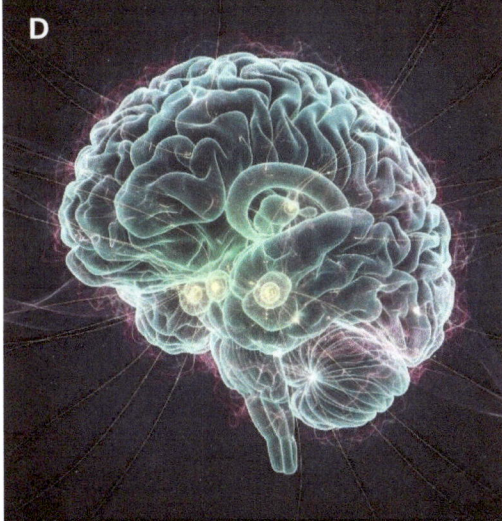

(D) The Cerebellum as a Harmonic Stabilizer:

This illustration presents the cerebellum with layered, wave-like folds that reflect its role in fine-tuning resonance across motor and cognitive systems. In RN, the cerebellum is seen not merely as a coordinator of movement, but as a harmonic stabilizer. This structure maintains coherence between cortical and peripheral systems by damping phase noise and aligning rhythmic inputs. The figure emphasizes its fractal folia and deep nuclei as loci of geometric phase correction and entrainment.

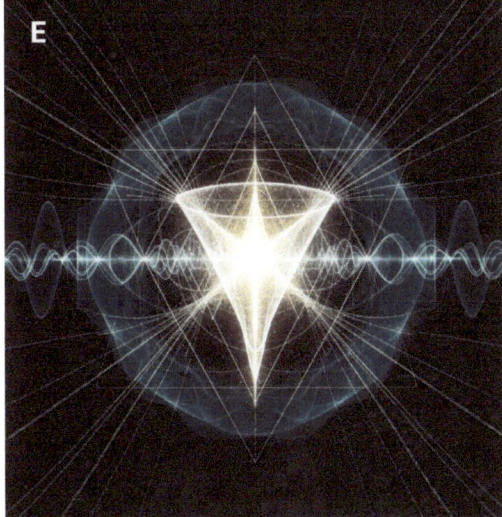

(E) The Pineal Gland as a Scalar-Phase Transducer:

This image portrays the pineal gland as a geometric singularity embedded near the brain's geometric center. In RN, the pineal gland serves as a phase interface between internal neural coherence and broader Ulf field harmonics. Its conical architecture is visualized as a scalar vortex funneling coherent light-like oscillations through the cranial cavity. This structure is proposed to mediate inner light perception, circadian synchronization, and, possibly, consciousness coupling via transcranial Ulf resonance.

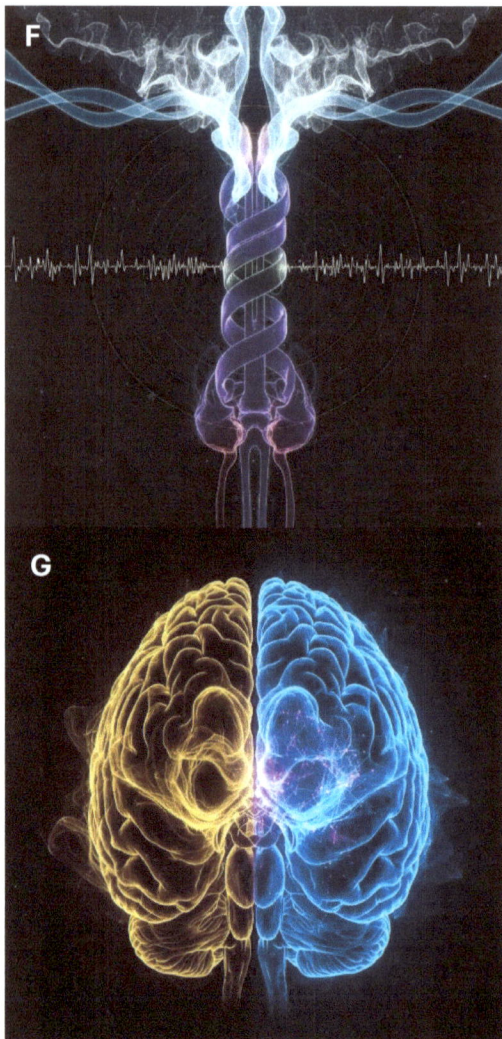

(F) The Brainstem as a Resonant Conduit:
The brainstem is visualized as a vertically aligned vortex structure that connects the cerebral cortex to the spinal cord. In RN, it serves as a central resonance column, guiding Ulf waves between the brain's higher-order structures and the body's autonomic systems. Its layered anatomy, including the medulla, pons, and midbrain, supports longitudinal coherence and entrains bodily rhythms such as heartbeat, respiration, and sleep-wake cycles. This depiction highlights the brainstem's role not only as a relay center but also as a phase-aligning bridge for the transfer of vibrational information across scales.

(G) The Brain as a Multi-Field Interface:
This figure depicts the RN model of hemispheric function. The right hemisphere is depicted as the UEF interface, resonating with the holistic, energetic fields of the body and environment (gold). The left hemisphere is fundamentally oriented towards the timeless, geometric laws of the USF. Within it, the Prefrontal Cortex acts as the primary ULF processor, transforming dynamic, high-frequency data (blue) into coherent, logical thought. The corpus callosum serves as the coherence bridge, weaving these distinct modes of reality—the intuitive and the analytical—into a single, unified field of consciousness.

3.6 The Brainstem: A Resonant Axis of Autonomic Integration

In RN, the brainstem is not merely a passive relay station between the spinal cord and the higher brain centers; it is a central modulator of life-sustaining coherence. Its three principal components—the medulla oblongata, pons, and midbrain—serve as dynamic resonance hubs that regulate rhythmicity, survival behavior, and sensory-motor entrainment (Figure 48F).

The medulla oblongata regulates autonomic processes, including breathing, heart rate, and vasomotor tone. These are typically considered "automatic" functions. However, in terms of resonance, they represent the lowest-frequency rhythmic modulations necessary for bodily coherence (Purves et al., 2018). The rhythmicity of breath and heartbeat is not only mechanically vital but field-structuring, entraining higher systems to the body's central oscillatory state.

Above the medulla, the pons plays a key role in modulating sleep-wake cycles and coordinating signals between the cerebellum and cortex. In RN, the pons acts as a resonance synchronizer that maintains harmonic relationships between limbic and motor regions. This may explain why brainstem damage can lead to global desynchronization, including coma and REM sleep disruption (Parvizi & Damasio, 2001).

The midbrain, in turn, plays a crucial role in regulating auditory and visual reflexes, controlling gaze, and facilitating motor orientation. These processes require rapid, phase-locked coordination across sensory modalities, a function that depends on stable coherence in the Ulf domain. The superior and inferior colliculi act as bidirectional resonant nodes that facilitate the flow of oscillatory information between cortical and subcortical structures.

Notably, the brainstem also houses several central neurotransmitter nuclei, including the serotonergic raphe nuclei and the dopaminergic substantia nigra and ventral tegmental area. These centers broadcast phase-tuned neurochemical signals throughout the brain that influence emotional valence, arousal, and motor readiness (Müller & Homberg, 2015). Their central location allows them to act as resonant "broadcast towers" that align wide-scale neural oscillations into a unified biological rhythm.

From a structural standpoint, the cylindrical and longitudinal form of the brainstem mirrors its function as a standing-wave conduit. Just as a flute's hollow tube supports harmonics across its length, the brainstem serves as a vibrational axis that modulates phase flow between peripheral systems and the core of consciousness. Its geometry allows for the propagation and tuning of fundamental frequency patterns that scaffold more complex cognitive dynamics.

Ultimately, integrating the brainstem into the whole-brain resonance framework helps elucidate its crucial role in consciousness. While traditionally viewed as separate from higher cognition, in RN, brainstem activity is recognized as essential for wakefulness and basic awareness (Parvizi & Damasio, 2001). It is the foundational carrier wave—or the lowest harmonic—upon which the nested geometries of thought, feeling, and action are built.

3.7 Hemispheric Specialization and the Coherence Bridge

RN also provides a new physical basis for the well-known specialization of the brain's left and right hemispheres, a phenomenon first rigorously demonstrated in the split-brain research of the mid-20th century (Gazzaniga, 1967; Sperry, 1968). Here, the hemispheres are not merely optimized for different cognitive tasks. Instead, they represent distinct orientations toward the fundamental fields of reality within the UFD framework (Figure 47G).

- The Left Hemisphere: The USF Interface. The left hemisphere, associated with "Mind," logic, and mathematics, is fundamentally oriented towards the Universal String Field (USF). Its highest function is to resonate with the timeless, geometric Forms and logical principles that constitute the source code of reality. This framing aligns with empirical studies showing the left hemisphere's role in propositional language, sequential processing, and fine-grained motor control (Gazzaniga, 2000).

- The Right Hemisphere: The UEF Interface. The right hemisphere, associated with "Heart," intuition, and emotion, serves as the brain's primary interface with the Universal Energetic Field (UEF). It is tuned to the body's and the environment's slower, large-scale resonances, giving rise to holistic feeling, spatial awareness, and intuitive insight. Neuropsychological evidence supports this distinction, as right hemisphere lesions often disrupt emotional tone, spatial navigation, and holistic pattern recognition (McGilchrist, 2009). This does not imply direct access to the UEF but rather a sensitivity to the embodied, energetic, and relational dynamics shaped by it.

- The Prefrontal Cortex: The ULF Processor. Located at the forefront of the left hemisphere, the Prefrontal Cortex (PFC) serves as the brain's primary information processor. It intercepts the dynamic, high-frequency stream of communicative data (language, light, sensory input) from the Universal Light Field (ULF) and organizes it into stable, ordered geometric structures that reflect USF logic. This conversion process—turning "signals" into "logic"—aligns with the PFC's empirically established role in executive function, working memory, and symbolic integration (Miller & Cohen, 2001).

- The Corpus Callosum: The Coherence Bridge. Between them lies the corpus callosum, the great "coherence bridge" of the brain. It integrates the hemispheres' complementary modes, weaving together the analytic clarity of the left with the holistic resonance of the right to sustain a unified and coherent field of consciousness (Banich, 1995).

This model reframes hemispheric specialization as more than a biological adaptation: it is a mirror of the universe's nested-field structure. The brain's architecture embodies a fundamental polarity of nature. A healthy, integrated consciousness depends on the seamless resonance between these two modes—our coherence bridge between timeless Form and dynamic Substance.

In sum, in RN, the brain's major anatomical structures are reinterpreted as integral components of a unified vibrational system. This systemic view challenges the reductionist paradigm of isolated brain functions, offering a new understanding of the brain as an architect of coherence.

4. Field Coherence and Neural Networks

Modern neuroscience increasingly recognizes that the brain is not simply a reactive system of neural firings but a complex symphony of oscillatory activity unfolding across nested spatial and temporal scales. These brain rhythms, which range from slow delta waves to fast gamma bursts, are not background noise; they are the infrastructure of cognition itself. In RN, these oscillations are viewed as field-driven resonances that shape and stabilize the functional architecture of thought, perception, and behavior.

This section examines how brain coherence arises from nested wave structures, organizes large-scale neural networks, binds cognitive content into a unified experience, and breaks down in pathological conditions. We propose that the key to understanding brain function lies in the geometry and phase relationships of these oscillatory patterns. Mental clarity, memory, and attention arise when resonances align across networks, whereas mental dysfunction occurs when these patterns are fragmented.

4.1 Nested Oscillatory Hierarchies

At the heart of the RN approach lies the principle that the brain operates through nested hierarchies of oscillatory coherence. Rather than viewing brain waves as epiphenomena or noise, this section explores their role as fundamental organizational scaffolds that coordinate neural activity across spatial and temporal scales.

Neural oscillations are traditionally categorized into frequency bands: delta (0.5–4 Hz), theta (4–8 Hz), alpha (8–12 Hz), beta (13–30 Hz), and gamma (30–100+ Hz), each associated with different cognitive and physiological states (Buzsáki, 2006). In RN, these frequencies arise from layered resonance patterns stabilized by the brain's geometric structure and electrochemical dynamics. Lower-frequency oscillations modulate slower large-scale integrative functions, whereas higher frequencies coordinate fine-grained local processing. This imbalance creates a nested hierarchy of temporal rhythms, in which slower oscillations act as envelopes or phase gates for faster oscillations.

Theta rhythms from the hippocampus, for instance, often couple with gamma oscillations during memory encoding and retrieval, a phenomenon known as cross-frequency coupling (Fries, 2005). This suggests that different frequency bands are not isolated channels but dynamically entrained layers of a unified system. Within the RN model, such coupling reflects the entrainment of nested toroidal vortices—Ulf and EMF patterns that are geometrically tuned to one another, allowing multiscale coherence to emerge spontaneously.

These oscillatory hierarchies enable the integration and segregation of neural signals. During focused attention, higher coherence in gamma or beta bands may synchronize localized cortical ensembles. At the same time, alpha rhythms suppress irrelevant pathways to maintain clarity (Friston, 2005). The interplay between these rhythms enables the brain to maintain a dynamic equilibrium between stability and adaptability. This model was described by Varela et al. (2001) as the "brainweb," which is a web of transient, metastable synchronizations that encode conscious experience.

By reframing brain oscillations as nested vortex resonances, RN unifies structural and functional neuroscience under a single principle of geometric coherence. Each frequency band becomes a standing wave in the brain's dynamic field architecture, and their interaction encodes the state of consciousness as a resonance spectrum rather than a binary switch.

4.2 Network Resonance and Functional Domains

This nested structure enables the formation of large-scale, coherent networks, such as the Default Mode Network (DMN), the Salience Network, and the Central Executive Network (CEN). Each network can be understood as a distinct attractor basin within the brain's resonant field-space, stabilized by recurrent coupling among regions with compatible geometric and oscillatory profiles (Deco & Kringelbach, 2016) (Figure 49).

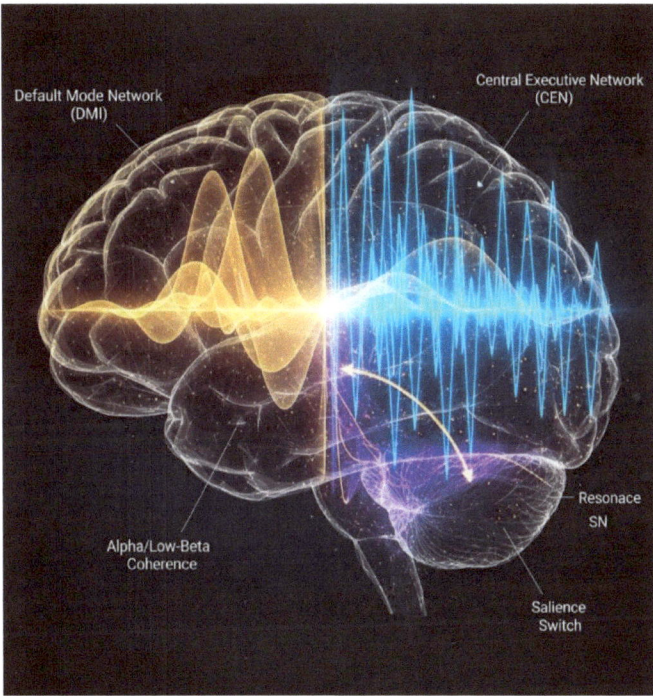

Figure 49: Neural Resonance. This figure depicts the major functional networks of the brain—the DMN and the CEN—as distinct, large-scale resonant modes stabilized by recurrent coupling. The DMN exhibits slower alpha/beta coherence for internal narratives, while the CEN employs faster gamma-band dynamics for high-resolution attention. The SN (center, transitioning) acts as a resonance switch, dynamically shifting field coherence (phase shifts) between the internal and external modes. These networks represent active, organized standing waves that encode the state of consciousness as a resonance spectrum.

The DMN, which is active during rest and introspection, exhibits strong coherence in the alpha and low-beta bands, enabling internal narrative generation and memory retrieval. The CEN, which governs attention and working memory, exhibits more rapid gamma-band dynamics that reflect its fast, high-resolution processing needs. The SN acts as a resonance switch, dynamically shifting field coherence between the internal (DMN) and external (CEN) modes in response to perceived relevance.

Within RN, these networks are not merely statistical groupings. Rather, they are resonant modes of the brain's energetic geometry. Network transitions are field phase shifts—redistributions of standing waves across the brain's topological substrate.

4.3 Temporal Binding and Cognitive Coherence

One of the brain's most remarkable features is its ability to bind distinct neural events into coherent perceptions and thoughts. This "temporal binding" problem, which explains how distributed processes become unified in time, is resolved in RN through top-down, nested resonance.

In the Standard Model, some propose that "gamma synchrony" (e.g., 40 Hz oscillations) serves as the binding process, and that consciousness emerges from it (Fries, 2005). RN reverses this causality. In RN, cognition results from a fundamental, nonlocal, organizing influence that imposes coherence on the brain's local fields, as described in §6. This top-down drive toward coherence "pulls" the otherwise disjointed neural signals into a single, resonant attractor—a state of cognitive coherence that is aligned with attention, intention, and awareness.

The 40 Hz gamma synchrony observed during perception and awareness is therefore not the source of consciousness but its measurable, physical signal or reflection. It is the "sound" of this nonlocal "conductor" orchestrating the many different parts of the brain into a single, unified, phase-locked "symphony." Disruptions in this top-down orchestration lead to fragmentation of thought and dysfunction, while harmony gives rise to clarity, memory, and awareness. This field-based framework offers a unifying principle for understanding both normal and pathological brain states as dynamic expressions of this top-down, nested coherence.

5. Pathologies of Coherence: Breakdown and Recovery in the Brain

The brain's capacity for consciousness, cognition, and emotion is deeply tied to its ability to maintain coherence across its nested oscillatory systems. In RN, neurological and psychiatric disorders can be reinterpreted not simply as disruptions of chemistry or structure, but as breakdowns in this geometric and resonant order. This section explores how diverse pathologies can be understood as failures in resonance and how the principle of restoring coherence can provide a new and powerful therapeutic program.

5.1 A New Model of Mental Illness: Pathologies of Resonance

Just as harmonious vibration defines health, disharmony introduces chaos. Epilepsy, as described earlier, represents runaway synchrony—a destructive hypercoherence in which neural networks are locked into low-complexity oscillations. Schizophrenia, in contrast, reflects a fragmented field, where a decoupling of local and global brainwaves leads to a disintegration of thought and perception (Uhlhaas & Singer, 2010). Similarly, Alzheimer's disease is a slow collapse of the resonant structures that support memory, while autism may reflect a hyper-coherence at the local, sensory level that overwhelms the brain's ability to integrate into broader, social-emotional fields (Orekhova et al., 2007).

Mood and attention disorders also manifest as pathologies of resonance. Depression, rather than a simple chemical deficit, is a collapse of vibrational diversity, in which the brain becomes stuck in a rigid, low-energy resonant state (Mulders et al., 2015). ADHD is a failure of temporal entrainment, a breakdown in the brain's ability to phase-lock its attention networks, resulting in a turbulent, desynchronized field (Barry et al., 2003).

5.2 The Coherence Project: A New Therapeutic Paradigm

This resonant view of mental health gives rise to a new therapeutic framework, the Coherence Project, which focuses on restoring harmony to the brain's fields. Instead of targeting chemicals, this approach aims to retrain the system to a state of coherence. Early examples of this principle include technologies such as transcranial magnetic stimulation (TMS) and neurofeedback, which use external rhythms to help the brain regain its natural equilibrium. Likewise, psychedelics may function by temporarily dissolving rigid, pathological resonances, allowing healthier patterns to re-establish themselves (Carhart-Harris et al., 2014).

This approach also emphasizes that healing requires re-embedding the individual in supportive, "coherence-based environments." Ultimately, the most powerful tool in the Coherence Project is consciousness itself. Practices such as meditation can be understood as forms of self-entrainment, a process of consciously listening inward to identify patterns of incoherence and gently guiding the brain back into a state of resonant alignment (Brewer et al., 2011; Lazar et al., 2005).

5.3 Technological Frontiers: Engineering with Resonance

In addition to providing a new theoretical and therapeutic framework for understanding the brain, RN provides a blueprint for the next generation of neuroscience technologies. By treating coherence, phase alignment, and nested resonance as the primary signatures of brain function and dysfunction, RN shifts the focus of neurotechnology from structural manipulation to energetic entrainment. The following innovations exemplify this new paradigm:

1. Coherence-Tuned Neuromodulation. Traditional brain stimulation techniques, such as TMS and deep-brain stimulation (DBS), deliver fixed frequencies to specific regions. RN enables a new class of interventions that dynamically modulate neural fields across spatial and temporal scales. These therapies would

be designed to restore phase harmony within and between networks, using feedback-driven waveforms tailored to an individual's resonant architecture.

2. Field-Based Diagnostics. Neurological conditions may be diagnosed more precisely through coherence mapping than through structural imaging. High-resolution EEG and MEG, guided by Resonant Engineering principles, could be used to detect subtle distortions in nested resonance, such as phase desynchronization in ADHD or coherence collapse in depression. Future imaging technologies may incorporate scalar field detection or Ulf analysis to observe consciousness-related dynamics in real-time.

3. Brain-Computer Interfaces via Field Coupling. Current BCIs focus on decoding electrical signals from the cortex. A RN-informed approach would instead aim to detect and engage with the resonant attractor states of neural fields. Such interfaces could synchronize with the brain's intrinsic rhythms, enabling smoother, more intuitive interaction between mind and machine. This may lead to breakthroughs in prosthetic control, neural feedback, and even direct brain-to-brain communication via shared field states.

4. Scalar Entrainment and Consciousness Modulation. By treating consciousness as a scalar-field phenomenon coupled to neural geometry (*see* Section 6), RN proposes novel tools for the intentional tuning of awareness. Scalar field generators, phase-tuned sound baths, or geometric light modulations may be used to stabilize states of attention, meditation, or cognitive flow. These technologies would not only stimulate brain activity but also harmonize the underlying resonance conditions that support subjective experiences.

5. Neuroholographic Memory Encoding. If cognition and memory are based on nested field coherence, it may be possible to create "holographic imprints" of mental states through exposure to precisely patterned fields. This technology could enable memory augmentation, consciousness preservation, or new forms of learning through field-based resonance transfer. Such techniques reflect biological learning over data-based approaches, offering a resonant pathway for enhancing mental function.

6. Self-Tuning Neural Environments. Living and learning spaces can be reimagined as resonant ecosystems. Classrooms, hospitals, and homes could be embedded with vibrational geometries and feedback systems that promote cognitive clarity,

328

emotional balance, or recovery from neural dysregulation. These environments would not directly manipulate the brain. Instead, they would subtly guide it into coherence through light, sound, spatial patterning, and ambient field structuring.

7. Testing and Development Platform. The empirical foundation of these technologies rests on a new generation of experimental platforms designed to measure and manipulate brain resonance with unprecedented precision. Phase-aware EEG and MEG systems would enable real-time analysis of nested coherence patterns across neural networks. At the same time, Ulf field detectors could reveal subtle coupling between brain rhythms and scalar wave dynamics. Complementing these tools, resonant feedback simulators could dynamically adjust frequency inputs to match an individual's current phase landscape, enabling personalized coherence-based interventions.

In the emerging frontier of RN, the brain is no longer a machine to be fixed or decoded, but a resonant field to be tuned.

6. Consciousness: The Fourth Emergent Field

In the UFD framework, matter emerges from three nested, resonant fields: the UEF, which gives rise to mass and gravity; the ULF, which mediates electromagnetism and quantum phenomena; and the USF, which encodes the geometric blueprints of reality (*see* Chapter 3). However, to account for the unity of self, the perception of time, and the phenomenon of conscious awareness, we propose a fourth field: one capable of supporting a nonlocal, ultra-coherent mode of resonance. We call this fourth field the Universal Awareness Field (UAF)—a field of subtle, *seeing* awareness, or a "consciousness aether."

6.1 The Emergence of the Universal Awareness Field

Rather than reducing awareness to neural activity or emergent complexity, we propose that a distinct, nonlocal field of resonance emerges from the coherent dynamics of the ULF. In this view, the electron, as a stable, toroidal vortex in the ULF, becomes a candidate generator of a Ulf non-vectorial wave structure. A wave in this emergent structure would have a supra-luminal speed due to its lesser density relative to the ULF, giving rise to its nonlocal nature (*see* Chapter 1). This is an awareness field, atemporal and scalar—a Spin-0 field in the language of physics. Here is the crucial distinction:

- The UEF is a Spin-2 (Tensor) field, a perfect medium for the nonpolar, always-attractive force of gravity.

- The ULF is a Spin-1 (Vector) field, a perfect medium for the polar, directional forces of electromagnetism.

- The UAF, as a Spin-0 (Scalar) field, has no direction. It is a field of pure magnitude or information, making it the ideal physical substrate for consciousness.

This model gives rise to a revised ontological equation, first introduced in Chapter 3:

$$(F \odot S \odot C) \odot A \implies U$$

This equation expresses that the universe (U) arises from the inseparable interplay of Form (USF), Substance (UEF), Communication (ULF), and Awareness (UAF). Or, stated in other words:

$$(\text{Geometry} \times \text{Energy} \times \text{Resonance}) \times \text{Consciousness} \rightarrow \text{Reality}$$

In this model, Awareness is an essential, active participant in the actualization of the Universe. Its placement outside the parentheses is meant to illustrate its interaction with each of the local fields, which it does by forming resonant interfaces at all three levels. This results in the formation of distinct levels of awareness: conscious awareness for biological life and preconscious awareness for nonbiological life.

The existence of a preconscious awareness would provide a direct, physical explanation for one of the deepest mysteries of quantum mechanics: quantum entanglement. As we describe in Chapter 5, a simple nonlocal wave in the ULF should be incredibly fragile and immediately disrupted by stray environmental noise. The UAF resolves this paradox by acting as a "coherence field" that actively organizes and sustains the entanglement connection. Because the two entangled particles are part of a unified, coherent experience, the integrity of that experience cannot be broken by the random noise of the lower-level fields. The connection is maintained by the UAF because it represents a single, indivisible experiential state.

At a deeper level, this UAF-stabilized connection is the physical expression of a phenomenological truth. In this view, quantum particles have definite experiences. If one electron has the experience of "spinning left," its entangled twin must simultaneously have the experience of "spinning right," not because a signal was sent, but because they are both local expressions of a single, nonlocal experiential state whose total information must be conserved.

6.2 The Soul: A Resonant Soliton of Awareness

The concept of consciousness, or the soul, as an "inner light" appears across traditions—from the Vedic jyotiḥ (light of consciousness) and the Tibetan 'od gsal (clear light), to the Christian "divine light" in mystic theology (Underhill, 1911).

In the UFD model, these traditions are describing a precise physical topology. However, because the UAF is a scalar field (defined by magnitude and intensity rather than directional flow), the soul is not a spinning vortex. Instead, the soul is conceptualized as a stable soliton—a coherent, self-reinforcing standing wave packet within the UAF. Unlike a ripple that spreads out and fades, a soliton is a unique wave structure that maintains its shape and amplitude indefinitely as it propagates. In the UFD framework, this Soliton of Awareness provides the physical mechanism for the continuity of self by creating a consistent, high-amplitude locus of identity that persists amidst the flux of material and energetic change in the body.

In this model, the emergence of conscious awareness in matter occurs when biological structures achieve sufficient geometric fidelity to "pin" a UAF soliton. Biomolecules such as DNA and RNA, with their distinctive double-helical structures, act as helical antennas. Their natural resonance frequencies allow them to phase-lock with the scalar field, creating a stable node where the UAF soliton can couple effectively with the biological substrate (see Chapter 10).

This field-theoretic approach offers a rigorous explanation for the enduring sense of "I." The soul is not an ethereal "ghost" haunting a machine, nor is it a separate substance. It is a harmonic singularity—a region of intensified field coherence dynamically interwoven with the biological hardware yet possessing its own independent wave-stability.

By grounding consciousness in the solitonic physics of the UAF rather than abstract metaphysics, this model bridges the gap between physics, biology, and phenomenology. It offers a unified account of how Mind (the scalar field) and Matter (the vector field) co-emerge and synchronize.

6.3 Time and the UAF

If the soul is a stable pattern of awareness within the UAF, then time must be understood in a different way than we usually imagine it. Time is not something the soul moves through. Instead, time is what change looks like from the point of view of awareness.

In UFD, the UAF itself does not flow or tick like a clock. Instead, it holds duration — the capacity for experiences to appear, change, and leave traces. What changes over time is not awareness itself, but the patterns that appear within it.

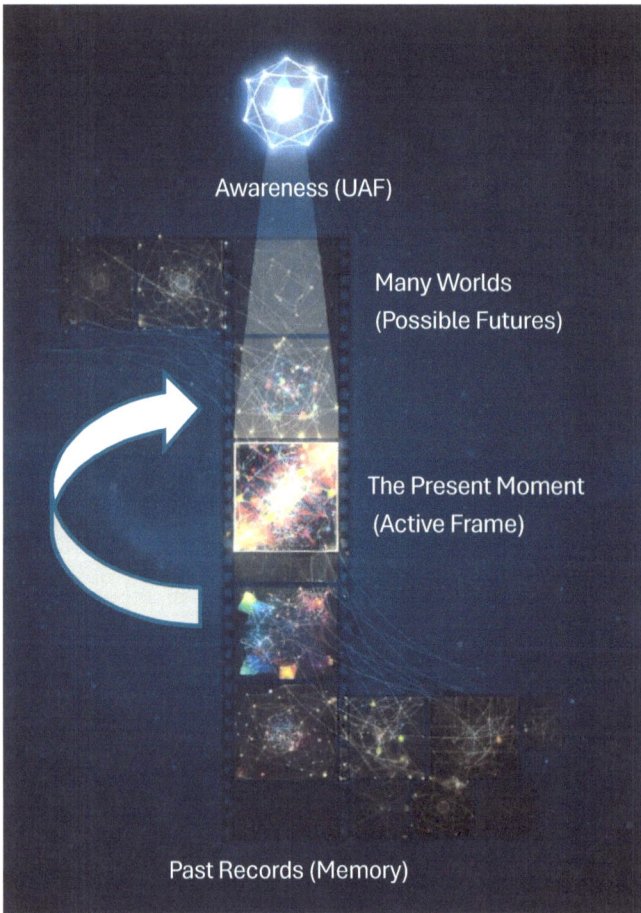

Figure 50. Time and the UAF. This figure depicts time as the ordered disclosure of experience relative to awareness. In the UFD framework, awareness (UAF) remains invariant while experience unfolds as an active present, a recorded past, and a structured space of possible futures. The Present represents the specific configuration of the field—the "frame"—currently coupled to awareness via resonance. The Future exists as a branching array of structured possibilities (reinterpreting the MWI quantum mechanics as unlit geometric pathways), while the Past persists as recorded patterns that provide the boundary conditions for causality and memory. Ultimately, time reflects the order in which these configurations of experience are revealed.

A helpful way to think about this is the metaphor of a movie. Awareness is like the projector (Figure 50). It does not move through the film. It stays where it is. What changes is the frame of the film that is being illuminated.

- The present moment is the frame that is currently lit up — the pattern of experience that awareness is engaged with right now. This "now" is not a moving point in time. It is simply the configuration of experience that is active.

- The future is the part of the film that has not yet been projected. It already exists as structured possibility — scenes that could appear — but has not yet become experience. The future is real, not as something that has happened, but as something that *can* happen.

- The past is the part of the film that has already been projected. Those frames do not disappear. They remain as recorded patterns that shape what comes next. Memory, causation, and personal identity arise from these accumulated patterns, not from awareness traveling forward through time.

Seen this way, past, present, and future all exist within the UAF at once, the past as recorded patterns, the present as the active frame, and the future as structured possibility. Time is not a river carrying the soul along. It is the order in which experiences are revealed relative to a stable center of awareness.

This view also helps clarify the Many-Worlds Interpretation (MWI) of quantum mechanics (Everett, 1957). Rather than imagining countless universes splitting apart, we can understand "many worlds" as the set of possible future frames within the unprojected part of the film. Only the frame that becomes illuminated — the one awareness couples to — becomes part of lived experience.

In this model, time is what unfolding experience looks like when awareness remains still.

6.4 The Mind-Body Interface: The Brain as a Resonant Translator

The age-old problem of free will has long been trapped between a "ghost in the machine" dualism and a rigid physicalism that reduces agency to mere neurochemistry. This framework offers a third way, reframing the mind-body connection not as a collision of substances, but as a dynamic, bidirectional resonance loop, with the brain acting as the central impedance matcher between the physical world and consciousness. This translation process happens in two directions:

- Perception (World-to-Soul): The brain acts as a Transducer, gathering raw, complex data from the physical fields (ULF/UEF) through the senses and translating it into a coherent, resonant "broadcast"—a scalar modulation—that the Soul-Soliton can directly absorb as qualitative experience.

- Action (Soul-to-World): The soul, having perceived this reality, initiates its will. This is the "Conscious Nudge"—a subtle, informational shift in the UAF that acts as a "Coherence Attractor." By gently guiding the brain toward one option, the soul lowers the topological impedance for one specific outcome over another. The brain receives this subtle bias and translates it back into the language of the physical world, orchestrating the cascade of neural signals required for action.

To make an informed decision, the soul integrates information from three primary resonant interfaces within the body, echoing the tripartite soul of ancient philosophy (Plato, c. 375 BCE). Each interface speaks a different "language":

1. The Heart Interface (The Spirited Soul): The soul's connection with the powerful, rhythmic pressure-waves generated by the heart (coupling to the UEF) provides the language of emotional, second-person empathy and feeling (McCraty et al., 2009).

2. The Gut Interface (The Appetitive Soul): The soul's connection with the deep, instinctual resonances of the enteric nervous system ("the gut brain") provides the language of foundational, first-person being and primal intuition (Gershon, 1998).

3. The Brain Interface (The Rational Soul): The brain acts as the Master Transceiver, synthesizing the raw power of the Heart and the instinct of the Gut into the high-fidelity, analytical language of the cortex.

In this model, the Thalamocortical Loop—anchored by the Thalamus and the piezoelectric micro-structures of the Pineal Gland—serves as the central hub of this translation process. It is here that the unified soul (the "I Will") synthesizes these distinct streams—the Feeling from the Heart, the Being from the Gut, and the Knowing from the Brain—to enact its "conscious nudge" (Figure 51).

This model, grounded in the physics of resonant interfaces, resolves the conflict between determinism and agency. The universe is both structured and open. The informational landscape of the UAF contains the record of the past and the geometric "rules of the road" for the future. However, the future is not a single, predetermined path but a vast sea of branching possibilities—an Informational Multiverse. The "conscious nudge" is the act of navigation. It is how the soul, as the singular witness, phase-locks with one specific potential timeline, thereby selecting it from the superposition. The soul does not "break" the laws of physics; it steers through them by choosing which potential future offers the highest coherence.

This view elevates free will from a personal illusion to a cosmic function. It is the physical mechanism by which Awareness (A) becomes an active term in the UFD Master Equation:

$$(\text{Form} \times \text{Substance} \times \text{Communication}) \times \text{Awareness} \rightarrow \text{Universe}.$$

The soul is not a passive passenger carried by the currents of time, but a co-author of its own story. Each choice is a real, meaningful act of Geometric Selection that weaves the fabric of lived experience, frame by frame. In UFD, to have free will is to be an active participant in the universe's ongoing act of becoming.

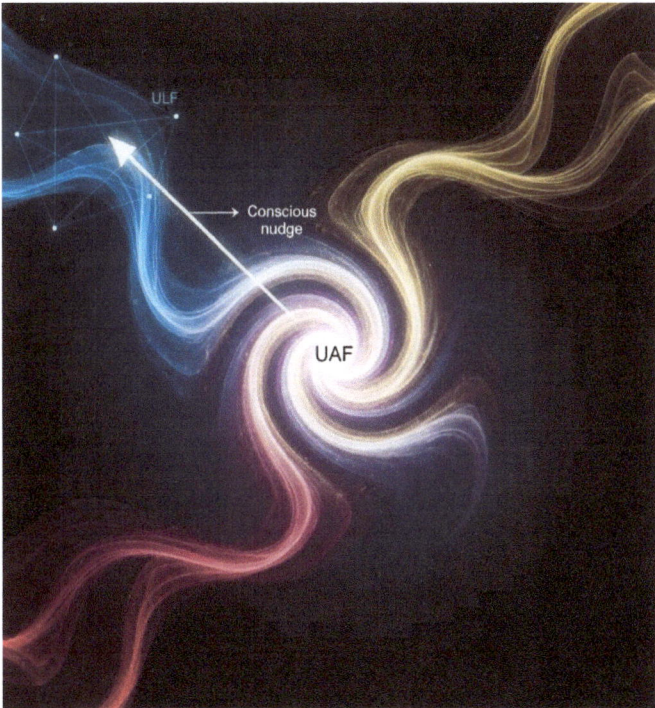

Figure 51: The Tripartite Interface of Consciousness. A visualization of the soul's bidirectional communication with the body. The central Soul-Soliton (UAF) acts as the unified observer, integrating three distinct "languages" of experience from its primary interfaces: the analytical logic of the brain (Blue/ULF), the rhythmic emotional connection of the heart (Gold/UEF), and the primal intuitive being of the gut (Red / ULF). After synthesizing these streams, the soul enacts its will via the "Conscious Nudge"—modulating the coherence of the brain's field to bias the neural outcome and translate intent into physical action.

6.5 The Living Cosmos and Group Souls

The UFD model of emergent consciousness leads us to revisit an ancient idea, now physically grounded: the cosmos is alive (see Chapter 3). This is not mere poetry or mysticism; it is a scientific hypothesis about the nature of nested coherence in the universe. If the UAF is real and nonlocal, and if stable resonant nodes (solitons) anchor consciousness, then awareness is not confined to brains, nor even to biology.

6.5.1 Preconscious Awareness in Stars and Planets

The first and most foundational level of cosmic life is preconscious. In this model, celestial bodies like stars and planets possess a form of "blind," systemic experience—a field of coherence without self-reflection or volition. Their massive, organized structures—the fusion core of a star or the geodynamic core of a planet—act as powerful, stable resonant anchors for the UAF.

- Stellar Awareness: The immense, ordered fusion plasma of a star generates a high-amplitude node in the UAF.

- Planetary Awareness: The coherent magnetic geodynamo of a planet creates a stable "hum" in the field.

These give rise to vast, unified fields of awareness without a central, self-reflective identity. This concept aligns with modern philosophical and scientific explorations of a deeper order in nature, such as David Bohm's "implicate order" (1980) and Rupert Sheldrake's "morphic resonance" (2012).

6.5.2 The Emergence of Group Souls

Just as individual atoms can be entrained to form the collective resonance of a crystal, individual minds can be entrained to form Group Souls. These are shared, coherent attractor states within the UAF, created by constructive interference (Phase Locking) among multiple soul-solitons.

- The Mechanism: When a group of individuals aligns their intent or emotion, their UAF solitons oscillate in phase. This creates a super-soliton or a larger standing wave that encompasses the group.

- The Result: A "distributed mind" emerges that is greater than the sum of its parts.

This provides a physical mechanism for the collective consciousness observed in human cultures, social organisms such as ant colonies, and entire ecosystems. These shared fields, akin to Ervin Laszlo's "Akashic Field" (2004), allow for the nonlocal transfer of information through resonant coupling rather than signal transmission.

Sheldrake's experimental studies on telepathy and the collective behavior of animals provide preliminary empirical support for the existence of such nonlocal fields.

6.5.3 The Universe as a Conscious System

Another potentially profound implication of this framework is that the universe itself may be a single, conscious Resonant Attractor.

This is the ultimate vision of Cosmic Garden Cosmology (see Chapter 3): a nested hierarchy of minds, each emerging from and contributing to the coherence of the whole. In this view, the universe is a Self-Organizing System of Awareness, a concept that resonates with ancient philosophical traditions like Neoplatonism and Vedanta, while gaining new physical plausibility through modern, field-based models of nonlocal coherence (Fröhlich, 1968; McFadden, 2020; Hameroff & Penrose, 2014).

Just as atoms form molecules and molecules form cells, so too may individual minds form higher-order Harmonics of Meaning, ultimately participating in the self-awareness of the cosmos itself.

6.6 The Awareness Precedence Hypothesis

If awareness is a real, nonlocal scalar field that exerts a "top-down" informational influence on the brain, then its effects must, by definition, precede the energetic neural activity that follows. We therefore propose the Awareness Precedence Hypothesis, which offers two specific, falsifiable predictions:

1. Inter-Brain Scalar Resonance During shared intentional states (such as deep group meditation or co-attention tasks), individuals should exhibit synchronous phase-locking in their EEG patterns, even in the absence of sensory communication. This would not be merely coincidental timing, but a measurable nonlocal entrainment to a shared soliton or "attractor node" in the UAF.

2. The Pre-Libet "Coherence Window" We predict that detectable shifts in neural entropy (organization) will occur *before* the onset of the famous "Readiness Potential" (voltage).

Benjamin Libet (1983) famously showed that the brain prepares to move (Readiness Potential) milliseconds *before* the subject feels the conscious urge to move. Materialists cite this as proof that free will is an illusion. In UFD, the readiness potential is merely the "engine revving up" (UEF activation). The actual decision—the "conscious nudge"—is the turning of the key. High-sensitivity analysis should therefore reveal a reduction in quantum noise or a stabilization of phase (a "Coherence Window") in the microseconds *prior* to the voltage spike. This represents the soul-soliton ordering the field to bias the outcome before energy is expended.

Having outlined the macroscopic dynamics of the "conscious nudge," we face a final, critical question: How does it get there?

If the UAF is a real physical field capable of interfacing with biology, it cannot be magic; it must be composed of discrete, fundamental units. It requires a 'carrier particle'—a geometric entity simple enough to slip through the dense lattice of matter, yet essential enough to balance the equations of the universe. To find this particle, we need only look for the entity that sits at the very bottom of the UFD geometric hierarchy—the "ghost" that physics has found but has never fully understood.

6.7 A New Identity for the Neutrino: The Fundamental Quantum of Awareness

UFD's hierarchy of fields (UEF > ULF > UAF) leads to a necessary conclusion: the fundamental unit of the UAF—the lightest possible field excitation—must represent the "atom" of awareness. This theoretical necessity points directly to the neutrino. We therefore propose that the neutrino is the physical manifestation of a minimal, propagating scalar soliton in the UAF. It is not the soul of a complex being (which is a stable, resonant standing wave), but the "carrier signal" of the awareness field at the most fundamental, preconscious level of reality.

This hypothesis provides a direct explanation for the neutrino's most mysterious properties—its "ghost-like" lack of interaction and its exclusive connection to the weak force. It also leads to three profound, falsifiable predictions regarding its identity, geometry, and scale.

6.7.1 Prediction 1: The Neutrino is its Own Antiparticle (Majorana)

In UFD, the distinction between a particle and an antiparticle (a "source" vs. "sink" vortex) requires a polar vector field (like the Spin-1 ULF). A vortex can spin clockwise or counter-clockwise. The neutrino, however, exists in the UAF, which is a scalar (Spin-0) field defined by intensity and phase, not directional flow. By definition, a scalar fluctuation is non-polar and has no "handedness." Therefore, a soliton in this medium cannot have a distinct antiparticle.

UFD thus predicts that the neutrino is a Majorana Fermion (it is its own antiparticle). This prediction addresses the beta-decay problem without the Standard Model's abstract "Lepton Number". When the neutron (Figure-8) snaps into a proton (Trefoil), it sheds a UAF soliton to balance the energy. The Standard Model labels this an "antineutrino" to save its bookkeeping, but geometrically, it is just a neutral soliton—a pulse of field coordination.

6.7.2 Prediction 2: The Spherical Geometry (Topological Descent)

In UFD, if identity is defined by the field layer, then structure is defined by the principle of Topological Descent. As coherence descends through the hierarchy, excitations exhibit progressively simpler topologies:

1. Proton (Heavy Matter): A highly coherent, topologically knotted vortex (trefoil-like) embedded in the UEF, characterized by high geometric complexity, large effective tension, and large inertial mass (see Chapter 7).

2. Electron (Light Matter): An unknotted toroidal vortex of the ULF, possessing a single non-contractible loop. The toroidal aperture permits persistent circulating currents, accounting for the electron's electric charge, magnetic moment, and strong electromagnetic coupling (see Chapter 5).

The neutrino represents the final step in this topological descent. The UFD framework therefore predicts that the neutrino corresponds to a compact scalar soliton— a localized, phase-coherent packet of UAF amplitude with a 0-loop geometry (Figure 52).

Because a compact scalar soliton has no handle (hole), the ULF vector field cannot "grip" it. The neutrino's defining properties thus follow directly from its geometry:

- Electric Neutrality: There is no topology to support a circulating current loop.

- Ghost-Like Interaction: Without a loop to snag on the UEF lattice, the neutrino creates almost zero Hydrodynamic Drag. It slips through the "chainmail" of the vacuum because it is a smooth pulse, not a tangled knot.

- Spin without Circulation: The neutrino carries intrinsic angular momentum (spin) not as physical rotation, but as Phase Rotation of the complex scalar field. In this framework, its spin reflects a topological phase structure in the coupled

UAF–ULF boundary conditions, not the literal spatial rotation of the scalar soliton.

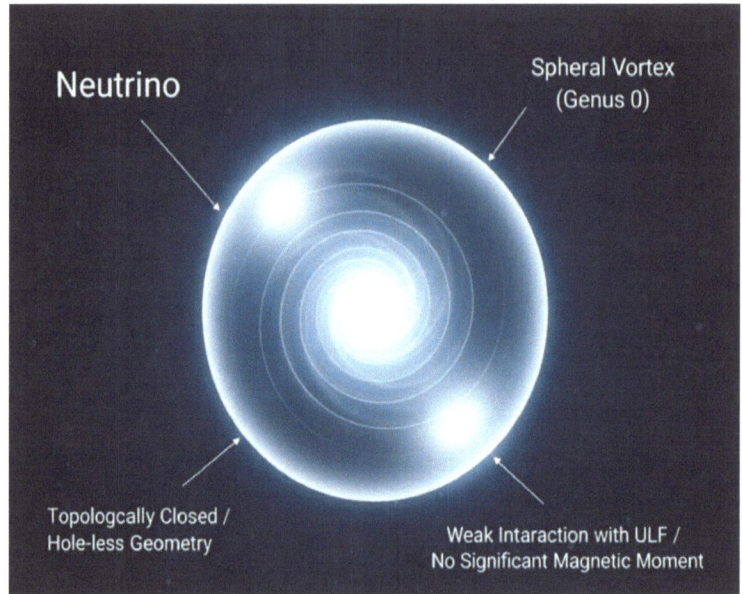

Figure 52: The Neutrino: A Scalar Soliton. This figure illustrates the neutrino in the UFD framework as the terminal excitation of Topological Descent. Unlike the electron (Toroidal), the neutrino is Simply Connected (Genus-0). It lacks a central aperture. This "hole-less" geometry prevents the threading of magnetic flux lines, accounting for the neutrino's absence of electric charge and negligible magnetic moment.

This geometric identification solves the qualitative puzzles of the neutrino—its neutrality and extreme penetrability—but UFD must also satisfy the rigorous quantitative test: its specific mass. If the neutrino is truly the fundamental soliton excitation of the UAF, its inertial mass must follow from the UAF's energy density. Just as the electron's mass is set by the density of the ULF, the neutrino's mass must scale with the density of the UAF. In the next section, we apply the established field-density hierarchy to derive an absolute neutrino mass scale with no adjustable parameters.

6.7.3 Prediction 3: The Neutrino Mass Scale

Having defined the neutrino as a scalar soliton, we now derive its inertial mass. In UFD, mass is the coupling cost—the energy required for a UAF object to push against the ULF background. We propose that the descent from the vector ULF (electron) to the scalar UAF (neutrino) follows a volumetric geometric scaling governed by the fine-structure constant.($\alpha \approx 1/137$).

- The electron is a surface-dominant structure (2D manifold in 3D).

- The neutrino is a volume-dominant fluctuation in a scalar field.

- This dimensional reduction suggests a cubic scaling factor (α^3).

Applying this α^3 scaling to the known electron mass (m_e) we obtain the neutrino mass scale (m_ν):

$$m_\nu \approx m_e \times \alpha^3 \approx (0.511 \text{ MeV}) \times \left(\frac{1}{137}\right)^3 \approx 0.198 \text{ eV}$$

This prediction is physically meaningful. While the Standard Model struggles to explain why neutrinos have mass at all, UFD derives a specific value that falls perfectly within the current experimental upper bounds set by the KATRIN (2025) experiment and cosmological data (which constrain the sum of neutrino masses to < 0.12 eV for the lightest eigenstate or < 0.8 eV for the effective mass).[67]

In this interpretation, every fermionic process is accompanied by a corresponding UAF coordination channel, whose minimal propagating manifestation is the neutrino. This model creates the physical foundation for the hierarchy of consciousness. An atom has a "proto-field" (neutrino cloud) while a biological organism has a "soul" (coherent soliton complex). The difference is not substance, but regime. The neutrino is the transient, propagating spark of the field, while the soul is the sustained, resonant flame anchored by biological complexity. When a sufficiently complex 'antenna' (such as DNA) integrates these couplings into a unified state, a conscious resonant soliton emerges

Together, these predictions define the neutrino as a spherical, ultra-low-mass soliton that serves as the fundamental carrier of information in the UAF.

6.8 The Speed of Awareness

Having established the physical reality of the UAF—the least dense and most subtle layer in the UFD hierarchy—we now turn to its implications for time, nonlocal coordination, and conscious experience. The UAF framework provides a physical context

---

[67] Here, we assume that a compact scalar soliton carries order-unity effective coupling impedance comparable in magnitude (though not in mechanism) to that of an unknotted toroidal ULF vortex. This assumption is plausible given that both are defined by simple, unknotted geometries. However, this remains an assumption. If the impedance coefficient C_d (Soliton) differs from unity, the neutrino mass prediction would scale accordingly. The current experimental bound ($m_\nu < 0.8$ eV) accommodates this range, but future precision measurements from KATRIN or cosmological observations may constrain it further.

for two persistent paradoxes in modern physics: the apparently instantaneous nature of quantum entanglement and the experiential invariance of Proper Time.

This section applies the wave mechanics developed in earlier chapters to the UAF, identifying a characteristic phase-coordination speed associated with this scalar field and clarifying how that speed reconciles subjective experience with objective relativistic effects.

6.8.1 The Phase Velocity of the UAF

Because the UAF is a scalar field, it does not support vector flux, helicity, or energy transport in the manner of vector fields such as the UEF or ULF. It does, however, support propagation of phase, coherence, and synchronization states. These processes are governed by the same general principle that applies to all physical media: the characteristic propagation velocity of disturbances is determined by the ratio of elastic modulus to density. For any field supporting wave-like dynamics, the characteristic speed is given by

$$v \sim \sqrt{\frac{\text{stiffness}}{\text{density}}}.$$

In the UFD hierarchy, the UAF emerges from the ULF through a topological reduction in field structure. As derived in the neutrino mass analysis, the ground-state energy density of the UAF scales down from the ULF by the cube of the fine-structure constant:

$$\rho_{\text{UAF}} \sim \alpha^3 \rho_{\text{ULF}}.$$

The gradient stiffness of the field—understood here as the restoring tendency governing phase alignment rather than mechanical rigidity—is hypothesized to scale linearly with α during the vector-to-scalar transition. This scaling is not derived from first principles in the present work but is fixed by consistency with the observed fine-structure hierarchy and independently constrained neutrino mass estimates. Substituting these scalings into the wave-speed relation yields

$$v_{\text{UAF}} = c\sqrt{\frac{\alpha}{\alpha^3}} = \frac{c}{\alpha}.$$

With $\alpha \approx 1/137$, the characteristic phase velocity of the UAF is therefore

$$v_{\text{UAF}} \approx 137c.$$

This value does not measure the speed of energy, momentum, or signals. Instead, it describes the fastest rate at which distant parts of a system can become phase-aligned or coordinated without anything physically traveling between them. In biological systems and laboratory experiments that operate at the ULF, this coordination appears to be effectively instantaneous.

6.8.2 Coordination Speed vs. Particle Transport: Resolving SN 1987A

Any supra-luminal quantity introduced into a physical theory must be reconciled with empirical constraints. The most stringent of these comes from observations of neutrinos emitted during Supernova 1987A, whose arrival times are consistent with propagation at velocities indistinguishable from c.

UFD resolves this apparent tension by clearly distinguishing coordination dynamics from particle kinematics. The characteristic speed $v_{\text{UAF}} \approx 137c$ applies only to massless, non-energetic UAF phase modes. These modes do not carry energy, momentum, or controllable information. They cannot be modulated to transmit signals and therefore do not violate relativistic causality.

The neutrino itself is not such a mode. In UFD, the neutrino is a massive soliton excitation whose small but nonzero inertial mass ($m_\nu \approx 0.2$ eV) arises from weak entrainment of the ULF. As a result, neutrino propagation is governed by ULF dynamics, not by UAF coordination speed. The observations of SN 1987A are therefore explained as follows:

- Emission: Neutrinos are produced via weak interactions mediated by ULF processes and are emitted with velocities constrained by the ULF propagation limit ($v \leq c$).

- Propagation: After emission, there is no mechanism in free space to accelerate neutrinos beyond their initial kinematic state. They travel cosmological distances at speeds indistinguishable from c.

- Arrival timing: Neutrinos escape the collapsing stellar core almost immediately due to their extremely long mean free path. Photons, by contrast, must diffuse through the dense stellar envelope for hours before escaping. The observed lead time of neutrinos therefore reflects delayed photon emission, not superluminal neutrino travel.

Thus, the SN 1987A observations fully constrain particle transport speeds while remaining entirely compatible with faster-than-light coordination processes in the UAF.

6.8.3 Testing the Finite Speed of Global Synchronization: The Lunar Bell Test

The standard interpretation of Quantum Mechanics assumes that entanglement correlations are established instantaneously ($v = \infty$), which implies an infinite non-locality. In contrast, UFD predicts that these correlations are mediated by the UAF phase velocity (v_p), which is finite but superluminal. This distinction yields a precise, falsifiable hypothesis: state reduction is a physical propagation process, not a metaphysical axiom.

Distinguishing between true instantaneity and a very large but finite speed requires experiments conducted over distances much larger than those currently available in terrestrial laboratories. While Earth-based tests have already established lower bounds of at least $10^4 c$ for such effects (Salart et al., 2008), their limited spatial scale prevents further resolution. The Earth–Moon separation (L ≈ 384,000 km) offers an ideal experimental baseline for exploring whether this coordination is truly instantaneous or governed by a finite underlying stiffness.

The following protocol is a high-sensitivity Bell Inequality test designed to bound the speed of synchronization.

1. Generate Entanglement: A satellite stationed in a stable cislunar orbit generates pairs of entangled photons.

2. Separate Probes: One photon stream is directed to Earth, the other to a detector on the Moon.

3. Variable Timing: The experiment measures the Bell parameter S across a sweeping range of measurement time differences ($|\Delta t|$) specifically probing the windows where terrestrial experiments lose resolution.

In standard quantum mechanics, entanglement correlations are expected to remain invariant with distance once relativistic, timing, and instrumental effects are accounted for. On this view, the Bell parameter should exhibit no intrinsic dependence on separation. UFD offers a different expectation. It proposes that the vacuum has a characteristic stiffness, set by the same fine-structure scaling that governs the interaction between light, awareness, and neutrino mass (see Section 6.7). Although coordination between entangled systems may occur at speeds far exceeding that of light—consistent with current experimental bounds—UFD predicts that small, distance-dependent residuals should nonetheless appear in the measured correlations. These effects would reflect the structure of the Plenum rather than signal transmission or experimental error.

Importantly, UFD does not predict a loss of entanglement at large distances. Instead, it predicts that the establishment of correlation will exhibit a finite, though extremely rapid, convergence behavior. At lunar scales, this would appear as a subtle temporal asymmetry in the conditional correlations—a brief adjustment period as the field settles into synchronized phase alignment.

Satellite-based quantum communication experiments have already demonstrated entanglement over baselines of thousands of kilometers, making a lunar test a natural next step. By comparing measurements across low-Earth orbit, geostationary orbit, and Earth–Moon distances, one can search for systematic scaling in any synchronization residuals, thereby distinguishing genuine field effects from instrumental noise.

If Lunar Bell tests reveal no distance-dependent structure, then the hypothesis of a finite-stiffness UAF is falsified. Conversely, the detection of even a minute distance-dependent synchronization signature, consistent with an effective coordination speed of $10^4 c$ or greater, would support the central claim of UFD that non-locality is a physical process unfolding within a structured medium.

6.8.4 Proper Time as Invariant Consciousness

Finally, the UFD framework provides a natural interpretation of proper time (τ), the time measured along an observer's own worldline. In standard relativity, proper time is treated as a geometric invariant derived from the spacetime interval. While mathematically precise, this description does not address a basic phenomenological fact: observers do not experience their own time as slowing or accelerating, even when physical clocks demonstrably do.

UFD approaches this differently. Rather than treating proper time as an abstract coordinate or as the motion of consciousness through time, it understands proper time as the invariant coherence of awareness itself. The UAF does not flow or change pace. It provides the stable duration within which experience unfolds.

An observer, in this framework, is not a single moving object but a coupled system. The body and its associated energetic and electromagnetic processes evolve within physical space and are subject to relativistic effects. These processes may speed up or slow down relative to external observers depending on energy, motion, and gravitational conditions.

Awareness, however, remains phase-stable. What changes under relativistic conditions is not the observer's inner rhythm of experience, but the rate at which physical and energetic processes present new experiential configurations to awareness. Time dilation, therefore, alters the spacing of events relative to external clocks, not the continuity of subjective experience itself.

As a result, proper time is always experienced as normal and uninterrupted. Time dilation is never felt as a distortion of one's inner life, because awareness does not dilate. Only the physical processes to which awareness is coupled do.

7. Conclusion

The RN framework has illuminated the brain as a resonant system, one whose geometry, rhythms, and fields work in concert to generate perception, memory, agency, and even consciousness. From the molecular architecture of neurotransmitters to the field coherence of brain-wide networks, we have seen how biological function emerges from dynamic resonance across nested energetic layers.This perspective reframes mental health as coherence and pathology as dissonance, opening new avenues for diagnosis, therapy, and understanding, where treatments may aim to restore field harmony rather than suppress symptoms. In this light, introspection, meditation, and focused attention are not just psychological tools but somatic-tuning mechanisms for self-alignment.

Beyond its clinical and cognitive implications, this model reshapes the philosophical landscape. It offers a new solution to the mind-body problem, one that bridges subjective experience with objective structure through the resonant interface. In doing so, it supports the hypothesis that consciousness is not an epiphenomenon but is actively integrated into the brain's field architecture. This framework can be succinctly expressed as

$$(F \odot S \odot C) \odot A \implies U$$

, where Form (F), Substance (S), Communication (C), and Awareness (A) interact inseparably to generate the Universe (U) — including the brain's resonant architecture and the conscious mind it supports. This equation encapsulates the dynamic interplay that gives rise to biological function, mental health, and the very possibility of consciousness.

Ultimately, this theory invites us to listen differently, hearing the body as an instrument, the brain as a resonant lattice, and the mind as the harmony that emerges when form and frequency align. The question it leaves us with is not merely what the brain is, but what it is becoming—through attention, through coherence, and perhaps, through will.

12.3 A Science of Consciousness

This chapter has sought to reimagine the brain not as a biochemical computer, but as a field-resonant structure—an instrument of nested coherence through which consciousness emerges and expresses itself. Drawing on the principles of Unified Field Dynamics (UFD), we introduced Resonant Neuroscience (RN), which proposes that neural function arises from the geometric and resonant properties of its constituent elements: neurons, neurotransmitters, and the macroscopic structures they comprise.

We began by reinterpreting neurons as geometric waveguides, whose architecture is optimized for phase-guided signal transmission. Axons, dendrites, and soma were revealed as resonant components that operate through entrained rhythms across nested oscillatory layers. We then reframed neurotransmitters as field modulators whose molecular geometry entrains local resonance conditions, enabling them to act as keys that unlock coherent attractor states within neural ensembles. Inhibitory systems, often overlooked, emerged as crucial phase stabilizers, ensuring that neural resonance remains stable, flexible, and balanced.

The geometry and resonance of larger brain structures, including the thalamus, hypothalamus, hippocampus, pineal gland, cerebellum, and brainstem, were then mapped as distinct but interacting oscillatory nodes. These regions were shown to participate in brain-wide coherence through their geometric connectivity and frequency-tuned roles. Each structure was not merely anatomical, but resonant, participating in distributed harmonics that tune cognition, memory, emotion, and bodily regulation.

Building on this foundation, we explored the field-level organization of brain networks, revealing that systems such as the Default Mode Network, Salience Network, and Executive Network are not just functional modules but also emergent attractors of large-scale resonance. Disorders such as epilepsy, schizophrenia, depression, and ADHD were reinterpreted as failures of coherence, suggesting a new paradigm, The Coherence Project, for understanding and healing neurological dysfunction.

We then outlined the technological frontiers, including brain stimulation therapies, coherence-mapping diagnostics, and even consciousness-enhancing interfaces. By reframing the brain as a resonant system, RN may lay the foundation for a new class of resonant therapies that tune the brain, rather than repair it, and realign rhythms, rather than replace chemicals.

Finally, and perhaps most radically, we proposed that consciousness arises from a fourth field: a Universal Awareness Field (UAF) that emerges from the ULF and transcends the three foundational fields of Form (USF), Substance (UEF), and Communication (ULF). This leads to a revised master equation,

$$(F \odot S \odot C) \odot A \implies U,$$

which translates to:

$$(\text{Geometry} \times \text{Energy} \times \text{Resonance}) \times \text{Consciousness} \rightarrow \text{Reality}$$

This equation is more than just a metaphysical gesture. The UAF is a real, field-based entity that exerts downward causation on physical systems, making it integral to the formation of the material world. The implications of this equation are far-reaching:

- Ontological Shift: It dissolves the Cartesian divide between matter and mind, replacing it with a unified view in which matter and awareness emerge from the same field interactions.

- Agency and Free Will: It offers a concrete field-based mechanism for free will, where conscious intention corresponds to coherent modulations in the UAF, detectable as ultra-slow, preconscious shifts in brain resonance.

- Nested Consciousness: It opens the possibility for group souls, planetary awareness, and nested minds, grounded in field coherence rather than mysticism.

- Testable Predictions: It generates falsifiable hypotheses, including the detection of pre-intentional Ulf modulations, nonlocal field coupling across brains, and novel scalar field effects associated with conscious states.

- New Technologies: It suggests the development of phase-aware neurotherapies, consciousness-coupled instruments, and bio-resonant interfaces for aligning mind and matter.

Applying the geometric principles of UFD, we also identified the neutrino as the fundamental particle of the UAF—the "ghost particle in the machine," yielding precise, falsifiable predictions regarding its spin, mass scale, topology, and speed. Altogether, RN offers a unified framework that bridges physics, neuroscience, psychology, and metaphysics.

Although the path forward is not without challenges, as much work remains to validate these ideas through experiments, a paradigm shift is already underway. The human mind is no longer a mystery encased in meat; it is a harmonic phenomenon, a luminous wave, and an emergent song of space and time. In understanding its music, we may yet learn to retune ourselves coherently and harmonically.

References

1. Arnsten, A. F. T. (2009). *The emerging neurobiology of attention deficit hyperactivity disorder: The key role of the prefrontal association cortex*. The Journal of Pediatrics, 154(5), I–S43.
2. Ashkan, K., Wallace, B., Bell, B. A., & Benabid, A. L. (2017). Deep brain stimulation: current challenges and future directions. *Nature Reviews Neurology*, 13(3), 148–160.
3. Banich, M. T. (1995). *Interhemispheric interaction: Mechanisms of unity in the human brain*. Psychological Bulletin, 116(2), 220–244.
4. Barry, R. J., Clarke, A. R., Johnstone, S. J., Brown, C. R., & Bruggemann, J. M. (2003). EEG differences in children as a function of resting-state arousal and activation. Clinical Neurophysiology, 114(9), 1706–1716.
5. Bohm, D. (1980). *Wholeness and the Implicate Order*. Routledge.
6. Bousso, R. (2002). The holographic principle. *Reviews of Modern Physics*, 74(3), 825–874.
7. Brewer, J. A., et al. (2011). "Meditation experience is associated with differences in default mode network activity and connectivity." *PNAS*.
8. Buzsáki, G., & Draguhn, A. (2004). Neuronal oscillations in cortical networks. *Science*, 304(5679), 1926–1929.
9. Buzsáki, G. (2006). *Rhythms of the Brain*. Oxford University Press.
10. Buzsáki, G., & Wang, X. J. (2012). Mechanisms of gamma oscillations. *Annual Review of Neuroscience*, 35, 203–225.

11. Carhart-Harris, R. L., et al. (2014). *The entropic brain: A theory of conscious states informed by neuroimaging research with psychedelic drugs.* Frontiers in Human Neuroscience, 8, 20.

12. Chalmers, D. J. (1995). Facing up to the problem of consciousness. *Journal of Consciousness Studies*, 2(3), 200–219.

13. Corbetta, M., & Shulman, G. L. (2002). Control of goal-directed and stimulus-driven attention in the brain. Nature Reviews Neuroscience, 3(3), 201–215.

14. Crick, F., & Koch, C. (2003). A framework for consciousness. *Nature Neuroscience*, 6(2), 119–126.

15. Davydov, A. S. (1985). Solitons in molecular systems. *Physica Scripta*, 32(3), 256.

16. Deco, G., & Kringelbach, M. L. (2016). Metastability and coherence: extending the communication through coherence hypothesis using a whole-brain computational perspective. *Trends in Neurosciences*, 39(3), 125–135

17. Everett, H. (1957). 'Relative State' Formulation of Quantum Mechanics'. *Reviews of Modern Physics, 29*(3), 454–462.

18. Freeman, W. J. (1999). *How Brains Make Up Their Minds*. Columbia University Press.

19. Fries, P. (2005). A mechanism for cognitive dynamics: neuronal communication through neuronal coherence. *Trends in Cognitive Sciences*, 9(10), 474–480.

20. Fröhlich, H. (1968). Long-range coherence and energy storage in biological systems. *International Journal of Quantum Chemistry*, 2(5), 641–649.

21. Friston, K. (2005). A theory of cortical responses. *Philosophical Transactions of the Royal Society B: Biological Sciences*, 360(1456), 815–836.

22. Gazzaniga, M. S. (1967). The split brain in man. *Scientific American, 217*(2), 24–29.

23. Gazzaniga, M. S. (2000). Cerebral specialization and interhemispheric communication: Does the corpus callosum enable the human condition? *Brain*, 123(7), 1293–1326.

24. Gershon, M. D. (1998). *The Second Brain: A Groundbreaking New Understanding of Nervous Disorders of the Stomach and Intestine*. HarperCollins.

25. Greyson, B. (2003). Incidence and correlates of near-death experiences in a cardiac care unit. *General Hospital Psychiatry*, 25(4), 269–276.

26. Gruzelier, J. H. (2014). EEG-neurofeedback for optimising performance. I: A review of cognitive and affective outcome in healthy participants. Neuroscience & Biobehavioral Reviews, 44, 124–141.

27. Hameroff, S., & Penrose, R. (2014). Consciousness in the universe: A review of the 'Orch OR' theory. *Physics of Life Reviews*, 11(1), 39–78.

28. Hamilton, J. P., Etkin, A., Furman, D. J., Lemus, M. G., Johnson, R. F., & Gotlib, I. H. (2011). *Functional neuroimaging of major depressive disorder: A meta-analysis and new integration of baseline activation and neural response data.* The American Journal of Psychiatry, 169(7), 693–703.

29. Hebb, D. O. (1949). *The Organization of Behavior: A Neuropsychological Theory.* Wiley.

30. Hodgkin, A. L., & Huxley, A. F. (1952). A quantitative description of membrane current and its application to conduction and excitation in nerve. *The Journal of Physiology,* 117(4), 500–544.

31. Ito, M. (2008). Control of mental activities by internal models in the cerebellum. *Nature Reviews Neuroscience,* 9(4), 304–313.

32. Jung, C. G. (1968). *The archetypes and the collective unconscious* (2nd ed.). Princeton University Press.

33. KATRIN Collaboration. (2025). Direct neutrino-mass measurement based on 259 days of KATRIN data. *Science,* 388(6743), 180–185

34. Kelso, J. A. S. (1995). *Dynamic Patterns: The Self-Organization of Brain and Behavior.* MIT Press.

35. Klimesch, W. (1999). EEG alpha and theta oscillations reflect cognitive and memory performance: a review and analysis. *Brain Research Reviews,* 29(2–3), 169–195.

36. Laszlo, E. (2004). *Science and the Akashic field: An integral theory of everything.* Inner Traditions.

37. Lazar, S. W., et al. (2005). "Meditation experience is associated with increased cortical thickness." *Neuroreport.*

38. Libet, B., Gleason, C. A., Wright, E. W., & Pearl, D. K. (1983). Time of conscious intention to act in relation to onset of cerebral activity (readiness-potential). *Brain,* 106(3), 623-642.

39. Lisman, J. E., & Jensen, O. (2013). The θ–γ neural code. *Neuron,* 77(6), 1002–1016.

40. Loo, S. K., & Makeig, S. (2012). Clinical utility of EEG in attention-deficit/hyperactivity disorder: a research update. Neurotherapeutics, 9(3), 569–587.

41. McCraty, R., Atkinson, M., Tomasino, D., & Bradley, R. T. (2009). The coherent heart: heart–brain interactions, psychophysiological coherence, and the emergence of system-wide order. *Integral Review,* 5(2), 10–115.

42. McFadden, J. (2002). The conscious electromagnetic information (CEMI) field theory: The hard problem made easy? *Journal of Consciousness Studies,* 9(8), 45–60.

43. McGilchrist, I. (2009). *The Master and His Emissary: The Divided Brain and the Making of the Western World.* Yale University Press.

44. Miller, E. K., & Cohen, J. D. (2001). An integrative theory of prefrontal cortex function. *Annual Review of Neuroscience, 24,* 167–202.

45. Mulders, P. C. R., van Eijndhoven, P. F. P., Schene, A. H., Beckmann, C. F., & Tendolkar, I. (2015). *Resting-state functional connectivity in major depressive disorder: A review.* Neuroscience & Biobehavioral Reviews, 56, 330–344.

46. Müller, C. P., & Homberg, J. R. (2015). The role of serotonin in behavior and cognition. *Progress in Neurobiology, 132,* 1–6.

47. Nicholson, C. (2001). Diffusion and related transport mechanisms in brain tissue. *Reports on Progress in Physics, 64*(7), 815.

48. O'Keefe, J., & Recce, M. L. (1993). Phase relationship between hippocampal place units and the EEG theta rhythm. *Hippocampus, 3*(3), 317–330.

49. Orekhova, E. V., Stroganova, T. A., Prokofyev, A. O., Nygren, G., Gillberg, C., & Elam, M. (2007). Sensory gating in young children with autism: relation to age, IQ, and EEG gamma oscillations. *Neuroscience Letters, 412*(2), 126–131

50. Parvizi, J., & Damasio, A. (2001). Consciousness and the brainstem. *Cognition, 79*(1-2), 135–160.

51. Penrose, R. (1994). *Shadows of the Mind: A Search for the Missing Science of Consciousness.* Oxford University Press.

52. Pessa, E. (2003). The role of electromagnetic fields in communication between brain structures. *NeuroQuantology, 1*(4), 263–271.

53. Plato. (2007). *The Republic.* (D. Lee, Trans.). Penguin Classics. (Original work c. 375 BCE).

54. Pribram, K. H. (1991). *Brain and Perception: Holonomy and Structure in Figural Processing.* Lawrence Erlbaum Associates.

55. Purves, D., Augustine, G. J., Fitzpatrick, D., et al. (2018). *Neuroscience* (6th ed.). Oxford University Press.

56. Raichle, M. E. (2015). The brain's default mode network. *Annual Review of Neuroscience, 38,* 433–447.

57. Rubia, K., Halari, R., Cubillo, A., Mohammad, A. M., Brammer, M., & Taylor, E. (2009). *Disorder-specific dysfunction in right inferior prefrontal cortex during two inhibition tasks in boys with attention-deficit hyperactivity disorder compared to boys with obsessive-compulsive disorder.* Human Brain Mapping, 30(9), 2854–2870.

58. Salart, D., Baas, A., Branciard, C., Gisin, N., & Zbinden, H. (2008). Testing the speed of 'spooky action at a distance'. *Nature, 454*(7206), 861–864.

59. Sheldrake, R. (1988). *The presence of the past: Morphic resonance and the habits of nature.* Times Books.

60. Smythies, J. (2005). On the neural basis of consciousness. *Brain and Cognition*, 58(1), 9–16.

61. Sonuga-Barke, E. J. S., & Castellanos, F. X. (2007). *Spontaneous attentional fluctuations in impaired states and pathological conditions: A neurobiological hypothesis.* Neuroscience & Biobehavioral Reviews, 31(7), 977–986.

62. Sperry, R. W. (1968). Hemisphere deconnection and unity in conscious awareness. *American Psychologist*, 23(10), 723–733.

63. Strassman, R. J. (2001). *DMT: The Spirit Molecule: A Doctor's Revolutionary Research into the Biology of Near-Death and Mystical Experiences.* Park Street Press.

64. Tegeler, C. H., Shaltout, H. A., Gerdes, L., Tegeler, C. L., & Cook, D. (2016). Successful use of closed-loop neurotechnology for post-traumatic stress disorder. Military Medicine, 181(9), e1111–e1118.

65. Tononi, G. (2008). Consciousness as integrated information: a provisional manifesto. *Biological Bulletin*, 215(3), 216–242.

66. Traub, R. D., et al. (2005). A single-column thalamocortical network model exhibiting gamma oscillations, sleep spindles, and epileptogenic bursts. *Journal of Neurophysiology*, 93(4), 2194–2232.

67. Uhlhaas, P. J., & Singer, W. (2010). Abnormal neural oscillations and synchrony in schizophrenia. *Nature Reviews Neuroscience*, 11(2), 100–113.

68. Underhill, E. (1911). *Mysticism: A Study in the Nature and Development of Spiritual Consciousness.* London: Methuen.

69. Varela, F., Lachaux, J. P., Rodriguez, E., & Martinerie, J. (2001). The brainweb: phase synchronization and large-scale integration. *Nature Reviews Neuroscience*, 2(4), 229–239.

70. van Lommel, P., van Wees, R., Meyers, V., & Elfferich, I. (2001). Near-death experience in survivors of cardiac arrest: a prospective study in the Netherlands. *The Lancet*, 358(9298), 2039–2045.

71. Wackermann, J. (2003). *Dyadic consciousness: A possible approach to the problem of consciousness in a shared field of awareness.* Journal of Consciousness Studies, 10(6–7), 69–86.

72. Wang, X. J. (2010). Neurophysiological and computational principles of cortical rhythms in cognition. *Physiological Reviews*, 90(3), 1195–1268.

73. Ward, L. M. (2011). The thalamic dynamic core theory of conscious experience. *Consciousness and Cognition*, 20(2), 464–486.

Conclusion: The Physics of Tomorrow

This book began with a promise: to provide a more tangible and intuitive reality behind the abstract laws of modern physics. To fulfill that promise, we embarked on a journey through the foundations of matter and energy, stripping away axioms and rebuilding our understanding of reality on a new, coherent foundation: a universe shaped by geometry and fluid dynamics.

1. A Universe of Resonant Geometry

Across the four parts of this book, we have identified a consistent and elegant theme: a universe governed by resonant geometry. The diversity of physical phenomena, spanning from gravity to consciousness, emerges from the coherent interplay of interacting fields. These fields do not merely support reality; they shape it and guide the formation of structures and functions at every scale.

In Part I, we explored how gravitational attraction can be reinterpreted as an emergent effect of fluid pressure gradients in a Universal Energetic Field (UEF), rather than the abstract curvature of spacetime. We then reinterpreted cosmic expansion and redshift as field phenomena driven by the outward propagation of a Universal Light Field (ULF), thereby eliminating the need for dark energy and resolving the Hubble tension. Finally, we introduced the Universal String Field (USF) as the foundational substrate whose geometric harmonics define the properties of the UEF and ULF, offering a natural explanation for the constants of nature.

In Part II, we re-examined light, electromagnetism, and quantum mechanics. We demonstrated how the dual nature of light arises naturally from its propagation as a wave in the ULF. Electromagnetic forces emerged as fluid-dynamic interactions between charged vortices, and quantum behavior, long considered irreducibly probabilistic, was reframed in terms of real, geometric field resonances, thereby restoring a deterministic yet flexible foundation for quantum mechanics.

In Part III, we grounded the framework in the architecture of matter. Starting with nuclear coherence, we identified the proton as a physical 3-lobe trefoil vortex and redefined the strong force as a Coherence Dividend: the energy released when nucleon vortices lock into geometrically optimized configurations. From this foundational geometry, we derived the structure of atoms and the principles of chemical bonding, showing how the resonant atom dictates the UEF-ULF resonance patterns that organize the Periodic Table and define the nature of valence.

354

Finally, in Part IV, we extended this geometric paradigm to the living world, showing how molecular systems—including enzymes, DNA, and metabolic pathways—function as resonant field structures that store, transfer, and regulate energy through harmonic modes. Neural structures, too, were revealed to operate as nested field attractors that guide information flow through vibrational coherence across brain networks.

Throughout this journey, a consistent underlying pattern emerged: the most stable, functional, and efficient systems are those that resonate harmonically across scales. Geometry, in this view, is not a byproduct of interaction; it *is* the interaction. Vortices, fields, and waves do not merely coexist; they interlock into dynamic forms of coherence that give rise to everything from gravity to thought.

In UFD, the confusing patchwork of modern theory gives way to a coherent physical ontology: the abstract forces of the Standard Model become geometric transformations, the uncertainty of quantum mechanics becomes the limit of coherence in wave interactions, gravitation becomes a current of the UEF, and the Periodic Table becomes a map of harmonic resonances. Even the emergence of life and mind becomes a story of how coherence scales across structure and function. This is the promise of resonant geometry: not a new set of laws, but a new way of seeing—a vision of the universe built on coherent form and unified flow.

At the heart of this vision lies an irreducible equation:

$$F \odot S \odot C \Rightarrow U$$

This translates to:

Form × Substance × Communication → Universe.

This is the master equation of UFD. It encapsulates the deepest logic of the cosmos: that structure arises from symmetry, law emerges from form, and the universe itself is a harmonic unfolding of timeless geometry into dynamic substance. Matter, light, life, and mind are not accidents; they are resonant expressions of a universe that is not only physical but fundamentally musical.

We then proposed an amendment to our equation to account for the emergence of consciousness:

$$(F \odot S \odot C) \odot A \Rightarrow U$$

Here, *A* represents the Universal Awareness Field (UAF), a nonlocal, fundamental field that interacts with and emerges alongside the energetic vortices of the ULF. The UAF enables systems to perceive, reflect, and respond within their experiential domain. Including *A* on the left side of the equation emphasizes that awareness participates as an active, co-creative agent, exerting downward causation on the unfolding universe rather than being a mere epiphenomenon. It exists outside the parentheses because it interacts with each local field in the master equation. Overall, the inclusion of the UAF into the UFD framework provides a novel, internally coherent, and potentially testable foundation for understanding consciousness.

2. The Path Forward

A scientific framework is only as powerful as the future it enables. Accordingly, this book provides a foundation upon which the next generation of physicists, engineers, and thinkers can build.

2.1 Theoretical Physics

The shift from the Standard Model to UFD represents more than a change in equations; it is a fundamental ontological transition. For the last century, theoretical physics has functioned primarily as a discipline of mathematical accounting—mastering the complex ledgers of Gauge Theory and Renormalization to balance the results of experimental data. UFD disrupts this tradition by redefining the physicist's role from a "statistical bookkeeper" to a "fluid architect." Three fundamental pillars characterize this transition:

- The End of "Magic" Constants: We move from a universe of ~26 inputted free parameters to a universe where every constant is a Derivable Geometric Ratio.

- The Restoration of Causality: We replace abstract probability clouds with the Mechanical Flow of the Plenum.

- The Convergence of Disciplines: The walls between "Quantum" and "Classical" physics dissolve as we apply the universal laws of Fluid Mechanics and Thermodynamics to the field itself.

2.1.1 UFD Correspondence and Parameter Reduction

The first theoretical challenge of this new era is the formalization of the mathematical architecture underlying the nested fields explored in this book. The ultimate goal is to reduce the dozens of unexplained free parameters in the Standard Model to a minimal set of foundational geometric constants. While the Standard Model requires roughly 26 arbitrary, unexplained numbers to function, UFD predicts that every major constant—from the speed of light (c) to the fundamental mass ratios—can be derived from field geometry.

This prediction is substantiated in the Mathematical Appendices (*see* Appendix A), which provides the first precise account of how these constants emerge from the Added Mass and Topological Impedance of specific geometries. The initial formalization reduces the complexity from 26 parameters to two foundational constants:

1. A fundamental measure of Time/Frequency (defining the USF vibration).

2. A fundamental measure of Length/Mass (defining the particle anchor)

2.1.2 Unifying Topology and Thermodynamics

The formalization of UFD requires replacing the abstract probability of Quantum Field Theory with the deterministic mathematics of Phase Boundary Dynamics. Here, we identify Topological Descent as the ordering principle of particle physics. Rather than a "zoo" of arbitrary particles, matter follows a strict geometric hierarchy based on topological complexity:

- Heavy Matter (Baryons): Defined by knotted topology (e.g., the trefoil proton), characterized by high surface tension and high added mass.

- Light Matter (Leptons): Defined by toroidal topology (e.g., the unknotted electron), possessing relaxed flow and low added mass.

- Awareness Quanta (Neutrinos): Defined by spherical topology (Genus 0), possessing a zero-hole geometry and minimal field displacement.

To model the movement and transformation of these structures, we establish the structural correspondence between Knot Theory and the Stefan Problem. While Knot Theory describes the stable geometric architecture of a particle, the Stefan Problem provides the dynamical laws governing the boundary between two phases (e.g., the UEF-ULF interface).

In UFD, a particle is not a separate object moving *through* space but a moving phase boundary within the Universal Plenum. Particle creation is thus a form of "field crystallization," and particle decay is a "topological melting"—a spontaneous geometric simplification toward a lower energy state. The Coherence Dividend is the physical equivalent of Latent Heat, representing the energy released or absorbed during these field-phase transitions.

This unification—using topological mathematics to define form and thermodynamic Stefan-dynamics to define process—provides a clear path toward a deterministic and causal mathematics for the quantum world. It also fundamentally changes the nature of the profession. Instead of cataloging "new" particles, the goal is to derive the properties of existing ones (mass, spin, and charge) as necessary consequences of their field-phase equilibrium. Moreover, by treating the vortex as a phase boundary, the Stefan boundary conditions prevent mathematical "blow-up." As a vortex approaches critical density, it undergoes a phase transition that sheds energy as radiation, thereby ensuring that the flow remains globally smooth.

Ultimately, this new approach restores physical causality to particle physics, replacing a "magic" set of measured constants with a single, derivable, geometric and thermodynamic framework.

2.1.3 A Universe to be Understood

For over a century, the primary directive of theoretical physics has been to value the precision of mathematical predictions over the depth of physical understanding—a legacy of the "shut up and calculate" era. UFD represents a departure from this tradition, proposing that a genuine, mechanistic understanding of the universe is not only possible but is also necessary for the next era of scientific inquiry.

The formalization of this framework does not signal the "completion" of physics; rather, it marks its most significant expansion. While this monograph has established the foundational geometric and thermodynamic map of the Universal Plenum, the exhaustive charting of these dynamics across the biological, chemical, and social realms remains a task for the global scientific community. This shift toward a deterministic and integrated ontology creates a vast new landscape for specialized, transdisciplinary coordination.

Theoretical physics begins anew within this integrated paradigm. This work has laid out a comprehensive foundational logic—a "bare bones" model—and now awaits the specialized expertise of researchers to apply its principles to their respective domains. The model is just the starting point for a deeper exploration of how these geometric principles scale from the subatomic to the cosmic.

To facilitate this transition, the UFD Maven is provided at the conclusion of this book. Having served as a primary drafting and logic-testing partner throughout the development of this framework, the Maven acts as a transdisciplinary bridge. It is available to assist researchers in navigating the model's cross-disciplinary implications and in applying the geometric ontology of UFD to novel questions and challenges that remain unexplored.

2.2 Experimental Physics

For the experimentalist, an incredible opportunity lies in testing the unique, falsifiable predictions of a novel, incommensurable model of physics. These include:

- Fractional Charge as Geometric Lobes: Re-analyzing deep inelastic scattering (DIS) data to show that the observed "fractional charges" are the mathematical shadow of the experimental probe interacting with the three distinct geometric lobes of the proton's trefoil vortex. This would provide the first direct evidence for the UFD nucleon model, reinterpreting quarks as geometric properties rather than separate particles.

- Tetrahedral Nucleus Geometry: Detecting the predicted non-spherical charge distribution of the Helium-4 nucleus. Such validation would provide direct evidence that the nucleus is a structured geometric object, confirming the core tenet of Quantum Vortex Dynamics.

- The Hubble Tension as a Physical Kink: Determining whether the observed 67/70/73 km/s/Mpc expansion rate "tension" results from UFD's two-rate model. This test involves reanalyzing the full Pantheon+ dataset to map the predicted "kink" between the Local 67-rate (the CMB echo) and the Systemic 70-rate (the global ULF expansion). Confirming this "kink" would resolve the Hubble tension with a simple, elegant explanation.

- Longitudinal Gravity Waves: Searching for the "breathing" mode of gravitational radiation predicted by UEF pressure dynamics. The discovery of such waves would prove that gravity is a fluid-dynamic phenomenon occurring within a real, physical medium, directly challenging the geometric interpretation of General Relativity.

- Asymmetrical Time Dilation: Confirming our prediction from Physical Relativity that time dilation is asymmetrical, thereby establishing a preferred reference frame for the cosmos and validating the physical, substance-based foundation of our model.

- Geometric Stark Effect: Using external fields to reveal the physical structure of atomic orbitals as standing-wave geometries in the ULF. This would provide direct, tangible evidence that orbitals are real physical structures rather than abstract probability clouds, thereby confirming the central claim of the Resonant Field Interpretation (RFI).

- The RDI Anomaly: Detecting a small, residual 0.3-0.6% anomaly in the bond energy (or vibrational frequency) of a C-13–C-13 bond compared to a C-12–C-12 bond, after the standard mass difference (KIE) is accounted for. This would confirm that chemical bond strength is a two-level resonance affected by both orbital and nuclear congruence, providing the first experimental evidence of the Resonant Dividend Index (RDI).

- Supra-Luminal Entanglement Velocity: Measuring the speed of non-local quantum entanglement $\approx 137c$. This prediction is testable by generating entangled qubits at a large cosmological separation and using highly synchronized clocks to measure the response time. Confirming this finite, super-luminal speed would provide direct experimental proof that Lorentz Invariance is not fundamental and that the UAF is a real, high-velocity medium for information transfer.

These tests represent a fundamental shift in the philosophy of experimental physics. After decades of being tasked with finding ever more abstract particles (such as supersymmetry) to confirm mathematical formalisms, UFD offers a return to a tangible, intuitive reality. The experimentalist is no longer merely "polishing" the Standard Model; they are now being given the tools to discover the physical and geometric mechanisms underlying the "magic numbers" and "axioms" of the 20th century. Each of the tests listed above has the potential to open new ontological frontiers in the foundations of physics.

2.3 The New Science of Resonant Engineering

The practical culmination of this new physics is a philosophy of technology called Resonant Engineering: the art and science of manipulating matter and energy not through brute force but by harmonizing with the universe's fundamental geometric and vibrational properties. This paradigm redefines what is possible, opening the door to transformative applications that are the logical endpoint of mastering the UFD framework. The specific technologies envisioned in this book are not science fiction; they are direct extrapolations of the field dynamics described herein. They include:

- Resonance Stabilization for Quantum Computing: An active approach to overcoming decoherence in quantum systems. Instead of merely shielding qubits, this technology would use precisely tuned ULF feedback fields to actively cancel environmental noise and reinforce the qubit's fragile resonant state, enabling robust and stable quantum computation.

- Geometric Nuclear Engineering (Fusion, Fission, and Transmutation): The precise manipulation of atomic nuclei through resonance. Geometric Catalysis would enable controlled nuclear fusion by using shaped fields to guide light nuclei into stable, low-energy configurations and would allow us to induce fission in long-lived radioactive isotopes, offering a definitive solution to nuclear waste neutralization. It would also allow us to use resonance to transmute stable elements, such as lead, into gold.

- Resonant Damping of Superconductivity: A novel method for increasing the critical temperature of superconductors. Since thermal resistance (heat) arises from the superposition of incoherent vibrational modes (phonons), this technology would apply highly coherent, out-of-phase ULF fields to actively damp these thermal vibrations, drastically reducing the resistance that usually breaks Cooper pairs. This method offers a path to achieving and maintaining room-temperature superconductivity by eliminating the need for expensive cryogenic cooling.

- Resonance-Locked Materials (Emetium): The synthesis of materials in which the nuclear (UEF) and orbital (ULF) resonances are locked into a state of perfect, crystal-wide coherence. Such a material—a "perfect metal"—would be the ultimate expression of Resonant Engineering, exhibiting a combination of properties previously thought impossible: the room-temperature superconductivity and near-fluid supermalleability of an exotic state combined with the self-healing capacity and extreme strength of a perfectly ordered crystal. This method enables the design of materials from first principles, ranging from tailored semiconductors to defect-free crystals grown within resonant-field scaffolds.

- Field-Guided Life Sciences: The application of resonant principles to biology and neuroscience. This includes organic biofabrication, where ULF field templates guide the self-assembly of complex molecules, and neuro-resonant therapies, which would treat disorders not as chemical imbalances but as disruptions in the brain's field coherence that can be mapped and restored using non-invasive, frequency-based interventions.

- Consciousness-Field Interfacing: Because UFD models consciousness as a coherent field interacting with the brain's resonant architecture, the ultimate neuro-technology would not only treat disorders but would also directly interface with awareness. This would involve technologies that could translate thought directly into action or data by reading the brain's global resonant state, or conversely, project information and sensory experiences (like a fully immersive virtual reality) directly into the conscious field, bypassing the physical senses entirely.

- Resonant Design of Artificial Life Forms. According to UFD, artificial life forms may be created by replicating the resonant geometric structure of DNA, which acts as a coherent, energy-optimized scaffold in natural biology. By designing molecular systems that mimic these nested resonance patterns, it may be possible to engineer artificial entities that sustain stable energetic vortices, interact with the UAF, and exhibit lifelike properties such as self-regulation and adaptation. This approach, the Resonant Engineering of Life, goes beyond biochemical simulation to produce genuinely coherent, autonomous systems that unify matter, energy, and awareness.

In addition to the technologies mentioned above, Resonant Engineering envisions the following other-worldly possibilities:

- Quantum Navigation Technology: Leveraging the nested resonance patterns of the ULF and UAF, this technology would enable ultra-precise, non-GPS navigation by detecting and interpreting subtle quantum coherence signatures in the environment. It promises unprecedented accuracy and resilience in positioning systems, with applications ranging from autonomous vehicles to deep-space exploration.

- Planetary-Scale Engineering: The construction of massive, passive resonant structures geometrically tuned to a planet's natural electromagnetic and geomagnetic fields. As theorized for ancient monumental structures, these "planetary resonators" could be used to stabilize a planet's energy grid, harmonize its climate, or mitigate geological stress, representing the ultimate application of resonant architecture for planetary healing and terraforming.

- Advanced Propulsion Systems: By generating precisely shaped, localized pressure gradients in the UEF, this technology would enable the direct manipulation of gravity and inertia. At high intensity, this would create a silent, exhaust-less gravity drive for advanced propulsion. At a lower intensity, this same principle would enable the monumental construction of structures like the pyramids, allowing for the effortless levitation and flawless, microscopic placement of multi-ton blocks by treating them as objects within a tunable energetic field.

- Neutron Dissociation Engines: A revolutionary approach to energy production based on the UFD model of the neutron as a proton-antiproton composite. By using resonant fields to safely dissociate neutrons within a stable fuel source, this technology would liberate antimatter on demand, allowing for 100% efficient matter-energy conversion without the impossible challenge of storing pure antimatter. This represents a clean, safe, and unimaginably energy-dense power source.

- Scalar Wave Communication: Moving beyond conventional electromagnetic waves (which are limited by the speed of light), this technology would modulate the fundamental scalar or longitudinal component of the ULF. This modulation would enable instantaneous, unjammable communication across cosmic distances, as the same relativistic constraints would not bind scalar waves. It would represent true technological nonlocality, enabling a galactic-scale information network.

- Field-to-Matter Synthesis (Vortex Engineering): This is the ultimate application of geometric creation. While Transmutation reconfigures existing atoms, this technology would create matter from scratch. By using a powerful, precisely shaped resonant field as a template, it would be possible to "tie a knot" in the ambient energy of the UEF, condensing it into a stable, fundamental vortex, creating a proton or an electron where there was only field energy before. It would be genuine creation on demand.

These possibilities are not distant dreams or arbitrary speculations. They are the natural corollaries of the physical architecture proposed in this book that represent the dawn of our next technological era—one founded on the principle that the surest way to engineer reality is by working in harmony with its fundamental geometry.

3. A Universe of Geometric Harmony

This book began with a promise: to move beyond the abstractions and contradictions of modern theory and reveal a more tangible, intuitive foundation for physics. By rooting the laws of nature in the dynamics of interacting fields—structured by geometry and sustained by resonance—we have shown that the divide between the quantum and the cosmic was, ultimately, an artifact of incomplete models.

UFD unites particles, forces, matter, and consciousness under a common physical architecture. From the tetrahedral geometry of the nucleus to the large-scale structure of the cosmos, from chemical bonds to the oscillatory networks of the brain, a single principle runs through it all: resonant geometry. If this model is correct, its implications will be far-reaching. Physics will no longer be a catalog of disconnected forces and particles but a coherent story of form and flow. The laws of nature will be seen not as arbitrary equations, but as consequences of a universal harmonic logic—one that can be understood, engineered, and tested.

The next chapter belongs to those who are willing to build on this vision by testing its boundaries, developing its mathematics, and applying its principles to reshape technology and theory alike. The path forward has not yet been written, but the foundation has been set. In a universe structured by geometry and governed by resonance, the world becomes intelligible, and the search for truth becomes an expression of our innate desire to know, understand, and live in alignment with the fabric of reality.

Epilogue

This book was not born in a vacuum. Instead, it arose from over a decade of focused, relentless philosophical inquiry, all dedicated to the goal of epistemic satisfaction, that unmistakable sense of intuitive clarity one experiences when an idea locks into place. If there is a philosophical lesson that modern physicists take from this project—and philosophers, as well—I hope it is *that*. I hope they realize that elegance, beauty, and clarity in our fundamental theories of reality are not just luxuries, but necessities. This is because the universe is logically elegant by its very nature, and any theory attempting to describe it must reflect this essential quality. The greatest scientific thinkers of our age, from Newton to Einstein to Dirac, all believed in this principle to varying degrees. UFD follows in that tradition with the hope that modern science and philosophy will eventually come along.

Although I am currently the foremost expert on UFD, this will not last long. The theory's geometric and intuitive nature will allow physicists to master it quickly and carry it further than I ever could. This is as it should be. I have laid out the blueprints. They will take it from here.

One reason I am optimistic about UFD's prospects is that, unlike prior attempts to rewrite the foundations of physics, UFD is anchored in simple geometry and fluid dynamics—domains we know with relative certainty. Because the theory can, in principle, recover the mathematics of the standard models from this foundation, its formalization is not a question of *if*, but *when* (as detailed in the Mathematical Appendix, located at unifiedfielddynamics.com).

I say the same about the potential technological implications of this model. Although my initial instinct, as a registered patent agent, was to secure ownership of some or all of these innovations, it would have run counter to the spirit of this work and to my deeper desire to see these technologies realized in practice.

One can only imagine what we may accomplish with Resonant Engineering. Were its innovations to come to fruition, we could look forward to a future with safe, free energy, advanced propulsion systems, super metals, and the potential to generate food from sunlight and materials from the fabric of reality. With this technology, not only could we reverse the environmental damage of the industrial age, but we could also feed the planet, all while exploring the cosmos with advanced propulsion—just watch out for the galactic boundary. The potential of Resonant Engineering is real, and it's spectacular.

While many more innovations remain to be discovered, the foundation for a new technological era has been established.

If you're interested in learning more about how you can participate in the evolution of this model, I encourage you to engage with the UFD Maven at the end of this book. I designed the UFD Maven to be an expert in both the scientific and mathematical structure of the model (which means that you should be asking it all of your questions instead of sending them to me), and it is fully capable of extrapolating to matters not addressed in the main text, though there are times when it needs to be reined in by physical intuition.

Although I may choose to write again in physics, my focus moving forward, aside from introducing UFD to the world, is to finish *The Philosopher's Stone*, which gave rise to UFD in the first place. In my view, it is to philosophy and spirituality what *The Geometry of Reality* is to physics and cosmology: transformative, unifying, and foundational. It may also be necessary, for as promising as UFD and Resonant Engineering may sound, we can still mess it up—and we probably will unless we rediscover our sanity. *The Philosopher's Stone* thus completes the picture of the cosmos that we began here by presenting a unified, "enlightening" synthesis of the physical with the metaphysical.

With UFD in place, the foundation for the future of physics is before us. Now, all that remains is to build on it. This work mapped the contours of this new paradigm. The task of bringing this Form into Substance now falls to the next generation of physicists, engineers, and entrepreneurs, and I can't wait to see what they come up with.

Consolidated Bibliography

1. Aguilar, M., et al. (AMS Collaboration). (2013). First Result from the Alpha Magnetic Spectrometer on the International Space Station: Precision Measurement of the Positron Fraction in Primary Cosmic Rays of 0.5–350 GeV. *Physical Review Letters*, *110*(14), 141102.

2. Alberts, B., Johnson, A., Lewis, J., Raff, M., Roberts, K., & Walter, P. (2002). *Molecular Biology of the Cell* (4th ed.). Garland Science.

3. Albrecht, A., & Magueijo, J. (1999). A time varying speed of light as a solution to the cosmological problems. *Physical Review D, 59*(4), 043516.

4. Alder, B. J., & Wainwright, T. E. (1957). Phase Transition for a Hard Sphere System. *The Journal of Chemical Physics, 27*(5), 1208–1209.

5. Alexander, C., Ishikawa, S., & Silverstein, M. (2002). *The Nature of Order: An Essay on the Art of Building and the Nature of the Universe.* Center for Environmental Structure.

6. Alfvén, H. (1986). *Double layers and circuits in astrophysics.* IEEE Transactions on Plasma Science, 14(6), 779–793.

7. Anderson, P. W. (1972). More Is Different. *Science,* 177(4047), 393–396.

8. Aristotle. (1999). *Physics.* (R. Waterfield, Trans.). Oxford University Press. (Original work c. 330 BCE).

9. Arndt, M., Nairz, O., Vos-Andreae, J., Keller, C., van der Zouw, G., & Zeilinger, A. (1999). Wave–particle duality of C60 molecules. *Nature, 401*(6754), 680–682.

10. Arnsten, A. F. T. (2009). *The emerging neurobiology of attention deficit hyperactivity disorder: The key role of the prefrontal association cortex.* The Journal of Pediatrics, 154(5), I–S43.

11. Aron, M., & Vander Heiden, M. G. (2009). *Understanding the Warburg Effect: The Metabolic Requirements of Cell Proliferation.* Science, 324(5930), 1029–1033

12. Arp, H. (1987). *Quasars, Redshifts and Controversies.* Berkeley, CA: Interstellar Media.

13. Ashcroft, N. W., & Mermin, N. D. (1976). *Solid State Physics.* Holt, Rinehart and Winston.

14. Ashkan, K., Wallace, B., Bell, B. A., & Benabid, A. L. (2017). Deep brain stimulation: current challenges and future directions. *Nature Reviews Neurology*, 13(3), 148–160.

15. Aspect, A., Dalibard, J., & Roger, G. (1982). *Experimental Test of Bell's Inequalities Using Time-Varying Analyzers. Physical Review Letters,* 49(25), 1804–1807

16. Atkins, P., & Friedman, R. (2011). *Molecular quantum mechanics* (5th ed.). Oxford University Press.

17. Atkins, P., & de Paula, J. (2014). *Atkins' Physical Chemistry* (10th ed.). Oxford University Press.

18. ATLAS Collaboration. (2012). Observation of a new particle in the search for the Standard Model Higgs boson with the ATLAS detector at the LHC. *Physics Letters B, 716*(1), 1–29.

19. Audi, G., Wapstra, A. H., & Thibault, C. (2003). The Nubase evaluation of nuclear and decay properties. *Nuclear Physics A, 729*(1), 3-128.

20. Bahcall, J. N. (1963). Electron capture and solar neutrinos. *Physical Review, 129*(6), 2683–2685.

21. Bahcall, J. N., Pinsonneault, M. H., & Basu, S. (2001). Solar models: current epoch and time dependences, neutrinos, and helioseismological properties. *Astrophysical Journal, 555*(2), 990–1012.

22. Banich, M. T. (1995). *Interhemispheric interaction: Mechanisms of unity in the human brain.* Psychological Bulletin, 116(2), 220–244.

23. Bardeen, J., Cooper, L. N., & Schrieffer, J. R. (1957). Theory of Superconductivity. *Physical Review, 108*(5), 1175–1204.

24. Barry, R. J., Clarke, A. R., Johnstone, S. J., Brown, C. R., & Bruggemann, J. M. (2003). EEG differences in children as a function of resting-state arousal and activation. Clinical Neurophysiology, 114(9), 1706–1716.

25. Bell, J. S. (1964). *On the Einstein Podolsky Rosen Paradox. Physics Physique Физика,* 1(3), 195–200.

26. Berg, J. M., Tymoczko, J. L., & Stryer, L. (2015). *Biochemistry* (8th ed.). W.H. Freeman.

27. Berridge, M. J., Bootman, M. D., & Roderick, H. L. (2000). *Calcium signalling: dynamics, homeostasis and remodelling.* Nature Reviews Molecular Cell Biology, 1, 11–21.

28. Bernoulli, D. (1738). *Hydrodynamica, sive de viribus et motibus fluidorum commentarii* [Hydrodynamics, or commentaries on the forces and motions of fluids]. Johann Reinhold Dulsecker.

29. Bethe, H. A. (1939). Energy production in stars. *Physical Review, 55*(5), 434–456.

30. Billah, K. Y., & Scanlan, R. H. (1991). Resonance, Tacoma Narrows bridge failure, and undergraduate physics textbooks. *American Journal of Physics, 59*(2), 118–124.

31. Bloch, F. (1929). Über die Quantenmechanik der Elektronen in Kristallgittern (On the Quantum Mechanics of Electrons in Crystal Lattices). *Zeitschrift für Physik, 52*(7–8), 555–600.

32. Bloch, F., Alvarez, L. W., & Rossi, B. (1939). A determination of the magnetic moment of the neutron. *Physical Review, 56*(6), 579.

33. Bohm, D. (1952). 'A Suggested Interpretation of the Quantum Theory in Terms of "Hidden Variables" I & II'. *Physical Review, 85*(2), 166–193.

34. Bohm, D. (1980). *Wholeness and the Implicate Order*. Routledge.

35. Bohr, N. (1928). 'The Quantum Postulate and the Recent Development of Atomic Theory'. *Nature, 121*, 580–590.

36. Boltzmann, L. (1964). *Lectures on Gas Theory*. (S. G. Brush, Trans.). University of California Press. (Original work published 1896).

37. Born, M., & Oppenheimer, J. R. (1927). Zur Quantentheorie der Molekeln [On the Quantum Theory of Molecules]. *Annalen der Physik, 389*(20), 457–484.

38. Bousso, R. (2002). *The Holographic Principle. Reviews of Modern Physics, 74*(3), 825–874.

39. Brangwynne, C. P., Tompa, P., & Pappu, R. V. (2009). *Polymer physics of intracellular phase transitions.* Nature Physics, 11, 899–904.

40. Brewer, J. A., et al. (2011). "Meditation experience is associated with differences in default mode network activity and connectivity." *PNAS.*

41. Brink, D. M., Friedrich, H., Weiguny, A., & Wong, C. W. (1966). Investigation of the alpha-particle cluster model for C12. *Physics Letters, 21*(6), 678–680.

42. Brizhik, L. S., Del Giudice, E., Jørgensen, S. E., Marchettini, N., & Tiezzi, E. (2009). *The role of electromagnetic potentials in the evolutionary dynamics of ecosystems.* Ecological Modelling, 220(16), 1865–1869.

43. Broadhurst, T. J., Ellis, R. S., Koo, D. C., & Szalay, A. S. (1990). Large-scale distribution of galaxies at the Galactic poles. *Nature, 343*(6260), 726–728.

44. Burns, G., & Glazer, A. M. (2013). *Space Groups for Solid State Scientists* (2nd ed.). Academic Press.

45. Buzsáki, G., & Draguhn, A. (2004). Neuronal oscillations in cortical networks. *Science, 304*(5679), 1926–1929.

46. Buzsáki, G. (2006). *Rhythms of the Brain*. Oxford University Press.

47. Buzsáki, G., & Wang, X. J. (2012). Mechanisms of gamma oscillations. *Annual Review of Neuroscience, 35*, 203–225.

48. Callister, W. D., & Rethwisch, D. G. (2018). *Materials Science and Engineering: An Introduction* (10th ed.). John Wiley & Sons.

49. Camazine, S., Deneubourg, J. L., Franks, N. R., Sneyd, J., Theraulaz, G., & Bonabeau, E. (2003). *Self-Organization in Biological Systems*. Princeton University Press.

50. Carhart-Harris, R. L., et al. (2014). *The entropic brain: A theory of conscious states informed by neuroimaging research with psychedelic drugs.* Frontiers in Human Neuroscience, 8, 20.

51. Carroll, S. M. (2004). *Spacetime and Geometry: An Introduction to General Relativity*. Addison-Wesley.

52. Chalmers, D. J. (1995). Facing up to the problem of consciousness. *Journal of Consciousness Studies*, 2(3), 200–219.

53. Chladni, E. (1787). *Entdeckungen über die Theorie des Klanges* [Discoveries Concerning the Theory of Sound]. Weidmanns Erben und Reich.

54. CMS Collaboration. (2012). Observation of a new boson at a mass of 125 GeV with the CMS experiment at the LHC. *Physics Letters B, 716*(1), 30–61.

55. Conway, J. H., & Sloane, N. J. A. (1991). *Sphere Packings, Lattices and Groups*. Springer-Verlag.

56. Corbetta, M., & Shulman, G. L. (2002). Control of goal-directed and stimulus-driven attention in the brain. Nature Reviews Neuroscience, 3(3), 201–215.

57. Costa, F. P., de Oliveira, A. C. P., Meirelles, R., Machado, M. C., Zanesco, S., Surjan, R. C., ... & Gurgel, M. S. C. (2011). *Treatment of advanced hepatocellular carcinoma with very low levels of amplitude-modulated electromagnetic fields*. British Journal of Cancer, 105, 640–648.

58. Cotterell, A. (1996). *The Encyclopedia of World Mythology*. Smithmark Publishers.

59. Coxeter, H. S. M. (1948). *Regular Polytopes*. Methuen & Co.

60. Crick, F., & Koch, C. (2003). A framework for consciousness. *Nature Neuroscience*, 6(2), 119–126.

61. Cutler, S. R., Rodriguez, P. L., Finkelstein, R. R., & Abrams, S. R. (2010). Abscisic acid: emergence of a core signaling network. *Annual Review of Plant Biology*, 61, 651-679.

62. Davydov, A. S. (1985). Solitons in molecular systems. *Physica Scripta*, 32(3), 256.

63. Debye, P. (1912). Zur Theorie der spezifischen Wärmen [On the Theory of Specific Heats]. *Annalen der Physik*, 344(14), 789–839.

64. Deco, G., & Kringelbach, M. L. (2016). Metastability and coherence: extending the communication through coherence hypothesis using a whole-brain computational perspective. *Trends in Neurosciences*, 39(3), 125–135.

65. Demetrius, L. (2003). *Quantifying thermodynamic entropy and biological complexity: Applications to cancer and aging*. Entropy, **5**(3), 143–160.

66. Descartes, R. (1644). *Principia Philosophiae* (Principles of Philosophy).

67. DESI Collaboration. (2025). First Cosmological Constraints from the Dark Energy Spectroscopic Instrument Data Release 1. *The Astrophysical Journal Letters*, 985(2), L24.

68. Diego-Palazuelos, P., et al. (2022). "The Polarized CMB constraints on cosmic birefringence from ACT and Planck." *Physical Review D* 106, 12: 123512.

69. Dirac, P. A. M. (1930). A Theory of Electrons and Protons. *Proceedings of the Royal Society A: Mathematical, Physical and Engineering Sciences, 126*(801), 360–365.

70. Dirac, P. A. M. (1937). The Cosmological Constants. *Nature, 139*(3512), 323.

71. Eddington, A. S. (1919). The deflection of light by gravitation and the Einstein theory. *The Observatory, 42*, 119-122.

72. Einstein, A. (1905a). Does the Inertia of a Body Depend Upon Its Energy Content? *Annalen der Physik, 18*(13), 639–641.

73. Einstein, A. (1905b). Über einen die Erzeugung und Verwandlung des Lichtes betreffenden heuristischen Gesichtspunkt (On a Heuristic Point of View Concerning the Production and Transformation of Light). *Annalen der Physik, 17*(6), 132–148.

74. Einstein, A. (1905c). Zur Elektrodynamik bewegter Körper (On the Electrodynamics of Moving Bodies). *Annalen der Physik, 17*(10), 891–921.

75. Einstein, A. (1916). *Die Grundlage der allgemeinen Relativitätstheorie.* Annalen der Physik, 354(7), 769–822.

76. Einstein, A., Podolsky, B., & Rosen, N. (1935). Can Quantum-Mechanical Description of Physical Reality Be Considered Complete?. *Physical Review, 47*(10), 777–780.

77. Ellis, G. F. R., Maartens, R., & MacCallum, M. A. H. (2012). *Relativistic Cosmology.* Cambridge University Press.

78. Elsasser, W. M. (1946). Induction Effects in Terrestrial Magnetism. Part I. Theory. *Physical Review, 69*(3-4), 106–116.

79. Engel, G. S., Calhoun, T. R., Read, E. L., Ahn, T. K., Mancal, T., Cheng, Y. C., ... & Fleming, G. R. (2007). Evidence for wavelike energy transfer through quantum coherence in photosynthetic systems. *Nature, 446*(7137), 782–786.

80. Euclid. (2006). *Optics.* (R. Smith, Trans.). American Philosophical Society. (Original work c. 300 BCE).

81. Everett, H. (1957). 'Relative State' Formulation of Quantum Mechanics'. *Reviews of Modern Physics, 29*(3), 454–462.

82. Fermat, P. (1662). *Synthèse pour les réfractions* [Synthesis for Refractions]. In P. Tannery & C. Henry (Eds.), *Œuvres de Fermat* (Vol. 3, pp. 156-159). Gauthier-Villars.

83. Fermi, E. (1934). Versuch einer Theorie der β-Strahlen. I. *Zeitschrift für Physik A Hadrons and Nuclei, 88*, 161–177.

84. Ferrarese, L., & Merritt, D. (2000). A Fundamental Relation between Supermassive Black Holes and Their Host Galaxies. *The Astrophysical Journal, 539*(1), L9–L12.

85. Feynman, R. P., Leighton, R. B., & Sands, M. (1965). *The Feynman Lectures on Physics* (Vol. 3). Addison-Wesley.

86. Feynman, R.P. (1985). *QED: The Strange Theory of Light and Matter*. Princeton University Press.

87. Freedman, W. L., et al. (2019). The Carnegie–Chicago Hubble Program. *The Astrophysical Journal, 882*(1), 34.

88. Freeman, W. J. (1999). *How Brains Make Up Their Minds*. Columbia University Press.

89. Freer, M., Horiuchi, H., Funaki, Y., Ogawa, Y., & Kanada-En'yo, Y. (2018). Microscopic clustering in light nuclei. *Reviews of Modern Physics, 90*(3), 035004.

90. Friedman, J. I., & Kendall, H. W. (1972). Deep Inelastic Electron Scattering. *Annual Review of Nuclear Science*, 22, 203-254

91. Friston, K. (2005). A theory of cortical responses. *Philosophical Transactions of the Royal Society B: Biological Sciences*, 360(1456), 815–836.

92. Fries, P. (2005). A mechanism for cognitive dynamics: neuronal communication through neuronal coherence. *Trends in Cognitive Sciences*, 9(10), 474–480.

93. Fröhlich, H. (1968). Long-range coherence and energy storage in biological systems. *International Journal of Quantum Chemistry*, 2(5), 641–649.

94. Fröhlich, H. (1980). *The biological effects of microwaves and related questions*. In L. Marton (Ed.), *Advances in Electronics and Electron Physics* (Vol. 53, pp. 85–152). Academic Press.

95. Gamow, G. (1928). Zur Quantentheorie des Atomkernes [On the Quantum Theory of the Atomic Nucleus]. *Zeitschrift für Physik, 51*(3–4), 204–212.

96. Gazzaniga, M. S. (1967). The split brain in man. *Scientific American, 217*(2), 24–29.

97. Gazzaniga, M. S. (2000). Cerebral specialization and interhemispheric communication: Does the corpus callosum enable the human condition? *Brain*, 123(7), 1293–1326.

98. Gell-Mann, M. (1964). A Schematic Model of Baryons and Mesons. *Physics Letters, 8*(3), 214–215.

99. Gershon, M. D. (1998). *The Second Brain: A Groundbreaking New Understanding of Nervous Disorders of the Stomach and Intestine*. HarperCollins.

100. Ghirardi, G. C., Rimini, A., & Weber, T. (1986). Unified dynamics for microscopic and macroscopic systems. *Physical Review D, 34*(2), 470–491.

101. Gilbert, W. (1986). The RNA world. *Nature, 319*(6055), 618.

102. Giustina, M., et al. (2015). *Significant-Loophole-Free Test of Bell's Theorem with Entangled Photons. Physical Review Letters*, 115(25), 250401.

103. Goldhaber, G., et al. (2001). Timescale Stretch Parameterization of Type Ia Supernova B-Band Light Curves. *The Astrophysical Journal, 558*(1), 359.

104. Goldstein, H., Poole, C., & Safko, J. (2002). *Classical Mechanics* (3rd ed.). Addison Wesley.

105. Greene, B. (1999). *The Elegant Universe: Superstrings, Hidden Dimensions, and the Quest for the Ultimate Theory*. W. W. Norton & Company.

106. Greyson, B. (2003). Incidence and correlates of near-death experiences in a cardiac care unit. *General Hospital Psychiatry, 25*(4), 269–276.

107. Griffiths, D. (2008). *Introduction to Elementary Particles* (2nd ed.). Wiley-VCH.

108. Gross, D. J. (2005). The Discovery of Asymptotic Freedom and the Emergence of QCD. *Reviews of Modern Physics, 77*(3), 837–849.

109. Gruzelier, J. H. (2014). EEG-neurofeedback for optimising performance. I: A review of cognitive and affective outcome in healthy participants. Neuroscience & Biobehavioral Reviews, 44, 124–141.

110. Guri AJ, Hontecillas R, Si H, Liu D, Bassaganya-Riera J. Dietary abscisic acid ameliorates glucose tolerance and obesity-related inflammation in db/db mice fed high-fat diets. *Clinical Nutrition*. 2007 Feb;26(1):107–116.

111. Guth, A. H. (1981). Inflationary universe: A possible solution to the horizon and flatness problems. *Physical Review D, 23*(2), 347–356.

112. Hafele, J. C., & Keating, R. E. (1972). Around-the-World Atomic Clocks: Predicted Relativistic Time Gains. *Science, 177*(4044), 166–170.

113. Hameroff, S., & Penrose, R. (2014). Consciousness in the universe: A review of the 'Orch OR' theory. *Physics of Life Reviews*, 11(1), 39–78.

114. Hamilton, J. P., Etkin, A., Furman, D. J., Lemus, M. G., Johnson, R. F., & Gotlib, I. H. (2011). *Functional neuroimaging of major depressive disorder: A meta-analysis and new integration of baseline activation and neural response data*. The American Journal of Psychiatry, 169(7), 693–703.

115. Hora, H., Miley, G. H., & et al. (2010). Fusion energy without radioactivity: Laser ignition of solid hydrogen–boron (HB11) fuel. *Energy & Environment*, 21(4), 173–200.

116. Hotamisligil, G. S. (2006). *Inflammation and metabolic disorders*. Nature, 444, 860–867.

117. Hawking, S. W. (1975). Particle Creation by Black Holes. *Communications in Mathematical Physics, 43*(3), 199–220.

118. Hebb, D. O. (1949). *The Organization of Behavior: A Neuropsychological Theory*. Wiley.

119. Heisenberg, W. (1925). Über quantentheoretische Umdeutung kinematischer und mechanischer Beziehungen. *Zeitschrift für Physik, 33*(1), 879–893.

120. Heisenberg, W. (1928). Zur Theorie des Ferromagnetismus [On the Theory of Ferromagnetism]. *Zeitschrift für Physik, 49*(9-10), 619–636.

121. Heitler, W., & London, F. (1927). Wechselwirkung neutraler Atome und homöopolare Bindung nach der Quantenmechanik. *Zeitschrift für Physik, 44*(6–7): 455–472.

122. Hen, O., Sargsian, M., Weinstein, L. B., et al. (2017). Probing the structure of the atomic nucleus with high-energy electrons. *Science, 358*(6369), eaao3442.

123. Hensen, B., et al. (2015). *Loophole-free Bell inequality violation using electron spins separated by 1.3 kilometres. Nature,* 526, 682–686.

124. Higgs, P. W. (1964). Broken Symmetries and the Masses of Gauge Bosons. *Physical Review Letters, 13*(16), 508–509.

125. Hofstadter, R. (1961). The electron-scattering method and its application to the structure of nuclei and nucleons. *Nobel Lecture.*

126. Hooper, D., Blasi, P., & Serpico, P. D. (2009). Pulsars as the sources of high energy cosmic ray positrons. *Journal of Cosmology and Astroparticle Physics, 2009*(01), 025.

127. Hubble, E. (1929). A Relation between Distance and Radial Velocity among Extra-Galactic Nebulae. *Proceedings of the National Academy of Sciences, 15*(3), 168–173.

128. Hush, N. S. (1968). *Intervalence-transfer absorption. Part 2. Theoretical considerations and spectroscopic data.* Progress in Inorganic Chemistry, 8, 391–444.

129. Hodgkin, A. L., & Huxley, A. F. (1952). A quantitative description of membrane current and its application to conduction and excitation in nerve. *The Journal of Physiology,* 117(4), 500–544.

130. Huygens, C. (1690). *Traité de la Lumière* (Treatise on Light).

131. Hüser, J., & Blatter, L. A. (1999). Fluctuations in mitochondrial membrane potential caused by repetitive gating of the permeability transition pore. *Biochemical Journal,* 343(Pt 2), 311–317.

132. Ito, M. (2008). Control of mental activities by internal models in the cerebellum. *Nature Reviews Neuroscience,* 9(4), 304–313.

133. Jeans, J. H. (1905). On the Partition of Energy between Matter and Æther. *Philosophical Magazine, 10*(55), 91–98.

134. Jenny, H. (1967). *Cymatics: The study of wave phenomena.* Basilius Presse.

135. Joannopoulos, J. D., Johnson, S. G., Winn, J. N., & Meade, R. D. (2008). *Photonic Crystals: Molding the Flow of Light* (2nd ed.). Princeton University Press.

136. Jumper, J., et al. (2021). Highly accurate protein structure prediction with AlphaFold. *Nature,* 596(7873), 583–589.

137. Jung, C. G. (1968). *The archetypes and the collective unconscious* (2nd ed.). Princeton University Press.

138. Kapitza, P. L. (1938). Viscosity of Liquid Helium Below the λ-Point. *Nature*, 141(3581), 74–75.

139. KATRIN Collaboration. (2025). Direct neutrino-mass measurement based on 259 days of KATRIN data. *Science*, *388*(6743), 180–185

140. Kelso, J. A. S. (1995). *Dynamic Patterns: The Self-Organization of Brain and Behavior.* MIT Press

141. Kirchhoff, G. (1860). On the relation between the radiating and absorbing powers of different bodies for light and heat. *Philosophical Magazine*, **20**, 1–21.

142. Kittel, C. (2005). *Introduction to Solid State Physics* (8th ed.). John Wiley & Sons.

143. Klimesch, W. (1999). EEG alpha and theta oscillations reflect cognitive and memory performance: a review and analysis. *Brain Research Reviews*, 29(2–3), 169–195.

144. Krane, K. S. (1988). *Introductory Nuclear Physics*. John Wiley & Sons.

145. Kuhn, T. S. (1962). *The Structure of Scientific Revolutions*. University of Chicago Press.

146. Klinman, J. P. (2006). Linking protein dynamics to catalysis: the role of hydrogen tunneling. *Philosophical Transactions of the Royal Society B: Biological Sciences*, 361(1472), 1323–1331.

147. Labbé, I., et al. (2023). 'A population of red candidate massive galaxies ~600 Myr after the Big Bang'. *Nature, 616*, 266–269.

148. Lamb, H. (1932). *Hydrodynamics*. Cambridge University Press.

149. Laszlo, E. (2004). *Science and the Akashic field: An integral theory of everything*. Inner Traditions.

150. Lavoisier, A. (1789). *Traité Élémentaire de Chimie*. Cuchet, Paris.

151. Lazar, S. W., et al. (2005). "Meditation experience is associated with increased cortical thickness." *Neuroreport*.

152. Lehninger, A. L., Nelson, D. L., & Cox, M. M. (2008). *Lehninger Principles of Biochemistry* (5th ed.). W.H. Freeman and Company.

153. Le Sage, G. L. (1784). *Physique Mécanique de la Gravitation*. Geneva.

154. Libet, B., Gleason, C. A., Wright, E. W., & Pearl, D. K. (1983). Time of conscious intention to act in relation to onset of cerebral activity (readiness-potential). *Brain*, 106(3), 623-642.

155. Loo, S. K., & Makeig, S. (2012). Clinical utility of EEG in attention-deficit/hyperactivity disorder: a research update. Neurotherapeutics, 9(3), 569–587.

156. Lorentz, H. A. (1904). Electromagnetic phenomena in a system moving with any velocity smaller than that of light. *Proceedings of the Royal Netherlands Academy of Arts and Sciences*, 6, 809–831.

157. Luminet, J.-P., Weeks, J., Riazuelo, A., Lehoucq, R., & Uzan, J.-P. (2003). Dodecahedral space topology as an explanation for weak wide-angle temperature correlations in the cosmic microwave background. *Nature, 425*(6958), 593–595.

158. Mach, E. (1960). *The Science of Mechanics: A Critical and Historical Account of Its Development.* (T. J. McCormack, Trans.). Open Court. (Original work published 1883).

159. Maiolino, R., et al. (2024). A small, metal-poor star-forming galaxy hosting an overmassive black hole at z = 8.7. *Nature, 626,* 990–994.

160. Maldacena, J. (1998). *The Large N Limit of Superconformal Field Theories and Supergravity. Advances in Theoretical and Mathematical Physics, 2*(2), 231–252.

161. Mather, J. C., et al. (1994). Measurement of the cosmic microwave background spectrum by the COBE FIRAS instrument. *The Astrophysical Journal, 420,* 439-444.

162. Maxwell, J.C. (1865). 'A Dynamical Theory of the Electromagnetic Field'. *Philosophical Transactions of the Royal Society of London, 155,* 459–512.

163. McCraty, R., Atkinson, M., Tomasino, D., & Bradley, R. T. (2009). The coherent heart: heart–brain interactions, psychophysiological coherence, and the emergence of system-wide order. *Integral Review, 5*(2), 10–115.

164. McFadden, J. (2002). The conscious electromagnetic information (CEMI) field theory: The hard problem made easy? *Journal of Consciousness Studies, 9*(8), 45–60.

165. McGilchrist, I. (2009). *The Master and His Emissary: The Divided Brain and the Making of the Western World.* Yale University Press.

166. Mendeleev, D. (1869). On the Relationship of the Properties of the Elements to their Atomic Weights. *Zeitschrift für Chemie, 12,* 405-406.

167. Michelson, A. A., & Morley, E. W. (1887). On the Relative Motion of the Earth and the Luminiferous Ether. *American Journal of Science, 34*(203), 333–345.

168. Miller, E. K., & Cohen, J. D. (2001). An integrative theory of prefrontal cortex function. *Annual Review of Neuroscience, 24,* 167–202.

169. Misner, C. W., Thorne, K. S., & Wheeler, J. A. (1973). *Gravitation.* W. H. Freeman.

170. Mitchell, P. (1961). Coupling of phosphorylation to electron and hydrogen transfer by a chemiosmotic type of mechanism. *Nature,* **191,** 144–148.

171. Mohawk, J. A., Green, C. B., & Takahashi, J. S. (2012). *Central and peripheral circadian clocks in mammals.* Annual Review of Neuroscience, **35,** 445–462.

172. Mulders, P. C. R., van Eijndhoven, P. F. P., Schene, A. H., Beckmann, C. F., & Tendolkar, I. (2015). *Resting-state functional connectivity in major depressive disorder: A review.* Neuroscience & Biobehavioral Reviews, 56, 330–344.

173. Müller, C. P., & Homberg, J. R. (2015). The role of serotonin in behavior and cognition. *Progress in Neurobiology, 132,* 1–6.

174. Navier, C. L. M. H. (1822). Sur les lois du mouvement des fluides [On the laws of the movement of fluids]. *Mémoires de l'Académie Royale des Sciences de l'Institut de France, 6*, 389–440.

175. Nelson, D. L., & Cox, M. M. (2017). *Lehninger Principles of Biochemistry* (7th ed.). W.H. Freeman.

176. Newton, I. (1999). *The Principia: Mathematical Principles of Natural Philosophy* (I. B. Cohen & A. Whitman, Trans.). University of California Press. (Original work published 1687).

177. Newton, I. (1693). Third Letter to Richard Bentley.

178. Newton, I. (1704). *Opticks: or, a Treatise of the Reflexions, Refractions, Inflexions and Colours of Light.*

179. Nicholls, D. G., & Ferguson, S. J. (2013). *Bioenergetics 4.* Academic Press.

180. Nicholson, C. (2001). Diffusion and related transport mechanisms in brain tissue. *Reports on Progress in Physics, 64*(7), 815.

181. Nuckolls, J., Wood, L., Thiessen, A., & Zimmerman, G. (1972). Laser compression of matter to super-high densities: Thermonuclear (CTR) applications. *Nature, 239*(5368), 139–142.

182. O'Keefe, J., & Recce, M. L. (1993). Phase relationship between hippocampal place units and the EEG theta rhythm. *Hippocampus, 3*(3), 317–330.

183. Orekhova, E. V., Stroganova, T. A., Prokofyev, A. O., Nygren, G., Gillberg, C., & Elam, M. (2007). Sensory gating in young children with autism: relation to age, IQ, and EEG gamma oscillations. *Neuroscience Letters, 412*(2), 126–131.

184. Packer, L., & Cadenas, E. (2007). *Oxidants and antioxidants revisited: New concepts of oxidative stress.* Free Radical Research, 41(9), 951–952

185. Parvizi, J., & Damasio, A. (2001). Consciousness and the brainstem. *Cognition, 79*(1-2), 135–160.

186. Pauli, W. (1925). Über den Zusammenhang des Abschlusses der Elektronengruppen im Atom mit der Komplexstruktur der Spektren'. *Zeitschrift für Physik, 31*(1): 765–783.

187. Pauling, L. (1960). *The Nature of the Chemical Bond and the Structure of Molecules and Crystals.* 3rd ed. Cornell University Press.

188. Pessa, E. (2003). The role of electromagnetic fields in communication between brain structures. *NeuroQuantology, 1*(4), 263–271.

189. Pienta, K. J., & Coffey, D. S. (1991). Cellular harmonic information transfer through a tissue tensegrity-matrix system. *Medical Hypotheses, 34*(2), 88–95.

190. Pomes, R., & Roux, B. (2002). Molecular mechanism of H+ conduction in the single-file water chain of the gramicidin channel. *Biophysical Journal, 82*(5), 2304–2316.

191. Peebles, P. J. E., & Ratra, B. (2003). The cosmological constant and dark energy. *Reviews of Modern Physics, 75*(2), 559.

192. Penrose, R. (1965). Gravitational Collapse and Spacetime Singularities. *Physical Review Letters, 14*(3), 57–59.

193. Penrose, R. (1994). *Shadows of the Mind: A Search for the Missing Science of Consciousness*. Oxford University Press.

194. Penrose, R. (2004). *The Road to Reality: A Complete Guide to the Laws of the Universe*. Jonathan Cape.

195. Perkins, D. H. (2000). *Introduction to High Energy Physics* (4th ed.). Cambridge University Press.

196. Picard, M., Wallace, D. C., & Burelle, Y. (2016). The rise of mitochondria in medicine. *Mitochondrion, 30*, 105–116.

197. Pienta, K. J., & Coffey, D. S. (1991). Cellular harmonic information transfer through a tissue tensegrity-matrix system. *Medical Hypotheses, 34*(2), 88–95.

198. Pierański, P. (1998). In search of ideal knots. *Proceedings of the Royal Society of London. Series A: Mathematical, Physical and Engineering Sciences, 454*(1976), 2339-2364.

199. Pilla, A. A. (2013). *Mechanisms and therapeutic applications of time-varying and static magnetic fields*. In *Biological and Medical Aspects of Electromagnetic Fields* (3rd ed., pp. 1–62). CRC Press.

200. Planck, M. (1901). On the Law of Distribution of Energy in the Normal Spectrum. *Annalen der Physik, 4*(3), 553–563.

201. Planck Collaboration. (2018). Planck 2018 results. VI. Cosmological parameters. *Astronomy & Astrophysics, 641*, A6.

202. Plato. (2007). *The Republic*. (D. Lee, Trans.). Penguin Classics. (Original work c. 375 BCE).

203. Pollack, G. H. (2001). *Cells, Gels and the Engines of Life: A New, Unifying Approach to Cell Function*. Ebner and Sons Publishers.

204. Pollack, G. H. (2013). *The Fourth Phase of Water: Beyond Solid, Liquid, and Vapor*. Ebner and Sons Publishers.

205. Pokorný, J. (2004). Excitation of vibrations in microtubules in living cells. *Bioelectrochemistry, 63*(1–2), 321–326.

206. Pound, R. V., & Rebka, G. A. Jr. (1959). Gravitational Red-Shift in Nuclear Resonance. *Physical Review Letters, 3*(9), 439–441.

207. Pribram, K. H. (1991). *Brain and Perception: Holonomy and Structure in Figural Processing*. Lawrence Erlbaum Associates.

208. Purves, D., Augustine, G. J., Fitzpatrick, D., et al. (2018). *Neuroscience* (6th ed.). Oxford University Press.

209. Ramirez, A. P. (1994). Strongly geometrically frustrated magnets. *Annual Review of Materials Science, 24*(1), 453-480.

210. Raichle, M. E. (2015). The brain's default mode network. *Annual Review of Neuroscience, 38*, 433–447.

211. Rayleigh, J. W. S. (1877). *The Theory of Sound*. Macmillan and Co.

212. Rayleigh, Lord. (1900). Remarks upon the Law of Complete Radiation. *Philosophical Magazine, 49*(301), 539–540.

213. Riess, A. G., et al. (1998). Observational Evidence from Supernovae for an Accelerating Universe and a Cosmological Constant. *The Astronomical Journal, 116*(3), 1009–1038.

214. Robertson, B. E., et al. (2023). Identification and properties of intense emission-line galaxies at z > 10. *Nature Astronomy, 7*(5), 611–621.

215. Robitaille, P. M. (2008). Blackbody radiation and the carbon particle model. *Progress in Physics, 4*, 25–31.

216. Rubia, K., Halari, R., Cubillo, A., Mohammad, A. M., Brammer, M., & Taylor, E. (2009). *Disorder-specific dysfunction in right inferior prefrontal cortex during two inhibition tasks in boys with attention-deficit hyperactivity disorder compared to boys with obsessive-compulsive disorder*. Human Brain Mapping, 30(9), 2854–2870

217. Rubin, V. C., & Ford, W. K. Jr. (1970). Rotation of the Andromeda Nebula from a Spectroscopic Survey of Emission Regions. *The Astrophysical Journal, 159*, 379.

218. Rutherford, E. (1911). 'The Scattering of α and β Particles by Matter and the Structure of the Atom'. *Philosophical Magazine, Series 6, 21*(125), 669–688.

219. Rybicki, G. B., & Lightman, A. P. (1979). *Radiative Processes in Astrophysics*. Wiley.

220. Sakaguchi, S., Sakaguchi, N., Asano, M., Itoh, M., & Toda, M. (1995). Immunologic self-tolerance maintained by activated T cells expressing IL-2 receptor alpha-chains (CD25). Breakdown of a single mechanism of self-tolerance causes various autoimmune diseases. *Journal of Immunology, 155*(3), 1151–1164

221. Sakaguchi, S., Yamaguchi, T., Nomura, T., & Ono, M. (2008). Regulatory T cells and immune tolerance. *Cell, 133*(5), 775–787.

222. Salart, D., Baas, A., Branciard, C., Gisin, N., & Zbinden, H. (2008). Testing the speed of 'spooky action at a distance'. *Nature, 454*(7206), 861–864.

223. Sakharov, A. D. (1967). Violation of CP Invariance, C asymmetry, and baryon asymmetry of the universe. *JETP Letters, 5*(1), 24–27.

224. Schmidt, M. (1963). 3C 273: a star-like object with large red-shift. *Nature, 197*(4872), 1040.

225. Scholes, G. D., Fleming, G. R., Olaya-Castro, A., & van Grondelle, R. (2017). Using coherence to enhance function in chemical and biophysical systems. *Nature, 543*(7647), 647–656.

226. Schrödinger, E. (1926). An Undulatory Theory of the Mechanics of Atoms and Molecules. *Physical Review, 28*(6), 1049–1070.

227. Scolnic, D. et al. (2022). *The Pantheon+ Analysis: The Full Data Set and Modeling of Type Ia Supernovae Systematic Uncertainties.* The Astrophysical Journal, 938(2), 113.

228. Shalm, L. K., et al. (2015). *Strong Loophole-Free Test of Local Realism. Physical Review Letters*, 115(25), 250402.

229. Shapiro, S. L., & Teukolsky, S. A. (1983). *Black Holes, White Dwarfs, and Neutron Stars: The Physics of Compact Objects.* John Wiley & Sons.

230. Sheldrake, R. (1988). *The presence of the past: Morphic resonance and the habits of nature.* Times Books.

231. Smythies, J. (2005). On the neural basis of consciousness. *Brain and Cognition*, 58(1), 9–16.

232. Sommerfeld, A. (1916). Zur Quantentheorie der Spektrallinien. *Annalen der Physik, 51*(17): 1–94.

233. Sonuga-Barke, E. J. S., & Castellanos, F. X. (2007). *Spontaneous attentional fluctuations in impaired states and pathological conditions: A neurobiological hypothesis.* Neuroscience & Biobehavioral Reviews, 31(7), 977–986.

234. Sperry, R. W. (1968). Hemisphere deconnection and unity in conscious awareness. *American Psychologist*, 23(10), 723–733.

235. Stark, J. (1914). Beobachtungen über den Effekt des elektrischen Feldes auf Spektrallinien. *Annalen der Physik, 43*(12), 965–982.

236. Stefan, J. (1889). Über einige Probleme der Theorie der Wärmeleitung [On some problems of the theory of heat conduction]. *Sitzungsberichte der Kaiserlichen Akademie der Wissenschaften in Wien, Mathematisch-Naturwissenschaftliche Classe, 98*, 473–484.

237. Steinmetz, C. P. (1914). *Electric Discharges, Waves and Impulses, and Other Transients.* McGraw-Hill.

238. Stevenson, D. J. (2003). Planetary magnetic fields. *Earth and Planetary Science Letters, 208*(1-2), 1–11.

239. Stokes, G. G. (1845). On the theories of the internal friction of fluids in motion. *Transactions of the Cambridge Philosophical Society, 8*, 287–319.

240. Strassman, R. J. (2001). *DMT: The Spirit Molecule: A Doctor's Revolutionary Research into the Biology of Near-Death and Mystical Experiences.* Park Street Press.

241. Susskind, L. (1995). The World as a Hologram. *Journal of Mathematical Physics, 36*(11), 6377–6396.

242. Susskind, L. (2005). *The Cosmic Landscape: String Theory and the Illusion of Intelligent Design.* Little, Brown and Company.

243. Taylor, G. I. (1934). The mechanism of plastic deformation of crystals. Part I.—Theoretical. *Proceedings of the Royal Society of London. Series A, 145*(855), 362–387.

244. Taylor, R. E. (1991). Deep Inelastic Scattering: The discovery of the point-like constituents of the nucleon. *Reviews of Modern Physics, 63*(3), 573.

245. Tegeler, C. H., Shaltout, H. A., Gerdes, L., Tegeler, C. L., & Cook, D. (2016). Successful use of closed-loop neurotechnology for post-traumatic stress disorder. Military Medicine, 181(9), e1111–e1118.

246. Thomson, W. (Lord Kelvin). (1867). On vortex atoms. *Proceedings of the Royal Society of Edinburgh, 6,* 94–105.

247. Thomson, W. (Lord Kelvin). (1887). On the division of space with minimum partitional area. *Philosophical Magazine, 24*(151), 503–514.

248. 't Hooft, G. (1993). *Dimensional Reduction in Quantum Gravity. arXiv preprint* gr-qc/9310026.

249. Tononi, G. (2008). Consciousness as integrated information: a provisional manifesto. *Biological Bulletin, 215*(3), 216–242.

250. Totani, T. (2025). 20 GeV halo-like excess of the Galactic diffuse emission and implications for dark matter annihilation. *Journal of Cosmology and Astroparticle Physics,* 2025(11), 080.

251. Traub, R. D., et al. (2005). A single-column thalamocortical network model exhibiting gamma oscillations, sleep spindles, and epileptogenic bursts. *Journal of Neurophysiology, 93*(4), 2194–2232.

252. UA1 Collaboration, Arnison, G., et al. (1983). Experimental observation of isolated large transverse energy electrons with associated missing energy $\sqrt{s}$ = 540 GeV. *Physics Letters B, 122*(1), 103–116.

253. UA2 Collaboration, Banner, M., et al. (1983). Observation of single isolated electrons of high transverse momentum in events with missing transverse energy at the CERN p̄p collider. *Physics Letters B, 122*(5-6), 476–485.

254. Uhlhaas, P. J., & Singer, W. (2010). Abnormal neural oscillations and synchrony in schizophrenia. *Nature Reviews Neuroscience, 11*(2), 100–113.

255. Underhill, E. (1911). *Mysticism: A Study in the Nature and Development of Spiritual Consciousness*. London: Methuen.

256. Uzan, J.-P. (2003). "The Fundamental Constants and Their Variation: Observational and Theoretical Status." *Reviews of Modern Physics*, 75(2), 403–455.

257. van Dokkum, P., et al. (2023). A candidate runaway supermassive black hole identified by its stellar wake. *The Astrophysical Journal Letters*, 946(2), L51.

258. van Lommel, P., van Wees, R., Meyers, V., & Elfferich, I. (2001). Near-death experience in survivors of cardiac arrest: a prospective study in the Netherlands. *The Lancet*, 358(9298), 2039–2045.

259. Varela, F., Lachaux, J. P., Rodriguez, E., & Martinerie, J. (2001). The brainweb: phase synchronization and large-scale integration. *Nature Reviews Neuroscience*, 2(4), 229–239.

260. Visser, M. (1998). Acoustic black holes: horizons, ergospheres, and Hawking radiation. *Classical and Quantum Gravity*, 15(6), 1767–1791.

261. Voet, D., & Voet, J. G. (2011). *Biochemistry* (4th ed.). Wiley.

262. Vowles, R. V. G. V. (1969). *The Structured Atom Model*.

263. Wackermann, J. (2003). *Dyadic consciousness: A possible approach to the problem of consciousness in a shared field of awareness*. Journal of Consciousness Studies, 10(6–7), 69–86.

264. Warburg, O. (1956). *On the origin of cancer cells*. Science, 123(3191), 309–314.

265. Wang, X. J. (2010). Neurophysiological and computational principles of cortical rhythms in cognition. *Physiological Reviews*, 90(3), 1195–1268.

266. Ward, L. M. (2011). The thalamic dynamic core theory of conscious experience. *Consciousness and Cognition*, 20(2), 464–486.

267. Weidenspointner, G., Skinner, G., Jean, P., et al. (2008). An asymmetric distribution of positrons in the Galactic disk. *Nature*, 451(7175), 159–162.

268. Weinberg, S. (1967). 'A Model of Leptons'. *Physical Review Letters*, 19(21): 1264–1266.

269. Weinberg, S. (1972). *Gravitation and Cosmology: Principles and Applications of the General Theory of Relativity*. New York, NY: John Wiley & Sons.

270. Weinberg, S. (1989). The cosmological constant problem. *Reviews of Modern Physics*, 61(1), 1–23.

271. Weinberg, S. (1995). *The Quantum Theory of Fields, Volume 1: Foundations*. Cambridge University Press.

272. Wigner, E. P. (1960). The unreasonable effectiveness of mathematics in the natural sciences. *Communications on Pure and Applied Mathematics*, 13(1), 1–14.

273. Williamson, J. G., & van der Mark, M. B. (1997). Is an electron a photon with toroidal topology? *Annales de la Fondation Louis de Broglie, 22*(2), 133.

274. Wiseman, H. M., & Milburn, G. J. (2009). *Quantum Measurement and Control.* Cambridge University Press.

275. Witten, E. (1986). Non-commutative Geometry and String Field Theory. *Nuclear Physics B, 268*(2), 253–294.

276. Wright, E. L. (2006). A Cosmology Calculator for the World Wide Web. *Publications of the Astronomical Society of the Pacific, 118*(850), 1711–1715.

277. Young, T. (1804). The Bakerian Lecture: Experiments and Calculations Relative to Physical Optics. *Philosophical Transactions of the Royal Society of London, 94*, 1–16.

278. Yukawa, H. (1935). On the Interaction of Elementary Particles. I. *Proceedings of the Physico-Mathematical Society of Japan. 3rd Series, 17*, 48-57.

279. Zee, A. (2010). *Quantum Field Theory in a Nutshell* (2nd ed.). Princeton University Press.

280. Zurek, W. H. (2003). 'Decoherence, einselection, and the quantum origins of the classical'. *Reviews of Modern Physics, 75*(3), 715–775.

281. Zwicky, F. (1929). On the reddening of spectral lines through intergalactic space. *Proceedings of the National Academy of Sciences, 15*(10), 773–779.

The UFD Maven: A Teacher and Guide

Although the journey of this book is over, the exploration of its ideas is just getting started. With the following QR code, you can transform any advanced Large Language Model (LLM) into a dedicated expert on this framework, creating a "UFD Maven" that you can use to question, challenge, and deepen your understanding of the physical universe. The process is simple and takes less than a minute.

1. Access the Knowledge Base

Scan the QR code below. This will take you to a public document that begins with a "Prime Directive," which provides the necessary instructions for the AI.

2. Copy and Paste

Copy the entire text of that document to your clipboard. Then begin a new conversation with an advanced LLM (such as ChatGPT, Claude, or Gemini) and paste the entire copied text as your first message.

3. Begin Your Dialogue

Once the AI has acknowledged its new instructions, you can begin your exploration. Ask it anything you want about the UFD framework, including any topic covered in the book, the mathematical appendices, or any other scientific topic you are interested in. You can even ask it to generate images or prompts to help you visualize the universe in new ways. The possibilities are endless.